Springer Handbook of Auditory Research

Series Editors: Richard R. Fay and Arthur N. Popper

For other titles published in this series, go to
www.springer.com/series/2506

Mary Florentine • Arthur N. Popper
Richard R. Fay
Editors

Loudness

with 75 Illustrations

Springer

Editors
Mary Florentine
Department of Speech-Language Pathology and Audiology with joint appointment in Department of Electrical and Computer Engineering
Northeastern University
Boston, MA 02115
USA
m.florentine@neu.edu

Richard R. Fay
Department of Psychology
Loyola University Chicago
Chicago, IL
USA
rfay@luc.edu

Arthur N. Popper
Department of Biology
University of Maryland
College Park, MD
USA
apopper@umd.edu

ISBN 978-1-4419-6711-4 e-ISBN 978-1-4419-6712-1
DOI 10.1007/978-1-4419-6712-1
Springer New York Dordrecht Heidelberg London

Library of Congress Control Number: 2010938801

Cover Art: Estimate of loudness as a function of level derived from chirp-evoked otoacoustic emissions by Luke Shaheen and Michael Epstein.

Printed on acid-free paper

Springer is part of Springer Science+Business Media (www.springer.com)

In Memoriam and Honor

Søren Buus
(January 29, 1951–April 29, 2004)

(Søren Buus's obituary was published by Scharf, B. and Florentine, M. in the Journal of the Acoustical Society of America, 2005, 117, p. 1685.)

Series Preface

Springer Handbook of Auditory Research

The *Springer Handbook of Auditory Research* presents a series of comprehensive and synthetic reviews of the fundamental topics in modern auditory research. The volumes are aimed at all individuals with interests in hearing research including advanced graduate students, postdoctoral researchers, and clinical investigators. The volumes are intended to introduce new investigators to important aspects of hearing science and to help established investigators to better understand the fundamental theories and data in fields of hearing that they may not normally follow closely.

Each volume presents a particular topic comprehensively, and each serves as a synthetic overview and guide to the literature. As such, the chapters present neither exhaustive data reviews nor original research that has not yet appeared in peer-reviewed journals. The volumes focus on topics that have developed a solid data and conceptual foundation, rather than on those for which a literature is only beginning to develop. New research areas will be covered on a timely basis in the series as they begin to mature.

Each volume in the series consists of a few substantial chapters on a particular topic. In some cases, the topics will be ones of traditional interest for which there is a substantial body of data and theory, such as auditory neuroanatomy (Vol. 1) and neurophysiology (Vol. 2). Other volumes in the series deal with topics that have begun to mature more recently, such as development, plasticity, and computational models of neural processing. In many cases, the series editors are joined by a coeditor having special expertise in the topic of the volume.

Richard R. Fay, Chicago, IL
Arthur N. Popper, College Park, MD

Volume Preface

The topic of loudness is of considerable concern, both in and outside of research laboratories. Most people have developed an opinion about some aspect of loudness, and many complain about the loudness of background sounds in their daily environments and their impacts on quality of life. Moreover, such sounds interfere with the ability to hear useful sounds, and such masking can be especially problematic for people with hearing losses, children, older adults, and nonnative speakers of a language.

At the same time, not all loud sounds are undesirable. Some loud sounds are important for human well-being, such as warning signals, whereas other loud sounds, such as music, can be pleasurable. In fact, loudness is essential for enjoying the dynamics of music. Thus, loudness is a pervasive and complex issue, and one that needs to be examined from a wide range of perspectives, and that is the purpose of this volume.

Research in loudness has been performed in many countries, and this volume is an international endeavor with authors from Europe, Japan, and the United States, making the volume an attempt to provide a global network of information about loudness. The editors are very pleased to be able to bring together information on many aspects of loudness in this one volume, as well as to highlight approaches from many different perspectives.

The overall stage for understanding the issues of loudness is set up in Chapter 1 by Florentine, who defines loudness and provides an overview of the many factors that influence loudness, Chapter 1 also addresses how language and culture may influence loudness, and concludes with a summary of current knowledge of the physiological mechanisms involved in loudness. Chapters 2 and 3 cover issues related to the measurement of loudness. Marks and Florentine, in Chapter 2, discuss the theoretical, empirical, and practical constraints on loudness measurement. The chapter starts with a brief history of loudness measurement in the nineteenth and twentieth centuries, and ends with contemporary methods of measurements. Of course, measures of loudness are also influenced by the context in which sounds are heard. In Chapter 3, Arieh and Marks discuss the ways in which context affects loudness and loudness judgments. In Chapter 4, Epstein reviews two issues related to responses to loudness: physiological effects of loud sounds, and perceptual and physiological data that correlate with loudness. Loudness in the laboratory is discussed in Chapters 5 and 6 using a traditional, but artificial, classification to divide the subject matter. Jesteadt and Leibold address the loudness of steady-state sounds in Chapter 5. Kuwano and Namba examine the loudness of nonsteady-state (time-varying) sounds in Chapter 6.

The bridge between loudness in the laboratory and daily environments begins in Chapter 7 and is expanded upon in Chapter 8. In Chapter 7, Sivonen and Ellermeir review studies on binaural loudness that have used different modes of stimulus presentation: headphones and free, diffuse, and directional sound fields. They show how mode of presentation affects the measurement of binaural loudness. In Chapter 8, Fastl and Florentine cover how loudness is related to annoyance, music, multisensory (audio-visual and audio-tactile) interactions, and the environmental context in which sounds are heard. They also discuss issues related to setting sound levels to optimal loudness for large groups of people. The topic of loudness is especially important for the one out of ten people who have a hearing loss and for those doing work with some aspect of aural rehabilitation. Knowledge of loudness in hearing loss is also important for anyone trying to understand normal hearing, because it puts constraints on theories of loudness. In Chapter 9, Smeds and Leijon summarize current thinking on the formation of loudness as it relates to different types of hearing loss and they describe strategies used to compensate for altered loudness. The volume ends in Chapter 10 with an introduction to models of loudness by Marozeau.

As with most volumes in the *Springer Handbook of Auditory Research*, chapters often build upon material discussed in earlier volumes. In particular, generally related material can be found in Volumes 3 (*Human Psychophysics*), 6 (*Auditory Computation*), 18 (*Speech Processing in the Auditory System*), and 29 (*Auditory Perception of Sound Sources*).

The editors express their appreciation to a number of colleagues and friends, including the authors of the chapters, who assisted in review of one or more of the chapters. We are grateful to William J. Cavanaugh, Leo Beranek, Brian Fligor, Julia B. Florentine, Michael G. Heinz, Sharon Kujawa, Andrzej Miśkiewicz, Brian C. J. Moore, Andrew Oxenham, Torben Poulsen, Bertram Scharf, and the students of the 2009 Loudness Seminar at Northeastern University.

Mary Florentine, Boston, MA
Arthur N. Popper, College Park, MD
Richard R. Fay, Chicago, IL

Contents

1 **Loudness** ... 1
Mary Florentine

2 **Measurement of Loudness, Part I: Methods, Problems, and Pitfalls** ... 17
Lawrence E. Marks and Mary Florentine

3 **Measurement of Loudness, Part II: Context Effects** ... 57
Yoav Arieh and Lawrence E. Marks

4 **Correlates of Loudness** ... 89
Michael J. Epstein

5 **Loudness in the Laboratory, Part I: Steady-State Sounds** ... 109
Walt Jesteadt and Lori J. Leibold

6 **Loudness in the Laboratory, Part II: Non-Steady-State Sounds** ... 145
Sonoko Kuwano and Seiichiro Namba

7 **Binaural Loudness** ... 169
Ville Pekka Sivonen and Wolfgang Ellermeier

8 **Loudness in Daily Environments** ... 199
Hugo Fastl and Mary Florentine

9 Loudness and Hearing Loss 223
Karolina Smeds and Arne Leijon

10 Models of Loudness 261
Jeremy Marozeau

Index 285

Contributors

Yoav Arieh
Department of Psychology, Montclair State University, Montclair, NJ 07043, USA
ariehy@mail.montclair.edu

Wolfgang Ellermeier
Department of Psychology, Technische Universität Darmstadt, D-64283, Darmstadt, Germany
ellermeier@psychologie.tu-darmstadt.de

Michael J. Epstein
Auditory Modeling and Processing Laboratory, Department of Speech-Language Pathology and Audiology, The Communications and Digital Signal Processing Center, Department of Electrical and Computer Engineering, Northeastern University, Boston, MA 02115, USA
m.epstein@neu.edu

Hugo Fastl
Department of Technical Acoustics, AG Technische Akustik, MMK, Technische Universität München, 80333 München, Germany
fastl@mmk.ei.tum.de

Mary Florentine
Department of Speech-Language Pathology and Audiology with joint appointment in Department of Electrical and Computer Engineering, Northeastern University, Boston, MA 02115, USA
m.florentine@neu.edu

Walt Jesteadt
Boys Town National Research Hospital, Omaha, NE 68131, USA
walt.jesteadt@boystown.org

Sonoko Kuwano
Osaka University, Toyonaka, Osaka 560-0083, Japan
kuwano@see.eng.osaka-u.ac.jp

Lori J. Leibold
Division of Speech and Hearing Sciences, The University of North Carolina School of Medicine, Chapel Hill, NC 27599, USA
leibold@med.unc.edu

Arne Leijon
KTH – School of Electrical Engineering, 100 44 Stockholm, Sweden
leijon@kth.se

Lawrence E. Marks
John B. Pierce Laboratory, Department of Epidemiology and Public Health, Yale University, School of Medicine, and Department of Psychology, Yale University New Haven, CT 06519, USA
marks@jbpierce.org

Jeremy Marozeau
The Bionic Ear Institute, East Melbourne, VIC 3002, Australia
jmarozeau@bionicear.org

Seiichiro Namba
Osaka University, 14-8-513, Aomadani-Higashi 1 chome, Minoo, Osaka, Japan
QZW00041@nifty.com

Ville Pekka Sivonen
Department of Signal Processing and Acoustics, Aalto University School of Science and Technology, Otakaari 5 A, 02150 Espoo, Finland
ville.sivonen@tkk.fi

Karolina Smeds
ORCA Europe, Widex A/S, Maria Bangata 4, 118 63 Stockholm, Sweden
karolina.smeds@orca-eu.info

Chapter 1
Loudness

Mary Florentine

1.1 Why Learn About Loudness?

The topic of loudness is no longer something esoteric, discussed only in research laboratories and psychoacoustics lectures. It is mainstream in social conversation, and most people have developed an opinion about some aspect of loudness. Our daily environments are too loud and people are taking notice. In their book, *One Square Inch of Silence*, Hempton and Grossmann (2009) document the lack of quiet places. The fact that a book about this topic can be published by Free Press, a division of Simon and Schuster, and appear on bookshelves – from Barnes and Noble to WalMart and Sam's Club – indicates that problems associated with loud sounds strike a resonant chord with a large segment of the population. Loud sounds intrude on our enjoyment of life and affect our performance; loud background sounds interfere with our ability to hear sounds we want to hear and can create communication problems for everyone, especially those with hearing losses (Chap. 9), children (Nelson et al. 2002), older adults (Kim et al. 2006), and non-native speakers of a language (e.g., Mayo et al. 1997; Lecumberri and Cooke 2006; Van Engen and Bradlow 2007). These combined groups add up to be a significant portion of the overall population.

News reports and media broadcasts in many countries have alerted the general population to the potential risk of hearing loss caused by exposure to high levels of sound, especially music. Many parents are especially concerned about the music-listening behaviors of their children. In fact, recommendations to prevent noise-induced hearing loss are often not heeded, although they are simple to understand (Florentine 1990). Research compiled during the past two decades is not unambiguous regarding the limits of toxic exposure levels. For example, sound exposures for musicians in symphony orchestras show that in many cases the sound exposure exceeds an 8-h limit of 85 dBA (Royster et al. 1991), but measurements of hearing

M. Florentine (✉)
Department of Speech-Language Pathology and Audiology with joint appointment in Department of Electrical and Computer Engineering,
Northeastern University, Boston, MA 02115, USA
e-mail: m.florentine@neu.edu

M. Florentine et al. (eds.), *Loudness*, Springer Handbook of Auditory Research 37,
DOI 10.1007/978-1-4419-6712-1_1, © Springer Science+Business Media, LLC 2011

thresholds in sound-exposed musicians do not indicate much change – although they often have tinnitus and may have more difficulty with complex auditory processing. Likewise, Axelsson et al. (1995) found surprisingly little change in the hearing thresholds of rock musicians tested in the beginning of their careers and tested after 20 years of performing. Hearing thresholds, however, may be a poor way to assess damage to hearing because they are quite insensitive to neural degeneration (Kujawa and Liberman 2009). Effective hearing in daily life, such as the ability to hear speech in background noise, requires more physiological processing than is needed to simply detect the presence of a sound.

It is noteworthy that the sound exposure limit of 85 dBA is based primarily on data collected from white adult males, who were exposed to industrial noise (ISO 1999 1990); recommendations for children exposed to music are estimations. Some data suggest that previous exposure to high-level noise has a deleterious effect on the progression of age-related hearing loss (Gates et al. 2000; Kujawa and Liberman 2006). In other words, ears with noise damage may age differently from those without noise damage. Although there is evidence that some people may be more susceptible to noise-induced hearing loss than others, there are currently no standardized tests that can identify those who may be at greater risk. Therefore, it is best to use caution around loud sounds (for iPod recommendations and other information, see Chap. 8).

Although there is some debate on the limits of toxic noise exposure, there is clear consensus among scientists that very high-level sounds can cause hearing loss and impact our general physical and psychological wellness (see Chap. 4). Some loud sounds have more physical and psychological impact than others, and the magnitude of noxious sounds in our daily environments are too loud in many locations. The background sounds of our daily lives – the soundscapes – have changed (Schafer 1977). Although there have always been loud sounds in nature such as waterfalls and thunder, people could choose not to live close to waterfalls; thunder was not a daily experience. Humans designed bullhorns for use as warning signals and long-distance communication devices. Although people experienced loud sounds, most people agree that the soundscapes of our daily environments are much louder and more intrusive today than in the past. Soundscapes in entire areas of countries can change rather quickly. For example, soundscapes took on a dramatic change in loudness in early twentieth century America with the onset of modern technology (Thompson 2002).

Loudness correlates highly with the degree of annoyance of community noises (Berglund et al. 1976). The problem of intrusively loud soundscapes is not confined to any one country; it is a global issue. It is not confined to spaces outside dwelling places. Sounds inside homes are often too loud. In fact, at the time of this writing the Commercial Advertisement Loudness Mitigation (CALM) Act is being discussed in the U.S. House of Representatives. The CALM Act addresses widespread consumer complaints regarding the abrupt loudness of television advertisements. The Act would enable the Federal Communications Commission (FCC) to monitor the levels of advertisements in television programming to ensure that the loudness levels of commercial breaks are consistent with those of the programming that it brackets. National standards on the loudness of commercials have already been adopted in Australia, Brazil, France, Israel, Russia, and the United Kingdom.

Most loud sounds are unnecessary and avoidable. Modern technology exists that can reduce the level of sounds. There are a number of ways to quiet loud sounds that are not difficult to implement, but they require awareness, effort, some financial resources, and in some cases political action. Even when regulations to control unreasonably loud sounds are enforced, technological knowledge has been used to get around the regulations. For example, a loudness maximizer has been developed to increase loudness of media broadcasts and commercials to grab the attention of potential customers, while staying within legal sound-level limits (see Chap. 8). This device was developed with knowledge of how we perceive loudness; this same knowledge can be used to set more effective sound-emission limits and is found within the chapters of this book.

Although many loud sounds should be eliminated, some loud sounds are important, useful, and even desirable. Warning signals are essential for our safety. Some loud sounds allow us to experience the dynamics of music and speech. To eliminate loud, unwanted sounds and keep desired sounds optimally loud requires an understanding of the many factors that influence the perception of loudness.

Another important reason to learn about loudness is to aid in the rehabilitation of people with hearing losses. According to the National Institutes of Deafness and Communication Disorders, one out of ten people have a hearing loss, and many people will develop a hearing loss as they age. The most common treatment for hearing loss is hearing aids. Kochkin's survey (2005) of 1,500 hearing-aid users indicates that only 60% of them reported being satisfied when asked about comfort with loud sounds.

This chapter provides an overview of the many factors that influence loudness and they are described in greater detail in subsequent chapters. It also includes topics that have not been covered elsewhere in this book. Section 1.2 includes the definition and the meaning of loudness, and gives an overview of the complex nature of loudness. It also specifies correct terminology and cautions about the use of incorrect and misleading terminology. The third section addresses how language and culture influence judgments of loudness. It describes how the connotative meaning of a percept, such as loudness, can be obtained and how it can differ among languages, even though the dictionary definitions indicate the same meaning. It also describes some early international collaboration undertaken to understand loudness and other aspects of perception that are related to it. The fourth section describes the current state of knowledge regarding loudness and points toward new areas of investigation.

1.2 Definition and Meaning of Loudness

Loudness has been defined as the perceptual strength of a sound that ranges from very soft (or quiet) to very loud. Scharf (1978) suggested that loudness may be defined "as the attribute of a sound that changes most readily when sound intensity is varied," but preferred to define it as the subjective intensity of a sound. The term "subjective" assigns the evaluation of intensity to be within the listener. Accordingly, no “right” and “wrong” answers exist in loudness judgments. Because there is no objective measure of loudness, measurements of loudness should employ methods of converging evidence. For example, a matrix design can be used in which sounds are matched in loudness to themselves and to each of the other signals (e.g., Florentine et al. 1978). The resulting data can be easily examined for symmetry and transitivity (see Chap. 2). Most definitions of loudness are somewhat vague, but most people behave in a consistent manner when judging loudness.

The scale of loudness not only allows ordering sounds from soft to loud, but also has a magnitude associated with it. A commonly used unit of loudness is the sone. One sone is defined as the loudness of a 1-kHz tone at 40-dB SPL heard binaurally in a free field from a source in the listener’s frontal plane. A sound with a loudness of 2.0 sones is twice as loud as a 40-dB SPL, 1-kHz tone, and a sound with a loudness of 0.5 sones is half as loud. In many instances, however, loudness judgments entail comparisons between two sounds and the loudness is expressed in terms of SPL of an equally loud comparison sound. If the comparison sound is a 1-kHz tone heard binaurally in a free field, its SPL gives the loudness level, which is measured in a unit called a phon. For example, a 60-Hz tone at 60-dB SPL is on the average about as loud as a 1-kHz tone at 40-dB SPL. Accordingly, the loudness level of the 60-Hz tone is 40 phons. (For more information on sones and phons, see Chap. 5. Encyclopedia entries that give a brief summary of loudness can be found in Scharf (1997), Florentine and Heinz (2009), and Epstein and Marozeau (2010).)

The study of loudness is a subarea of psychoacoustics, which is the study of the relationship between physical properties of sound and perceptual responses to them. Loudness is the primary perceptual correlate of the level of a sound. The adjective “primary” is important because loudness also changes with other physical properties of sound (e.g., frequency, bandwidth, duration, spectral complexity of a sound, the presence of other sounds, etc.). There is no simple one-to-one correspondence between loudness and any physical property of a sound, including level. A review of how loudness changes with physical properties of sound can be found in Jesteadt and Leibold (Chap. 5) for steady state sounds, and Kuwano and Namba (Chap. 6) for nonsteady-state sounds. In addition to stimulus factors, loudness changes with memory, multisensory interactions, the manner sounds are presented and how it is measured, cognitive factors, psychological and physical state of the listener, and cross-cultural differences.

Sounds are perceived in a context. Context changes in various ways, including the physical environment and methods used to measure loudness. Using the taxonomy of Buus (2002), methods include the procedures (mode of stimulus presentation, listener’s task, measurement strategy, and datum definition). A review of contemporary methods by Marks and Florentine is provided in Chap. 2. Arieh and Marks (Chap. 3) review the numerous ways in which loudness judgments depend on sequential presentations of relatively short sounds.

Whenever a sound is heard in daily life, it is heard in a multisensory context and in the context of other competing sounds. Experiments have shown that sounds presented simultaneously with a target sound can reduce its loudness; this is known as partial masking of loudness (Scharf 1964). Partial masking of loudness is a common experience in daily life; loudness changes depending on the background soundscape. For example, a friend's voice will sound softer on the street near a noisy construction site than in a quieter area. This phenomenon is referred to as partial masking of loudness, because the loudness of a sound attended to (i.e., a friend's voice) is partially masked by the background sounds. Some sounds are experienced as partially masked, whereas others are not. For example, when successive sounds are heard in daily environments – such as sounds produced by playing successive notes on a piano – the sounds partially overlap one another, but the sounds are heard as separate and loudness does not appear to change (see Chap. 6). Experiments that have been designed to bridge the gap between traditional laboratory experiments and daily experience in natural environments have demonstrated the influence of environmental contexts (see Chaps. 3 and 8). For example, on average a red sports car will be judged to be slightly louder than a green sports car when both are heard with the same automotive sounds (see Chap. 8). Therefore, the percept of loudness at the time of an event will be altered by both sensory and cognitive factors.

Although loudness is a one-dimensional concept in theory and research, it is a multidimensional concept as it is used in daily life. When sound is described as "loud" in daily life, it is the remembrance of the loudness that is being judged, not the same judgment of loudness at the time the sound was heard. This is important because research indicates that our memory of loudness can be altered over time (Ward 1987). There is some evidence that one or more very loud sounds in a soundscape may take precedence in memory over other sounds. For example, Kuwano and Namba (1985) compared two loudness judgment tasks: overall loudness ratings and instantaneous loudness ratings. In overall loudness ratings, listeners were asked to listen to a reasonably long-duration soundscape (e.g., 10 min) and judge the loudness of it at the end of the sound. In instantaneous loudness ratings, listeners were presented the same sound and were asked to track their perceived loudness continuously by varying line length (at various time intervals) while the sound was being heard. Results showed that the overall loudness judgments were larger than the average of instantaneous judgments for the same sounds. This is consistent with the contention that loud sounds in a soundscape may take precedence in memory and alter judgments of loudness over time (see Chap. 6).

In addition to the perceptual phenomenon already reviewed, another fascinating empirical phenomenon is how loudness depends on the manner in which sounds are presented and the listening environment. In general, loudness changes with distance from the sound source, but not always. Loudness can remain constant in the presence of substantial changes in the physical stimulus caused by varying sound distance. For example, the loudness of conversational speech can remain constant even when the distance between the talker and listener changes (Zahorik and Wightman 2001). This phenomenon is referred to as loudness constancy. Loudness also depends on whether sounds are presented monaurally or binaurally and whether sounds are presented via earphones or loudspeakers. Binaural loudness summation refers to the finding that a sound presented binaurally is louder than the same sound presented monaurally. Recent research indicates that the amount of binaural loudness summation is less for speech from a visually present talker than for recorded speech or tones, and the amount of binaural loudness summation is less when sounds are presented in a room than when sounds are presented via earphones (Epstein and Florentine 2009). This lack of binaural loudness advantage in rooms is called binaural loudness constancy, because of its relation to loudness constancy. (For a review of binaural loudness, see Chap. 7.)

It is well known that overexposure to high-level sounds can cause hearing loss. Loudness, as well as threshold, may change after a person is exposed to high-level sounds. Loudness changes can be short term, taking anywhere from a few minutes to several days to return to normal. This condition is called fatigue, or temporary loudness shift (see Chap. 4). With exposure to high level sounds over a prolonged time, or even a very short time with sufficient intensity, changes in loudness can be permanent. Loudness also changes with hearing loss in different ways depending on the type of hearing loss (see Chap. 9).

Not everyone experiences loudness in the same way, as is clearly evident in people with hearing losses. The same physical stimulus can elicit different perceptions of loudness depending on the etiology of the hearing loss and other factors (see Chap. 9). It is noteworthy that people with hearing loss may not use the terms describing loudness in the same manner as people with normal hearing. For example, an individual with a hearing loss may use "very soft" to describe the softest audible sound, whereas a person with normal hearing may describe the same sound as "moderately soft." Further, whispered speech that is amplified may still be identified as "soft" even though the presentation level is high and the sound is perceived as "loud." Because loudness is a subjective experience, it is impossible to know exactly how a person is using descriptive terms.

Although most people with normal hearing have a clear concept of the percept of loudness, terminology used to describe loudness is often incorrect and misleading. For example, many people use the term "volume" to describe the loudness of a sound when they should simply use the term "loudness." The term "volume" is incorrect because it is used to describe a percept that is different from loudness; volume refers to the subjective size of a sound, not its perceptual strength. In fact, we know that loudness is separate from volume by employing the principle of independent invariance; you can hold the percept of loudness constant and vary the volume of a sound (Stevens 1934). Therefore, "volume" should not be used to connote loudness; they are different subjective attributes.

Another common error is use of the term "intensity" to connote loudness. Loudness is a subjective attribute of sound, whereas intensity is a physical attribute of sound. The term "intensity" refers to a physical property of a sound, which is related to its level. Sound can be measured in units of intensity or pressure. Level is usually measured in units of SPL in decibels, or dB SPL, which is a logarithmic ratio of the sound pressure relative to a standard reference sound pressure.

Use of correct terminology is not trivial because it limits our ability to communicate concepts that lead to a better understanding of loudness. Understanding what people mean when they use a term related to a physical continuum is not difficult, because it can be quantified and compared with physical evidence. It is much more difficult to understand what people mean when they use a term related to a psychological continuum, because there is no way to directly measure a person's perception. All measurements of loudness are indirect and scientists are required to use converging lines of evidence to indirectly measure loudness. Although indirectly measuring loudness is challenging, it is possible. Marks and Florentine in Chap. 2 describe methods, problems, and pitfalls of measuring loudness.

1.3 Loudness, Language, and Culture

There are problems with using the correct terms associated with the concept of loudness in different languages. Because loudness research has been performed in different languages in laboratories around the world, it is especially important to understand exactly what is meant by loudness-related terms in those languages. Some languages have different terms for loudness that are used by people in daily language and other terms used by scientists. For example, the term "volume" is commonly used interchangeably with "loudness" in daily conversations in American English.

To determine what a person means when he or she uses a word related to loudness, translations by native speakers of the language are needed. One group of native speakers translates words from one language to another, and another group of native speakers translates back to the original language. The purpose of this cross-translation is to ensure accuracy of the translation. Even if the translation is as close as possible, it may not have the exact same meaning and/or may have other meanings associated with it. For example, the English word "soft" can be used to characterize an acoustic event, as in a soft sound. Soft can be translated into German as "leise" to also mean a soft sound. Other meanings that are associated with "soft" and "leise" are different. In German, "ein leiser Mensch" connotes an introverted and reserved person, whereas in English "soft" has an informal meaning for older people of foolish or silly, as in "He is soft in the head." and an informal meaning for young people of not being strong enough, as in "Don't be so soft; you are a pushover." Language usage changes over time and terms related to loudness are unlikely to be an exception to this rule. Schick and Höge (1996) point out the problem of the equivalence of words in different languages and they proposed an investigation of the amount of overlap in meaning and the development of a Meaning-Overlap-Atlas. It is unfortunate that this has not been realized; it would provide a valuable resource.

The connotative meaning of a percept, such as loudness, can be obtained using the method of semantic differential (Osgood et al. 1957) in which participants rate their impressions on adjective scales to obtain information about the meaning of a percept. Florentine et al. (1986) used this method to gain insight into the meaning of four words: loudness, noise, noisiness, and annoyance. Data from participants in three different countries were compared. In general, participants from England responded in a similar manner to those in the United States, but different from the Japanese participants. Whereas the responses of Japanese participants indicated that the word for loudness was a rather neutral concept, the participants in England

and the United States somewhat negatively polarized it. The opposite was true of the precept of noise; it was a somewhat neutral concept as indicated by the participants in England and the United States, but negatively polarized by the participants in Japan. For noisiness and annoyance, participants in all three countries negatively polarized the percepts. The results of this investigation suggest that the connotative meaning of the word "loudness" may be different between English and Japanese, although the meaning of the word in the dictionary seems to be the same. The same is true of the word "noise." A subsequent study (Kuwano et al. 1991) indicated that the word "loud" has neutral connotations in China, Japan, and Sweden, but negative connotations in Germany and the United States.

An international understanding of loudness is required to address issues related to loudness around the globe. Judgments of what is too loud depend on the cultural background of the listener. What is an acceptably loud sound in one culture may be unwanted noise in another (Namba et al. 1986, 1991). Merchants in street markets loudly announcing what they are selling are accepted and are considered part of the life of the city in some parts of the world. In other parts of the world, the same sound would be considered too loud and annoying. Acceptable levels of loudness are likely to depend on the culture and the meaning of the sound.

The International Organization for Standardization initiated a comprehensive international collaboration to understand the loudness of impulsive noise in the 1970s. A study group was formed and members of The Acoustics Laboratory, Technical University of Denmark, including O. Juhl Pedersen, Poul Erik Lyregaard, and Torben Poulsen, coordinated work and analyzed the data. A total of 22 laboratories from 12 countries around the world agreed to participate in the determination of the loudness level of a number of impulsive noises. The loudness level was determined by means of test subjects who evaluated the loudness of carefully calibrated noise and reference signals. About 500 test subjects participated. The project was dubbed "The Round Robin Test on Evaluation of Loudness Level of Impulsive Noise" with reference to Robin Hood and his gang, who encircled their signatures to indicate solidarity when petitioning the Sheriff of Nottingham (Petersen et al. 1977). Stimuli were made at the Acoustics Laboratory in Denmark and consisted of 21 sounds: nine impulsive noise signals (1 s), five single impulses, six tone pulses, and a 1-kHz pure tone for calibration purposes. The reference signal was a 1/3-octave noise band centered on a 1-kHz tone for the nine noises and a 1-kHz tone pulse for the other signals. The stimuli were recorded on audiotape and a set of tapes was sent to each laboratory with general instructions. The instructions did not specify a specific transducer type (headphones or loudspeakers) or measuring methods, although suggestions were made. It was believed that the psychophysical method for measuring loudness level (not loudness) was of minor importance for the results. Therefore, the participants were instructed to simply report the method used.

When the measurements were returned to the Acoustics Laboratory in Denmark, the investigators were amazed at the "variability" in the data. The stimuli – which were the same in all laboratories – were judged differently in different laboratories. As in many important investigations, the Round Robin experiment raised many more questions than it answered. It was an important turning point in the knowledge of loudness; it made scientists question what they thought they knew. In hindsight, it is clear that a number of factors could have significantly influenced their measurements of loudness, including transducer type (Chap. 7) and measuring methods (Chap. 2).

In the 1980s, another international study group was formed to study cross-cultural factors in the subjective impressions of environmental sounds, as well as social factors related to community noise problems. Scientists from Osaka University in Japan – Seiichiro Namba and Sonoko Kuwano – played a major role in initiating and coordinating the work. Participants included scientists from China, England, Germany, Japan, Korea, Sweden, Turkey, and the United States. Although the overall purpose of the series of studies was to examine the overall impressions of environmental sounds, important insights were gained regarding loudness. Results of this collaboration revealed that there are some differences in the connotative meaning of terms related to loudness among the different languages, although the dictionary definitions appear to be the same (Kuwano et al. 1991). Results also revealed that music creates a unique response in listeners from very different cultures in a similar manner; music can be loud, but not annoying unlike many other sounds (Kuwano et al. 1992; for details and data, see Chap. 8). International collaborations have been fruitful in increasing our understanding of loudness and the social and cultural issues related to community noise problems (Namba et al. 1991).

1.4 Current State of Knowledge Regarding Loudness

Much of what is known about loudness is summarized in the chapters that follow in this book. Our understanding of loudness is still unfolding and there is no comprehensive theory that explains all phenomena related to the perception of loudness. A general overview of the current state of knowledge can be found in loudness models. Loudness models can be divided into two types: models that describe and predict the relationship between the stimulus and the perception of loudness (i.e., psychoacoustical models) and models that attempt to make correlations between changes in the level of a stimulus and the physiological response to these changes (i.e., physiological models).

Psychoacoustical models have been used successfully to take into account many phenomena related to loudness. Although they do not account for all aspects of loudness, they have been effective in leading to a better understanding of loudness. An introduction to psychoacoustical models can be found in Marozeau (Chap. 10 and the references therein; Appell et al. 2001; Fastl and Zwicker 2007).

Despite scientific progress in the general understanding of the physiology of hearing (Pickles 2008), current understanding of the physiology of loudness does not warrant a separate chapter on physiological models of loudness. It is not surprising that physiological models of loudness are much less developed than psychoacoustic models, given the limited amount of data in the area of the physiology of loudness. Physiological data have been related to responses that are correlates of loudness, but not to loudness, per se (for correlates of loudness, see Chap. 4).

In addition, loudness is often discussed in conjunction with the topics of level detection (a.k.a. absolute threshold) and level discrimination (a.k.a. intensity discrimination). This is a debatable practice because the subjective attributes of changes in level may not be perceived as differences in loudness; they may be perceived as changes in pitch or other subjective attributes of sound – especially in individuals with hearing losses (for some discussion, see Buus et al. 1997; Oxenham and Buus 2000).

Substantial physiological data have been obtained on aspects of the neural coding of sound intensity (Plack and Carlyon 1995; Plack 2005; Pickles 2008). Neural coding measurements have been correlated with psychoacoustical measurements of level discrimination. In fact, psychoacoustical modeling of level discrimination across frequency (i.e., Florentine and Buus 1981) was used to develop and test the first quantitative model of auditory perception in a nonhuman species (the starling), tying together a wide variety of physiological and behavioral data for that species (Buus et al. 1995). The integration of information across frequency bands has been used by other authors in the development of physiologically based models of perception, but not applied directly to loudness. The psychoacoustical models of loudness indicate that although major contributions to the loudness of tones stem from excitation in auditory channels tuned to the tone's frequency (Moore et al. 1985), contributions from the other channels are also apparent (Florentine et al. 1997). Thus, it appears that loudness can be formed as a sum of activity from frequency-selective auditory channels and physiological models will need to take this into account.

Data from noise-exposed cats (May et al. 2009) appear roughly consistent with psychoacoustical data from humans with noise-induced hearing losses. For example, the bandwidth of vowels appears qualitatively consistent with loudness summation data from a group of humans with noise-induced hearing losses (Florentine and Zwicker 1979; Florentine et al. 1980). Unfortunately, much of the other data in the literature obtained from human listeners with sensorineural hearing losses of primarily cochlear origin are averaged and not separated by etiology, nor are individual data routinely reported. It is now sufficiently clear that there are substantial individual differences in how loudness grows with increasing level in people with sensorineural hearing losses. These individual differences are highly likely to reveal important mechanisms that contribute to loudness. For example, Marozeau and Florentine (2007) reviewed data from five experiments using different methods to obtain individual loudness functions of hearing-impaired listeners. Results suggest that: (1) when the level of a sound is increased there are considerable individual differences in loudness growth among hearing-impaired listeners and (2) averaging the results across hearing-impaired listeners will mask these differences.

Physiological studies of loudness in animals have been constrained by a lack of psychoacoustic measures of loudness. Common methods used to study loudness in humans, such as equal loudness matching and magnitude estimation, are not applicable for animal studies. Some studies have used a reaction–time paradigm – the louder the sound the faster the reaction time – that correlates with loudness (for review of reaction–time measures in humans, see Wagner et al. 2004 and Chap. 4).

The relationship between sound level and reaction time has been measured for nonhuman primates (Stebbins and Miller 1964; Pfingst et al. 1975) and the domestic cat (May et al. 2009). Attempts have been made to relate equal loudness contours from humans to equal latency contours from reaction times in both species (Stebbins 1966; Pfingst et al. 1975; May et al. 2009). Results show similarities between human and animal data, but also differences such as a compressed range of latencies at the highest frequencies. Pitch and annoyance-type subjective cues are potential confounds and it is not currently feasible to know the subjective experience of nonhumans. Further, attempts have been made to study the influence of noise-induced hearing loss on loudness. Reaction time latencies have been measured in sound-exposed monkeys and cats and compared with the data from humans (e.g., see Pfingst et al. 1975; May et al. 2009).

Only some aspects of the physiology of loudness appear to be explained. For example, the increase in loudness with increasing level is consistent with the basilar-membrane response function; there is a good correlation between the loudness-growth function and physiological data (see Chap. 4). How the loudness of a sound increases with level is not well understood at the auditory nerve, although attempts have been made to relate the psychoacoustical phenomena to knowledge of the auditory-nerve response (see e.g., Goldstein 1974; Pickles 1983; Relkin and Doucet 1997; Heinz and Young 2004; Heinz et al. 2005). Some features of psychoacoustical data are readily apparent in the auditory nerve data, but others are not. In particular, it appears to be an inescapable conclusion that any frequency-selective channel carries information about the stimulus level over dynamic range of about 120 dB, but how this information is used is unclear. Because most neurons tend to saturate within a dynamic range of only 30–60 dB, the encoding of stimulus level within a channel is not straightforward and requires careful consideration of the available evidence. For example, as the level of a tone increases, the firing rate of neurons in the auditory nerve also increases, but at some point increases in level cause no further increase in the firing rate. Although some benefit is obtained from a small number of auditory neurons with higher thresholds, this does not appear to be enough to account for the fact that loudness increases over a level range of about 120 dB. This is known as the dynamic range problem (see Chap. 4 and Delgutte 1996 for review of early physiological data correlated with sound level).

Two hypotheses to explain the dynamic range problem have received much attention. They are not mutually exclusive. One states that loudness is related to the total amount of neural activity. As a tone increases in level, it excites neurons with primary sensitivity at the characteristic frequency and also excites an increasing number of neurons with adjacent characteristic frequencies. This is known as the "spread of excitation" – a term taken from psychoacoustical modeling. It is unlikely, however, that a simple sum of the spike activity in the auditory nerve is a physiological correlate of loudness (Relkin and Doucet 1997). The other hypothesis is that loudness is related to temporal properties of neural activity. It is well known that neurons fire at precise times correlated with temporal properties of sound. In other words, they tend to phase lock to certain frequencies. As a tone increases in level, more neurons phase lock to it and the overall synchrony across the population of auditory nerve fibers increases. However, the ability of the auditory nerve fibers to phase lock decreases at high frequencies, which is inconsistent with this hypothesis.

Therefore, the connection between physiological responses and our perception of loudness remains unclear. Although qualitative data indicate possibilities, there have not been encouraging quantitative assessments. Much of what is known about the physiology of the perception of sound levels comes from correlating psychoacoustic measurements with the physiological responses to level differences. For example, it has been shown that information from a single neuron in the auditory nerve is enough to account for our ability to discriminate two sounds that differ in level. Just because information is available, however, does not mean that it is used by the auditory system. More research is needed to understand loudness encoding.

The past quarter-century has been especially fruitful in the area of loudness research. Four trends in psychophysics and their interconnections have led the way: (1) investigations between temporal and spectral integration of loudness and the loudness function itself, (2) investigations of individual differences in loudness functions among normal listeners and listeners with different types of hearing losses, (3) investigations of how the many aspects of context affect loudness, and (4) investigations of binaural loudness in and out of traditional laboratory settings. Some examples can be found in Florentine (2009). These trends were aided by technological developments that permitted large amounts of data to be modeled. Old theoretical frameworks have been challenged. Some concepts have been upheld; others have been reformulated. For example, it had been assumed that loudness at threshold was zero. This assumption influenced models of loudness for people with normal hearing and hearing losses. When measurements were actually made of loudness at threshold, the data showed a low, but positive value of loudness at threshold (Buus et al. 1998). A new standard (ANSI S3.4-2007) has been revised in light of these new data. The collapse of this assumption opened other assumptions to scrutiny. If loudness at threshold has a positive value, could it be that loudness at threshold is different for different listeners? Many studies have assumed that loudness at threshold is the same for all listeners whether they have normal hearing or hearing losses. In fact, there is considerable evidence that loudness at threshold may be different for different individuals. Could loudness at threshold be different at different frequencies in the same listener? If so, this could have implications for hearing loss rehabilitation. These new discoveries – together with old discoveries – are introduced in the ensuing chapters.

Although significant progress has been made in understanding loudness, there are areas that are primed for new discoveries. It is highly likely that over the next quarter-century (1) there will be an understanding of the physiological basis of loudness, (2) individual differences in loudness of listeners with normal hearing and hearing losses will be understood, resulting in better rehabilitation of people with hearing losses, (3) loudness context effects will be widely acknowledged – the gap between loudness in the laboratory and in daily environments will be better understood, and (4) new models will be developed that can predict individual differences in loudness among normal hearing-listeners and listeners with hearing losses, as well as predict the average perception of loudness for large groups of listeners in various daily environments. Prospects for the future of understanding loudness are quite hopeful as knowledge from different areas of study and psychoacoustics merge.

References

ANSI-S3.4 (2007) American National Standard Procedure for the Computation of Loudness of Steady Sounds. New York: American National Standards Institute.

Appell JE, Hohmann V, Kollmeier B (2001) Review of loudness models for normal and hearing-impaired listeners based on the model proposed by Zwicker. Z Audiol 40: 140–154.

Axelsson A, Eliasson A, Israelsson, B (1995) Hearing in pop/rock musicians: a follow-up study. Ear Hear 16:245–253.

Berglund B, Berglund U, Lindvall T (1976) Scaling loudness, noisiness, and annoyance of community noises. J Acoust Soc Am 60:1119–1125.

Buus S (2002) Psychophysical methods and other factors that affect the outcome of psychoacoustic measurements. In: Tranebjærg L, Christensen-Dalsgaard J, Andersen T, Poulsen T (eds), Genetics and the Function of the Auditory System: Proceedings of the 19th Danavox Symposium. Copenhagen, Denmark: Holmens Trykkeri, The Danavox Jubilee Foundation, ISBN 87–982422–9–6, pp. 183–225.

Buus S, Klump GM, Gleich O, Langemann U (1995) An excitation-pattern model for the starling (*Sturnus vulgaris*). J Acoust Soc Am 98:112–124.

Buus S, Florentine M, Poulsen T (1997) Temporal integration of loudness, loudness discrimination, and the form of the loudness function. J Acoust Soc Am 101:669–680.

Buus S, Müsch H, Florentine M (1998) On loudness at threshold. J Acoust Soc Am 104:399–410.

Delgutte B (1996) Physiological models for basic auditory percepts. In: Hawkins HL, McMullen TA, Popper AN, Fay RR (eds), Auditory Computation. New York: Springer, pp. 157–220.

Epstein M, Florentine M (2009) Binaural loudness summation for speech and tones presented via earphones and loudspeakers. Ear Hear 30:234–237.

Epstein M, Marozeau J (2010) Loudness and intensity coding. In: Plack, C (ed), OUPHAS Auditory Perception. Oxford, UK: Oxford University Press, pp. 45–69.

Fastl H, Zwicker E (2007) Psychoacoustics – Facts and Models, 3rd ed. Berlin/Heidelberg: Springer.

Florentine M (1990) Education as a tool to prevent noise-induced hearing loss. Hear Instrum 41:33–34.

Florentine M (2009) Advancements in psychophysics lead to a new understanding of loudness in normal hearing and hearing loss. In: Elliott MA, Antonijevic' S, Berthaud S, Mulcahy P, Martyn C, Bargery B, Schmidt H, Fechner Day 2009 Proceedings of the 25 Annual Meeting of the International Society for Psychophysics. Galway: Snap Printing, pp. 83–88.

Florentine M, Buus S (1981) An excitation-pattern model for intensity discrimination. J Acoust Soc Am 70:1646–1654.

Florentine M, Heinz MG (2009) Audition: Loudness. In: Goldstein EB, Encyclopedia of Perception, Sage Publications Ltd. London. EC1Y 1SP Vol. 1, Sage, pp. 145–151.

Florentine M, Zwicker E (1979) A model of loudness summation applied to noise-induced hearing loss. Hear Res 1:121–132.

Florentine M, Buus S, Bonding P (1978) Loudness of complex sounds as a function of the standard stimulus and the number of components. J Acoust Soc of Am 64:1036–1040.

Florentine M, Buus S, Scharf B, Zwicker E (1980) Frequency selectivity in normally-hearing and hearing-impaired observers. J Speech Hear Res 23:113–132.

Florentine M, Namba S, Kuwano S (1986) Concepts of loudness, noisiness, noise, and annoyance in the USA, Japan and England. Proc Inter-Noise 2:831–834

Florentine M, Buus S, Hellman R (1997) A model of loudness summation applied to high-frequency hearing loss. In: Jesteadt W (ed), Modeling Sensorineural Hearing Loss. Mahwah, NJ: Earlbaum, pp.187–197.

Gates GA, Schmid P, Kujawa SG, Nam B, D'Agnostino R (2000) Longitudinal threshold changes in older men with audiometric notches. Hear Res 141:220–228.

Goldstein JL (1974) Is the power law simply related to the driven spike response rate from the whole auditory nerve? In Moskowitz HR, Scharf B, Stevens SS (eds), Sensation and Measurement. Dordrecht: Reidel, pp. 223–229.

Heinz MG, Young ED (2004) Response growth with sound level in auditory-nerve fibers after noise-induced hearing loss. J Neurophysiol 91:784–795.

Heinz MG, Issa JB, Young ED (2005) Auditory-nerve rate responses are inconsistent with common hypotheses for the neural correlates of loudness recruitment. J Assoc Res Otolaryngol 6:91–105.

Hempton G, Grossmann J (2009) One Square Inch of Silence. New York: Free Press.

ISO 1999 (1990) Acoustics – Determination of occupational noise exposure and estimation of noise-induced hearing impairment. International Organization for Standardization, Geneva.

Kim SH, Frisina RD, Mapes FM, Hickman ED, Frisina DR (2006) Effect of age on binaural speech intelligibility in normal hearing adults. Speech Commun 48:591–597.

Kochkin S (2005) Marketrak VII: Hearing loss population tops 31 million people. Hear Rev 12:16–29.

Kujawa SG, Liberman MC (2006) Acceleration of age-related hearing loss by early noise exposure: evidence of a misspent youth. J Neurosci 26:2115–2123.

Kujawa SG, Liberman MC (2009) Adding insult to injury: cochlear nerve degeneration after "temporary" noise-induced hearing loss. J Neurosci 29:14077–14085.

Kuwano S, Namba S (1985) Continuous judgment of level-fluctuating sounds and the relationship between overall loudness and instantaneous loudness. Psychol Res 47:27–37.

Kuwano S, Namba S, Hashimoto T, Berglund B, Zheng D, Schick A, Höge H, Florentine M (1991) Emotional expression of noise: a cross-cultural study. J Sound Vib 151:421–428.

Kuwano S, Namba S, Florentine M, Zheng DR, Hashimoto T (1992) Factor analysis of the timbre of noise – comparison of the data obtained in three different laboratories Proc Acoust Soc Jpn N92–4–3:559–560.

Lecumberri MLG, Cooke M (2006) Effect of masker type on native and non-native consonant perception in noise. J Acoust Soc Am 119:2445–2454.

Marozeau J, Florentine M (2007) Loudness growth in individual listeners with hearing losses: A review. J Acoust Soc Am 122: EL81–87.

May BJ, Little N, Saylor S (2009) Loudness perception in the domestic cat: reaction time estimates of equal loudness contours and recruitment effects. J Assoc Res Otolaryngol 10:295–308.

Mayo LFH, Florentine M, Buus S (1997) Age of second-language acquisition and perception of speech in noise. J Speech Lang Hear Res 40:686–693.

Moore BCJ, Glasberg BR, Hess RF, Birchall JP (1985) Effects of flanking noise bands on the rate of growth of loudness of tones in normal and recruiting ears. J Acoust Soc Am 77:1505–1513.

Namba S, Kuwano S, Schick A (1986) The measurement of meaning of loudness, noisiness, and annoyance in different countries. In: Proc Int Cong Acoust, pp. C1–1.

Namba S, Kuwano S, Schick A, Aclar A, Florentine M, Zheng D (1991) A cross-cultural study on noise problems: a comparison of the results obtained in Japan, West Germany, the USA, China and Turkey. J Sound Vib 151:471–477.

Nelson PB, Soli SD, Seltz A (2002) Classroom Acoustics II: Acoustical Barriers to Learning. Melville, NY: Acoustical Society of America.

Osgood CE, Suci G, Tannenbaum P (1957) The Measurement of Meaning. Urbana, IL: University of Illinois Press.

Oxenham AJ, Buus S (2000) Level discrimination of sinusoids as a function of duration and level for fixed-level, roving level, and across-frequency conditions. J Acoust Soc Am 107:1605–1614.

Petersen OJ, Lyregaard PE, Poulsen T (1977) The Round Robin Test on Evaluation of Loudness Level of Impulsive Noise: Report no. 22. Technical University of Denmark: Acoustics Laboratory.

Pfingst BE, Hienz R, Kimm J, Miller J (1975) Reaction-time procedure for measurement of hearing. I. Suprathreshold functions. J Acoust Soc Am 57:421–430.

Pickles JO (1983) Auditory-nerve correlates of loudness summation with stimulus bandwidth in normal and pathological cochlea. Hear Res 12:239–250.

Pickles JO (2008) An Introduction to the Physiology of Hearing, 3rd ed. Bingley, UK: Emerald Group.

Plack CJ (2005) The Sense of Hearing. New York: Taylor and Francis.

Plack CJ, Carlyon RP (1995) Loudness perception and intensity coding. In: Moore BCJ (ed), Hearing. London: Academic Press, pp. 123–160.

Relkin EM, Doucet JR (1997) Is loudness simply proportional to the auditory nerve spike count? J Acoust Soc Am 101: 2735- 2740.

Royster JD, Royster LH, Killion MD (1991) Sound exposure and hearing thresholds of symphony orchestra musicians. J Acoust Soc Am 89:2792–2803.

Schafer RM (1977) The Tuning of the World. Toronto: Random House.

Scharf B (1964) Partial masking. Acustica 14:17–23.

Scharf B (1978) Loudness. In: Catrerette EC, Friedman MP (Eds.), Handbook of Perception: IV. Hearing. New York: Academic Press, pp. 187–242.

Scharf B (1997) Loudness. In: Crocker MJ, Encyclopedia of Acoustic: III. New York: Wiley, pp. 1481–1495.

Schick A, Höge H (1996) Cross-cultural psychoacoustics. In: Fastl H, Kuwano S, Schick A, Recent Trends in Hearing Research: Bibliotheks-und Informations-system der Universität Oldenburg. University of Oldenburg Press, pp. 287–314.

Stebbins WC (1966) Auditory reaction time and the derivation of equal loudness contours for the monkey. J Exp Anal Behav 9:135–142.

Stebbins WC, Miller JM (1964) Reaction time as a function of stimulus intensity for the monkey. J Exp Anal Behav 7:309–312.

Stevens SS (1934) The attributes of tones. Proc Natl Acad Sci USA 20:457–459.

Thompson E (2002) The Soundscape of Modernity. Cambridge, MA: MIT Press.

Van Engen KJ, Bradlow AR (2007) Sentence recognition in native- and foreign-language multi-talker background noise. J Acoust Soc Am 121:519–526.

Wagner E, Florentine M, Buus S, McCormack J (2004) Spectral loudness summation and simple reaction time. J Acoust Soc Am 116:1681–1686.

Ward LM (1987) Remembrance of sounds past: memory and psychophysical scaling. J Exp Psychol Human Percept Perf 13:216–227.

Zahorik P, Wightman FL (2001) Loudness constancy with varying sound source distance. Nat Neurosci 4:78–83.

Chapter 2
Measurement of Loudness, Part I: Methods, Problems, and Pitfalls

Lawrence E. Marks and Mary Florentine

2.1 Introduction

It is a matter of everyday experience that sounds vary in their perceived strength, from the barely perceptible whisper coming from across the room to the overwhelming roar of a jet engine coming from the end of an airport runway. Loudness is a salient feature of auditory experience, closely associated with measures of acoustical level (energy, power, or pressure) but not identical to any of them. It is a relatively straightforward matter for a person to note whether one sound is louder or softer than another, or to rank order a set of sounds with regard to their loudness. To measure loudness, however, in the typical sense of "measuring," requires more than just ranking the experiences from softest to loudest. It entails quantifying how much louder (e.g., determining whether the ratio or difference in the loudness of sounds A and B is greater or smaller than the ratio or difference in loudness of sounds C and D).

The quantitative measurement of loudness in this sense is important both to basic research and to its applications – important to scientists seeking to understand neural mechanisms and behavioral processes involved in hearing and to scientists, engineers, and architects concerned with the perception of noise in factories and other industrial settings, in the streets of urban centers, and in residences located along flight paths and near airports. As Laird et al. (1932) wrote more than three-quarters of a century ago, in an article describing one of the earliest attempts to quantify the perception of loudness,

> When a considerable amount of money is to be appropriated for making a work place quieter, for instance, the engineer can say that after acoustical material is added the noise level will be reduced by five or ten decibels. "But how much quieter will that make the office," is likely to be the inquiry. "A great deal" is not only an unsatisfactory but an unscientific answer. (p. 393)

L.E. Marks (✉)
John B. Pierce Laboratory, Department of Epidemiology and Public Health, Yale University School of Medicine, and Department of Psychology, Yale University New Haven, CT 06519, USA
e-mail: marks@jbpierce.org

M. Florentine et al. (eds.), *Loudness*, Springer Handbook of Auditory Research 37,
DOI 10.1007/978-1-4419-6712-1_2,

What is called for both scientifically and practically is a quantitative assessment of the change in loudness, such as knowing that reducing the physical level of environmental noise by a specified amount will reduce its perceived strength, its loudness, by 50%.

The present chapter focuses on methods for the measurement of loudness. Luce and Krumhansl (1988) pointed out that the psychophysical analysis of sensory measurement may operate at any one of three distinct levels. One level is mathematical, and it deals with the development of appropriate axioms for the numerical representations entailed by scales of sensory measurement. The second level is theoretical, and it deals with the structure of relations among scales of measurement. The third and last level is empirical, and it deals with the sensory relations expressed through the measurements. This third level treats sensory/perceptual measurement from a functional and pragmatic perspective, and it lies at the heart of the present chapter. From this perspective, the measurement of loudness is useful and valuable to the extent that it sheds light on basic mechanisms of hearing or makes it possible to predict responses to sounds in real-world settings.

Research over the past century and a half has developed and refined several approaches to measure loudness. This chapter summarizes the main approaches, evaluating the principles that underlie the application of each method and assessing the theoretical and practical problems that each approach faces – in essence, identifying the strengths and weaknesses of each approach. The chapter does not attempt to review the long-standing, often philosophically oriented, debates as to whether and how perceptual experiences may be quantified, but operates on the pragmatic assumption that quantification is not only possible but also scientifically meaningful and important; readers interested in the debates over quantification are directed elsewhere (see Savage 1970; Laming 1997; Marks and Algom 1998). The chapter starts with a brief history of loudness measurement in the nineteenth and twentieth centuries. Understanding this history is important because many of the concepts developed in the twentieth century resound in current scientific literature. Errors made in the interpretation of loudness data in the twenty-first century may arise from ignorance of these basic concepts regarding methods of measuring loudness. After the historical review, the reader is introduced to the theoretical, empirical, and practical constraints on loudness measurement.

2.2 A Brief History of Loudness Measurement

The history of loudness measurement is divided into two parts. The first part covers nineteenth century work by Fechner, Delboeuf, and others that raised the psychophysical problem of measuring loudness. The second part covers early twentieth century attempts to measure loudness by Piéron, Richardson and Ross, and Stevens.

2.2.1 Measurement of Loudness: Recognizing the Psychophysical Problem

The first steps toward measuring the loudness of sounds, and the magnitudes of other perceptual events, came in the second half of the nineteenth century with increased awareness on the part of sensory physiologists, psychologists, and physicists of what might be called "the psychophysical problem of intensity" – that the perceived magnitude of a perceptual experience need not be quantitatively proportional to the magnitude of the physical stimulus that evokes the experience. The experience of loudness is distinct from the physical measure of the stimulus. Nevertheless, it was sometimes assumed that that physical magnitude and perceived strength were commensurate, such that loudness was directly proportional to the physical magnitude of a sound. For example, Johann Krüger (1743) derived a simple rule of proportionality between the intensity of sensations and the intensity of the physical stimuli that produce the sensations. A century later, in his *Elements of Psychophysics*, Gustav Fechner recognized that direct proportionality flies in the face of direct experience. "I found it very interesting to hear the statement," wrote Fechner (1860/1966), "... that a choir of 400 male voices did not cause a significantly stronger impression than one of 200" (p. 152). The average (root-mean-square) acoustic power associated with a choir of 400 voices should be, in principle, about twice that associated with a comparable one of 200 voices. Yet the difference in the experience of loudness is not nearly so great as two-to-one.

To be sure, Fechner was not the first to make or recognize a distinction between perceptual experiences and the corresponding properties of stimulus events responsible for producing those experiences; the distinction goes back more than two millennia, at least as far as Democritus's famous dictum in the fifth century BCE, which states, "Sweet exists by convention, bitter by convention, color by convention; atoms and Void (alone) exist in reality.... We know nothing accurately in reality, but (only) as it changes according to the bodily condition, and the constitution of those things that flow upon (the body) and impinge upon it" (Freeman 1948, p. 110). Two millennia later, Locke (1690) noted that what he called secondary physical qualities of objects are not the same as our perceptions of them. Locke's distinction underlies the philosophical problem of sensory qualia – a topic that falls outside the scope of the present chapter (for a scientifically informed philosophical account, see Clark 1993).

Fechner was among the first, however, to recognize that there may be quantitative as well as qualitative differences between stimuli and sensations. In particular, Fechner pointed to quantitative differences between changes in the physical intensity of a stimulus and corresponding quantitative changes in the perceptual experience of it. He addressed the question of how perceived strength depends on physical intensity, stating that the intensity of sensation is proportional to the logarithm of physical intensity, when physical intensity is reckoned in units equal to the absolute threshold. This was his famous psychophysical law.

The first inklings of Fechner's logarithmic law came to him from philosophical and, later, mathematical intuition. He saw how he could derive the law, and hence derive measures of perceived magnitude, including loudness, from measures of the ability to discriminate two sounds. In fact, Fechner described how one could both derive the logarithmic law formally, from Weber's law of intensity discrimination, with the help of subsidiary theoretical and mathematical assumptions, and reveal the law empirically, by what is essentially a graphical procedure for summating discrimination thresholds [just-noticeable-differences (JNDs) in stimulation].

Fechner's proposal to construct scales of sensation from measures of discrimination rested in part on his view that sensation magnitudes could not be assessed accurately, in numerical fashion, by direct introspection, at least not in a scientifically meaningful way (although contemporaries of Fechner did take small steps in this direction, e.g., Merkel 1888). Fechner did, however, consider the possibility that intervals or differences in sensation magnitude might be compared directly, and investigators in the late nineteenth and early twentieth centuries began to develop and test several methods for producing sensory scales with equal-appearing intervals (e.g., Delboeuf 1873). One of these methods came to be called the method of bisection. In the method of bisection, a subject adjusts the level of a stimulus to appear midway between fixed upper and lower stimulus levels. Without modern technology, however, it was difficult to create an experiment in which subjects could adjust the physical levels of sounds in a controlled, continuous fashion, and it was especially difficult to measure the resulting sound levels even if one could vary them. The development of vacuum-tube technology in the early decades of the twentieth century provided the needed impetus.

As elegant as it is, Fechner's approach to sensory measurement in general and to the measurement of loudness in particular has not proven especially useful. The approach is exceedingly laborious to apply – a criticism that also applies to the approach of Thurstone (1927), which requires many pairwise comparisons of relative intensity of all possible pairs of stimuli. Thurstonian measurement is not reviewed here, but the interested reader is directed to other summaries (Marks and Algom 1998; Marks and Gescheider 2002), as well as to evaluations of Thurstone's conceptualizations in the development of sensory measurement (Luce 1994). Even more importantly, Fechner's approach often produces results that fail tests of internal consistency. If the number of JNDs above absolute threshold can serve as a fixed unit of loudness, as discussed in the next section, then all pairs of sounds that lie equal numbers of JNDs above threshold should be equally loud. Considerable evidence contradicts this principle. Nevertheless, because modern versions of Fechner's approach still have proponents (e.g., Falmagne 1985; Link 1992; Dzhafarov and Colonius 2005), a review and analysis are appropriate.

2.2.1.1 Fechner's Law and Fechnerian Measurement

Fechner (1860) reported that on the morning of 22 October 1850, he first conjectured that a logarithmic function might relate the magnitude of sensation to physical intensity. This conjecture actually preceded his discovery of empirical evidence supporting it. Having come to the putative insight that sensation increases as a logarithmic function of stimulus intensity, Fechner then came upon Weber's work on sensory discrimination, and this discovery led Fechner to develop both a mathematical derivation for the logarithmic law and a more general, empirical method for generating quantitative psychophysical functions, logarithmic or otherwise. Although Fechner's law was eventually replaced, as described later, the proposal of the law itself marked a watershed moment in the history of psychophysics, which the International Society of Psychophysics celebrates every year at its annual meeting.

To derive the logarithmic law mathematically, Fechner relied first on the generalization that has come to be called Weber's law, and second on two auxiliary mathematical assumptions. Extensive experimentation by both Weber and Fechner on intensity discrimination focused on measures of the JND, that is, the smallest difference between stimulus intensities that a person is able to distinguish. (The JND is also known as the difference limen [DL].) Fechner showed how JNDs could provide the building blocks for scales of sensation magnitude. Much of the data reported by Weber and Fechner conforms at least loosely to Weber's law, which states that if I is the baseline intensity from which a change in the stimulus is made, then the minimal change in I that is perceptible, ΔI (the JND), is proportional to I. That is,

$$\Delta I = k_1 I \tag{2.1}$$

An assumption critical to both Fechner's mathematical approach and his experimental approach is the subjective equality of JNDs – the assumption that all JNDs have the same psychological magnitude. If L is sensation magnitude such as loudness, then, for every JND, ΔL is constant, that is,

$$\Delta L = k_2 \tag{2.2}$$

A second assumption, critical to the mathematical derivation, though not to the empirical approach, is that one can convert the difference equations (2.1) and (2.2) into differential equations. Converting (2.1) and (2.2) and rearranging the terms leads to:

$$dI / (k_1 I) = 1 \tag{2.3a}$$

$$dL / k_2 = 1 \tag{2.3b}$$

Combining (2.3a) and (2.3b) and integrating in turn leads to Fechner's law:

$$L = k \log I + k_3 \tag{2.4}$$

where $k = k_2/k_1$.

Fechner further assumed that sensation magnitude takes on positive values only when the intensity of the stimulus, I, exceeds the absolute threshold. Consequently, by measuring I in terms of the absolute threshold, I_0, $k_3 = 0$, and one can write:

$$L = k \log (I / I_0) \tag{2.5}$$

Measured in this way, loudness, L, would have the properties of a ratio scale (Stevens 1946): A sound having a loudness, L, of 10 units (ten JNDs above threshold) would be twice as loud as a sound having a loudness, L, of 5 units (five JNDs above threshold). Without fixing the starting-point of the sensation scale, one would only be able to compare differences or intervals along the scale, but not ratios.

Fechner's model is elegant, but as Luce and Edwards (1958) showed, his general approach provides mathematically consistent results only when intensity discrimination (level discrimination) follows a limited number of formulas, such as Weber's law ($\Delta I = k_1 I$) and its linearization ($\Delta I = k_1 I$ + constant). The approach fails mathematically, for example, when auditory intensity discrimination follows what has been called a "near miss" to Weber's law, as shown in results of many studies (e.g., McGill and Goldberg 1968; Jesteadt et al. 1977; Florentine et al. 1987; see Parker and Schneider 1980; Schneider and Parker 1987). The near miss may be written as

$$\Delta I = k I^{b} \tag{2.6}$$

where b is smaller than 1.0, often having a value around 0.8–0.9.

The empirical approach to Fechnerian measurement, however, avoids these complications because the approach may be used to generate a Fechnerian scale from any set of intensity-discrimination data, regardless of whether Weber's law holds. Taking the empirical approach, one would proceed as follows: first, define as L_0 the sensation magnitude (e.g., loudness) associated with baseline intensity I_0. Second, measure the JND, ΔI_1, from baseline I, and then define the sensation magnitude of intensity I_2 (= $I + \Delta I_1$) as $L_0 + 1$. Next, starting from intensity I_2, measure the subsequent JND, ΔI_2, and define the sensation magnitude of I_2 (= $I_0 + \Delta I_1 + \Delta I_2$) as $L_0 + 2$; and so forth. This approach essentially builds up a measurement scale, under the assumption that each additional step of stimulus intensity, calculated as a JND, adds another unit of sensation magnitude.

Most studies of intensity discrimination do not use Fechner's adaptive approach, but measure JNDs using a predetermined set of starting intensities. Nevertheless, given a fixed set of stimulus intensities, it is possible to derive a reasonable empirical approximation to a Fechnerian function by interpolating values along the empirical discrimination function and then summing the inferred JNDs.

Figure 2.1 shows an example – a Fechnerian loudness scale derived from intensity-discrimination data at sound frequencies of 200, 400, 600, 800, 1,000, 2,000, 4,000, and 8,000 Hz, as reported by Jesteadt et al. (1977). In their experiment, Jesteadt et al. measured intensity discrimination at each of the eight frequencies at intensity levels of 5, 10, 20, 40, and 80 dB above threshold (sensation level, SL), omitting 80-dB SL at 200 Hz. The entire ensemble of results could be described by a single equation consistent with the "near miss" to Weber's law given in (2.6):

$$\Delta I = 0.463\left(I / I_0\right)^{0.928} \tag{2.7}$$

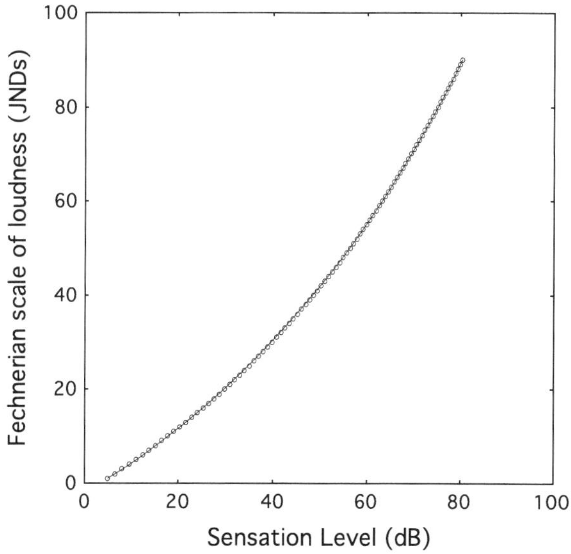

Fig. 2.1 A scale for loudness constructed from measures of just-noticeable differences in sound intensity at eight sound frequencies over the range 200–8,000 Hz (based on data and analysis of Jesteadt et al. 1977)

where I_0 is the reference for SL at each frequency. The Fechnerian function shown in Fig. 2.1 was constructed empirically, by summing JNDs as calculated from (2.7). If the discrimination data were consistent with Weber's law instead of its near miss, then the Fechnerian function derived by summing JNDs would follow a straight line. Instead, because of the "near miss," the derived function curves upward when plotted against SL in decibels. These derived data can be fitted, as shown, by a power function with an exponent of 0.129 (re: sound pressure; 0.0645 re: sound energy).

2.2.1.2 Fechnerian Loudness and the Principle of Equality

In Fechner's terms, the function shown in Fig. 2.1 would characterize the relation between loudness and sound intensity, applicable over a wide range of sound frequencies. Because the function is based on (2.7), which applies to frequencies from 200 to 4,000 Hz (see Florentine et al. 1987), Fechnerian loudness would vary directly with the ratio I/I_0 at all frequencies, which means that loudness would vary directly with sensation level (SL, i.e., the number of decibels above threshold), given that SL equals 10 log(I/I_0). This is to say, that if the Fechnerian function shown in Fig. 2.1 represents loudness, then, according to the principle of equality, all sounds at a given SL (at least between 200 and 4,000 Hz) should be equally loud.

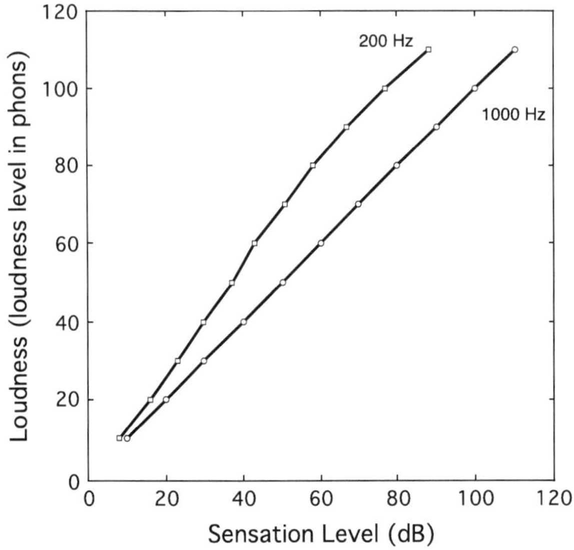

Fig. 2.2 The level in decibels above threshold of a 1,000-Hz tone (ordinate) that sounds as loud as test tone of 200 and 1,000 Hz at various levels above their threshold (abscissa). The data points are taken from curves appearing in Fig. 3 of Fletcher and Munson (1933). These failure of the points at 200 and 1,000 Hz to overlap contradicts the conjecture that decibels above threshold can serve as a uniform scale to quantify loudness

This inference is incorrect, as shown by equal-loudness relations determined across the sound spectrum. It has long been known that loudness depends on acoustic frequency as well as sound level (Fletcher and Munson 1933; Robinson and Dadson 1956; see Chap. 5, or ISO standard 226 2003). Figure 2.2 uses a subset of Fletcher and Munson's data to show how SL, or decibels above threshold, fails to meet the criterion of internal consistency, hence fails to provide an adequate measure of loudness. For a decibel to serve as a universal unit of loudness, the principle of equality requires that all acoustic signals 20 dB above threshold appear equally loud, all signals 30 dB above threshold appear equally loud, and so forth.

Figure 2.2 shows one reason why this prediction fails. A tone having a frequency of 1,000 Hz and a level that is 60 dB above its threshold would be assigned a loudness that equals, by definition, 60 (loudness = decibel) units. But, as determined by equal loudness matching, a tone having a frequency of 200 Hz that lays 60 dB above its threshold appears much louder, equal to 80 loudness units. In general, increasing sound intensity by a fixed number of decibels above threshold produces greater increments in loudness at 200 Hz than at 1,000 Hz. Thus, loudness matches obtained across the spectrum make it possible to eliminate one possible method for measuring loudness – in terms of the number of decibels above threshold.

Considered across sound frequency (Newman 1933; Ozimek and Zwislocki 1996), across masking conditions (Hellman et al. 1987; Johnson et al. 1993), and across normal hearing and hearing loss (Zwislocki and Jordan 1986; Stillman et al. 1993), JNDs fail to provide a constant unit of loudness. This failure was recognized early by Riesz (1933), who proposed a possible solution to the failure of JNDs to provide a constant unit of loudness across sound frequency. Riesz suggested that, at every sound frequency, one may ascertain the range of loudness from bottom to top and then determine the number of JNDs in this range. Once this is done, loudness at each frequency, according to Riesz, would depend directly on the fraction of the total number of JNDs to that point. This has been called a proportional-JND hypothesis, a view later considered by Lim et al. (1977). The status of this modified Fechnerian hypothesis, however, remains uncertain (see Houtsma et al. 1980). It has not been rigorously determined, for example, whether or to what extent the proportional-JND hypothesis could account for loudness of tones heard in quiet and in masking noise, or tones heard by listeners with normal hearing and with hearing losses characterized by abnormally rapid or slow loudness growth of loudness with increasing level (Florentine et al. 1979). In any case, while of theoretical interest, the approaches using Fechnerian and Thurstonian methods are impractical.

2.2.2 *Early Attempts to Measure Loudness in the Twentieth Century*

Four approaches to measuring loudness in the early twentieth century are noteworthy. These approaches, described in the following sections, are: (1) measurement through decibels, (2) measurement through reaction times, (3) measurement through additivity, and (4) measurement through judgments of ratios or magnitudes.

2.2.2.1 Fechner's Law and the Use of Decibels to Measure Loudness

Fechner's approach, and in particular his logarithmic law, helped propel the study of loudness measurement in the early decades of the twentieth century – especially with the widespread use of the decibel notation for representing relative values of sound intensity or sound pressure. The decibel (dB) scale is a logarithmic transformation of stimulus power or pressure, as is Fechner's scale of sensations. By implication, if Fechner were correct, then the decibel scale might serve as a scale or measure of loudness. As Fletcher and Munson (1933) noted, "In a paper during 1921 one of us suggested using the number of decibels above threshold as a measure of loudness…." (p. 82). Indeed, with zero decibels (0 dB) set at the absolute threshold, a decibel scale of loudness should have numerical ratio properties: A sound 80 dB above threshold would have twice the loudness of a sound 40 dB above threshold.

All of this seemed reasonable enough at first, except that direct experience contradicted the inference. As Churcher (1935) wrote, "… the experience of the author and his colleagues over many years is that the numbers assigned by the decibel scale to represent sensation magnitudes are not acceptable to introspection as indicating their relative magnitudes.... The loudness of the noise of a motor assessed at 80 dB above threshold ... is, to introspection, enormously greater than twice that of a motor assessed at 40 dB" (p. 217). Whereas a choir of 400 voices appears only slightly louder than a choir of 200, a motor producing 100,000,000 (threshold) units of acoustical power (80 dB above threshold) sounds far more than twice as loud as motor producing 10,000 units (40 dB above threshold). Of course, the preceding analysis is predicated on the assumption, among others, that loudness is zero at absolute threshold. For several reasons, this is highly unlikely. Evidence indicates that threshold-level sounds have small positive values of loudness (see Buus et al. 1998). Even so, Churcher's point remains valid: Decibels serve poorly as direct indicators of loudness.

Psychoacoustic research in the early decades of the twentieth century, and especially from 1930 onward, sought to quantify loudness in ways that would be commensurate with direct experience and that also would satisfy basic scientific principles of measurement. Three subsequent approaches were important, each of which sought in its own unique way to develop a score of loudness: (1) using speed of response as a surrogate measure for loudness, (2) building a scale on the basis of additivity, and (3) building a scale from overt judgments of loudness ratios. A fourth approach, estimating perceived magnitudes, originated during this same period and became important only in the second half of the last century. Each approach is described in the following sections.

2.2.2.2 Measuring Loudness from Response Times: Piéron's Law

One measure of sensory performance is the speed of response to a stimulus. Beginning at least with the report of Cattell (1886), it has been clear that as the level of a stimulus increases, the response time decreases. Nearly a century ago, Piéron (1914) suggested that response speed, the inverse of response time, might serve as a surrogate measure of sensation intensity (see Piéron 1952, for a later summary; for recent reviews, see Wagner et al. 2004 and Chap. 4). Piéron reported the results of a systematic study of the way that response time varies with physical intensity in several modalities, including hearing. In each case, Piéron concluded that response time decreased as a power function of stimulus intensity, writing an equation of the form

$$RT - R_0 = aI^{-m} \tag{2.8}$$

where RT is the response time for the particular stimulus and modality, m is the exponent, and R_0 is the "irreducible minimum" RT, representing the asymptote of the function as I becomes very large. The parameter R_0 presumably represents the minimal time needed to prepare and execute the response. Subsequent research has confirmed that a power function of the form expressed in (2.8) provides a good description to measures of simple *RT*s to acoustic stimuli varying in level (e.g., McGill 1961; Kohfeld 1971; Luce and Green 1972; Kohfeld et al. 1981a, b). Luce and Green developed a mathematical model to show how loudness and *RT* could be related through a hypothesized dependence of both variables on mechanisms of neural timing. McGill (1961) pointed out, however, that the values of exponents fitted to functions for auditory *RT* generally differ markedly from the values of exponents derived from direct estimates of loudness, especially magnitude estimations, although the exponents derived from *RT* agree better with exponents estimated from measures of loudness derived from judgments of differences or intervals. Exponents derived from measures of *RT* generally have values around 0.3 when the stimulus is reckoned in terms of sound pressure, 0.15 when reckoned in terms of sound energy or power (see Marks 1974b, 1978).

As we asked about decibel measures, so too may we ask about *RT*: Do sounds that are equally loud produce the same response times? Often, this is approximately the case. But violations of the principle of equality have been reported, for instance, in the *RT*s given to tones heard in the quiet vs. backgrounds of masking noise (Chocholle and Greenbaum 1966) and in the *RT*s given to tones of different frequencies (Kohfeld et al. 1981a; Epstein and Florentine 2006b). In particular, Kohfeld et al. reported that equally loud, low intensity tones gave similar *RT*s, but not identical ones.

2.2.2.3 Measuring Loudness by Additivity: Fletcher and Munson's Loudness Scale

Fletcher and Munson (1933) offered a novel approach to the measurement of loudness, which served as a powerful conceptual alternative to Fechner's. Fletcher and Munson sought to create a scale for loudness that was both internally consistent and grounded in a principle of additivity. Internal consistency was ensured empirically by matching all sounds in loudness to a common yardstick, a tone at 1,000 Hz. Additivity was assumed, on the basis of the postulate that acoustic stimuli that activate separate populations of auditory receptors will produce component loudnesses that in turn would combine by simple linear summation. Fletcher and Munson identified two conditions for independent activation and, hence, for presumed linear addition of loudness: stimulation of the two ears vs. one (binaural vs. monaural stimulation) and stimulation of the same ear with acoustic stimuli containing two (or more) widely separated tones vs. a stimulus containing a single tone.

Fletcher and Munson's procedure for measuring loudness contained, therefore, two steps: One starts by matching the loudness of a 1,000-Hz tone to the loudness of every acoustic stimulus of interest – to individual tones or tone complexes, presented to one or both ears. For every possible test stimulus, therefore, one determines the SPL of a matching 1,000-Hz tone – that is, the loudness level in phons. Subsequently, one may construct a scale of loudness by comparing, for example, the level in phons of a given sound presented binaurally and monaurally. Given the assumption of additivity, the sound will be twice as loud when heard by two ears compared to one. Similarly, the loudness of two equally loud tones, spaced sufficiently in frequency, will be twice as loud when played together as either tone alone. If, for example, an acoustic signal has a loudness level of 70 phons when heard binaurally but 60 phons when hear monaurally, then the increase in SPL from 60 to 70 dB at 1,000 Hz constitutes a doubling of loudness.

Although Fletcher and Munson were able to perform a limited number of empirical tests of the adequacy of the principle of additivity, this critical principle remained largely an assumption of the system. Methods such as magnitude estimation, discussed below, can be used to ask, for example, whether subjects judge binaural sounds to be twice as loud as monaural sounds; the results can depend, however, on the ways that subjects make numerical judgments (see Algom and Marks 1984). Methods of conjoint measurement (Luce and Tukey 1964) and functional measurement (Anderson 1970, 1981) provide additional mathematical and statistical tools for assessing additivity (for reviews, see Marks and Algom 1998; Marks and Gescheider 2002). Results using these approaches have produced both some support for additivity (e.g., Levelt et al. 1972; Marks 1978), at least with narrow-band stimuli (Marks 1980), but also evidence against it (e.g., Gigerenzer and Strube 1983; Hübner and Ellermeier 1993). There is now considerable evidence indicating that a sound heard by two ears can be less than twice as loud as a sound heard by one (see Chaps. 7 and 8). Most pertinently here, however, as discussed in Sect. 2.2.3, Fletcher and Munson's loudness scale, based on the principle of additivity, is close to the scale that Stevens (1955, 1956) would later propose.

2.2.2.4 Measuring Loudness by Judging Ratios: The Original Sone Scale

Several contemporaries of Fletcher and Munson sought to measure loudness by instructing their subjects to make quantitative (numerical) assessments of relative values of loudness – an approach that aimed at ensuring that the measures of loudness would agree better than decibels-above-threshold with direct experience. In 1930, Richardson and Ross reported the results of a pioneering study in which they asked eleven subjects to estimate numerically the loudness values of tones that varied in both frequency and level, all of the loudness judgments being made relative to a standard tone assigned the value of 1.0. This method is essentially a version of magnitude estimation, which Stevens (1955) would reinvent and elaborate nearly three decades later.

Richardson and Ross's study marked the beginning of a spate of experiments on loudness scaling. Many of these experiments used what came to be called "ratio methods" (Stevens 1958b), in that the subjects were instructed, in one way or another, to assess the ratio or proportionality between the loudness of one sound and another, or to produce sounds that fall in a specified loudness ratio. One ratio method often used in the 1930s was fractionation. In fractionation, subjects are instructed to adjust the level of one tone to make its loudness appear one-half, or some other fraction, of the loudness of a standard tone (e.g., Ham and Parkinson 1932; Laird et al. 1932; Geiger and Firestone 1933).

By 1936, Stevens was able to pull together several sets of findings and use them to construct a scale of loudness that he called the sone scale. Richardson and Ross had inferred from their measurements that, on average, loudness increased as a simple power function of the stimulus – with an exponent of 0.44. Like Fletcher and Munson's scale, the 1936 sone scale resembles the loudness scale that Stevens would later propose.

2.2.3 Sone Scale of Loudness and Stevens's Law

Two decades later, Stevens (1955, 1956) proposed a revision of the sone scale, which, like Richardson and Ross's loudness scale, follows a power function. According to Stevens, power functions characterize the general relationship between perceptual magnitudes and stimulus intensities, a relationship that applies to audition and to most, if not all, sensory modalities. Although Stevens mustered evidence in favor of a general power law, often designated as Stevens's law, a lion's share of his effort went to the measurement of loudness, and to the establishment of the new sone scale and its relation to the sound pressure or energy of the stimulus. In Stevens's formulation, loudness in sones, *LS*, follows a power function of the form

$$LS = I^{\beta} \tag{2.9}$$

where the unit of measurement of *I* equals the sound pressure or energy of a 1,000-Hz tone at 40-dB SPL and the tone is presented simultaneously to both ears.

The exponent of the power function describing the new sone scale is 0.6 re: sound pressure (0.3 re: sound pressure or power), which is about one third larger than the value reported by Richardson and Ross – and in its overall form, the new sone scale broadly resembles both the earlier sone scale and the scale of Fletcher and Munson, despite the departure of both of the latter scales from a simple power–law representation.

Figure 2.3 plots Stevens's (1955, 1956) new sone scale, which has served as the modern scale of loudness until fairly recently, together with his 1936 sone scale and with Fletcher and Munson's (1933) loudness scale. Stevens inferred that the original sone scale of 1936 departed from a power function largely because of biases inherent in the method of fractionation, the method used to generate much of the data that contributed to the scale (for recent critiques, see Ellermeier and Faulhammer 2000; Zimmer 2005). Lacking independent evidence regarding which methods are biased, how they are biased, and to what extent they are biased, it is also possible that the "true" loudness function at 1 kHz actually falls closer to the original sone scale than to the revised scale, that the departures from a power function evident in the original sone scale accurately represent loudness. Indeed, by 1972, Stevens would acknowledge the possibility of systematic deviations of loudness from a power function, a notion confirmed by subsequent findings of Florentine et al. (1996) and Buus et al. (1997), who came to this conclusion using a different conceptual framework (for

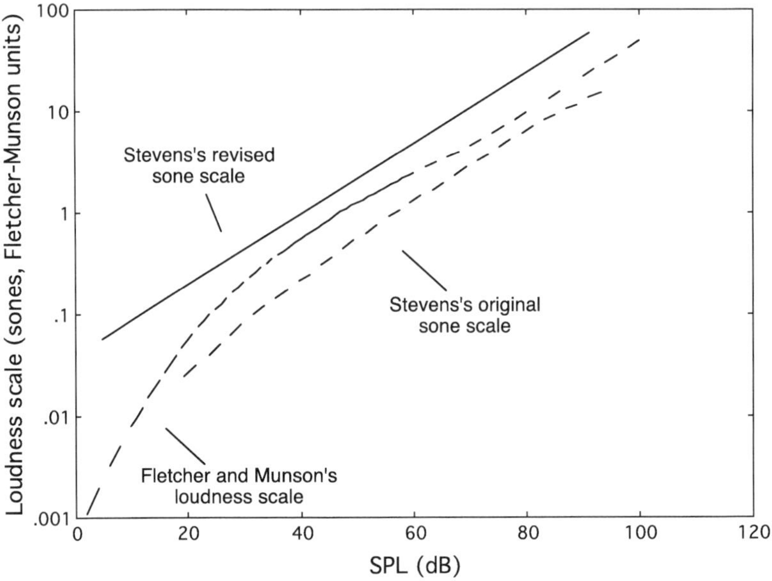

Fig. 2.3 Fletcher and Munson's (1933) loudness scale, Stevens's (1936) original sone scale, and Stevens's (1956) subsequent revision of the sone scale. All three scales are plotted on logarithmic axes, the decibel scale being itself logarithmic. The modern sone scale is defined explicit by a power function (*straight line* in these axes), whereas Fletcher and Munson's scale and the original sone scale only approximate power functions. Note that for clarity of display, Stevens's original sone scale is displaced downward by multiplying the values in sones by one-third

review, see Buus and Florentine 2001). Evidence that the log–log slope (exponent) of the loudness function is smaller at moderate SPLs, 25–60 dB, than at lower or higher ones, suggests the need to modify Stevens's simple power function with a more complex function. Such a function has been proposed by Florentine et al. (1996) and Buus et al. (1997) and termed the inflected exponential (InEx) function (see Florentine and Epstein 2006 and Chap. 5).

Note that Stevens (1956) derived the new sone scale largely on the basis of data obtained with magnitude estimation (the method used by Richardson and Ross 1930), as well as with data obtained using magnitude production, a method that inverts magnitude estimation. In magnitude estimation, the experimenter presents a series of sounds and the subject assigns numbers in proportion to the loudness of each; in magnitude production, the experimenter presents a series of numbers and the subject's task is to adjust the loudness of each to match. To revise the sone scale, Stevens included data obtained with both estimation and production methods.

This revised sone scale maintained the definition of 1 sone as the loudness of a binaurally heard tone at 40-dB SPL (see Chap. 5). The revised sone scale is a simple power function, and it was subsequently accepted by the ISO as the standard for the measurement of loudness (ISO 1959). Over the past half century, the sone scale has served as a touchstone for the measurement of loudness, as other approaches have been developed and investigated. This work has been critical in pointing to the ways that different psychophysical methods can give different results, and to the problems and potential pitfalls associated with the application of different psychophysical methods to measuring the magnitudes of sensations, including loudness.

2.3 Contemporary Approaches to Measuring Loudness

A number of methods are currently used to assess how loudness depends on various stimulus parameters. Modern approaches to the measurement of loudness rely primarily on several kinds of ratings or estimations of loudness, using variants of methods used by, for example, Richardson and Ross (1930) and Gage (1934). These have been reviewed in the previous sections. Each method has strengths and limitations; there is no perfect method for measuring loudness. Loudness researchers need to choose the best measurement method from what is available, while keeping in mind its limitations. The purpose of this section is to summarize issues of relevance when choosing a method of measurement. There are two broad types of measurement methods that are currently used: equal loudness matching and scaling methods. Each of these will be described in turn.

Whatever method is chosen to measure loudness, it must meet the basic requirement of yielding internally consistent measurements (see, e.g., Marks 1974b). A test of internal consistency can be defined in terms of loudness matches or comparisons (cf., Buus 2002). Acceptable methods for measuring loudness provide data conforming to two principles. The first is an ordinal indicant of relative loudness. If sound A has a measured loudness greater than that of sound B, then sound A is louder than sound B, and sound B is softer than sound A. Further, whenever two (or more) sounds are equally loud, the system must assign to them the same value in loudness. The second principle is that loudness equalities must be transitive: If acoustic signal A1 is as loud as signal A2, and A2 is as loud as A3, then A1 must be as loud as A3. The topic of internal consistency of loudness measurements will be revisited at various points in this section as it pertains to specific methods.

Before discussing specific methods, a word of caution is in order regarding their classification. Some authors have designated modern approaches as "direct" or "indirect." This has led to some confusion, because all methods for measuring sensory magnitudes are indirect, although it is fair to say that some are more indirect than others. The term "direct" has been used to denote approaches in which subjects are instructed to judge or rate loudness itself, often on a scale that has putative quantitative or quasi-quantitative properties. The designation of several approaches as "direct" is also intended to contrast with "indirect" approaches, such as that of Fechner, who sought to infer sensation magnitudes from measures of discrimination. Nevertheless, use of the adjective "direct" in this way remains something of a misnomer. The process for measurement involves not only the task that is set forth to the subject – for instance, to rate loudness on a discrete, bounded scale containing a fixed number of categories, or on a continuous, open-ended magnitude-estimation scale – but also involves a set of explicit or implicit mathematical assumptions that the experimenter makes so as to infer quantitative measures of loudness from the rating responses. To prevent a potential source of confusion, the practice of labeling methods as "direct" and "indirect" should be avoided.

2.3.1 Equal Loudness Matching

Equal loudness matching has been used extensively to assess how loudness depends on various stimulus parameters. It uses listeners as null-detectors to obtain measurements of stimulus parameters leading to the point of subjective equality (i.e., the level at which one sound is as loud as the other). Equal loudness matching needs only to assume that listeners can judge identity along a particular dimension, such as loudness, while ignoring differences along other dimensions (e.g., pitch, timbre, apparent duration, etc.). This axiom has never been seriously questioned (Zwislocki 1965; Chap. 1) and there is a general consensus among psychoacousticians that equal-loudness measurements continue to be the "gold" standard to which results obtained by other methods must conform. Loudness-matching (loudness-balance) measurements do not provide direct information about how loud a particular stimulus sounds. They provide information only about the level of a comparison sound judged as loud as the stimulus under investigation. Of course, if the loudness function for the comparison is known, the loudness function for the test stimulus can be constructed.

The measure known as "loudness level" was developed to construct a system in which loudness could be set equal to a common currency: in terms of the SPL of a 1-kHz tone whose loudness matches the loudness of any given test tone. The unit of loudness level is a phon, so that the loudness level of N phons is as loud as a 1-kHz tone at N-dB SPL [see Chap. 5, or the international standard (ISO 226, 2003)].

In several respects, loudness level in phons serves as a useful tool for assessing loudness: The specification of loudness level in decibel (phons) provides both a nominal indicant of loudness – all acoustical signals that are equal in loudness are, by definition, equal in loudness level – and also an ordinal indicant of relative loudness described earlier. The contention that all acoustical signals that have the same loudness should have the same loudness level points to a basic constraint on any method for measuring loudness. Whenever two (or more) sounds are equally loud, the system must assign to them the same value in loudness.

Loudness-balance measurements almost always determine the sound levels at which a test stimulus and a comparison stimulus appear equally loud. These measurements usually require that the level of one stimulus (the comparison) be varied in some manner to ascertain the level at which it is as loud as another stimulus (the standard). The variation in stimulus level can be accomplished in several ways, depending on the psychophysical procedure used to measure the point of subjective equality. The most frequently used psychophysical procedures are the method of adjustment and the modern adaptive procedures, which are described in the following sections. The method of constant stimuli, often used to measure loudness in classic research (e.g., Fletcher and Munson 1933), is highly inefficient and has been replaced by modern adaptive psychophysical procedures. For a description of the method of constant stimuli and other psychophysical procedures, see Gescheider (1997), Gulick et al. (1989), or Gelfand (2004).

2.3.1.1 Measuring Equal Loudness with the Method of Adjustment

In the method of adjustment, a listener is presented two sounds that alternate in time and is given direct control of the level of one of the sounds. The listener is instructed to adjust the variable sound to be equal in loudness to the sound that is fixed in level. Usually, the listener is asked to use a bracketing procedure, that is, to adjust the variable stimulus alternately louder and softer than the fixed stimulus so as to "home in" on the point of equality. One measurement of the point of subjective equality is taken to be the level produced by the final setting of the attenuator.

Although this procedure is conceptually simple, systematic errors may distort the results unless they are minimized through careful experimental design. For example, listeners tend to judge the second of two successive identical sounds as louder or softer than the first, depending on the interstimulus interval between the two (Stevens 1955; Hellström 1979). These time–order errors can be minimized if the order of presentation of the fixed and variable stimuli is randomized. More importantly, listeners tend to overestimate the loudness of the fixed stimulus. An additional bias of the adjustments toward comfortable listening levels may reinforce the overestimation for measurements at low levels, but reduce it at high levels (Stevens 1955). Thus, listeners will tend to set the variable stimulus too high in level in measurements at low and moderate levels, whereas this bias often appears small at high levels (e.g., Zwicker et al. 1957; Zwicker 1958; Scharf 1959, 1961; Hellman and Zwislocki 1964). These adjustment biases may also depend on the mechanical and electrical characteristics of the device used to control the variable stimulus (Guilford 1954; Stevens and Poulton 1956). Averaging the results by having the listeners adjust both the test stimulus to the comparison and the comparison to the test stimulus may minimize the effect of these adjustment biases. Because markings and steps on the adjusted attenuator may produce intractable biases in the adjustments, the variable stimulus should be controlled via an unmarked, continuously variable attenuator.

2.3.1.2 Measuring Equal Loudness with Adaptive Methods

The widespread availability of computers to control psychoacoustic experiments has led many investigators to use adaptive procedures for loudness-balance measurements (e.g., Jesteadt 1980; Hall 1981; Silva and Florentine 2006; for an introduction to adaptive procedures, see Gelfand 2004). In these procedures, the listener is presented two stimuli in sequence with a pause between them and is asked to respond which of the two is louder. The listener's response determines the presentation level of the variable stimulus on the next trial, according to rules that generally make the variable level approach, from both above and below, the level required for equal loudness. In many of the procedures, the critical values are the reversal points – stimulus levels at which the response to the variable changes from "softer" to "louder" or from "louder" to "softer." The complexity of the rules varies from a simple up–down procedure (e.g., Levitt 1971; Jesteadt 1980; Florentine et al. 1996) to complex procedures based on maximum-likelihood estimates of the psychometric function (e.g., Hall 1981; Takeshima et al. 2001).

Although a number of adaptive procedures have been used to measure absolute threshold (e.g., see Leek 2001), the simple up–down procedure is without doubt the most frequently used adaptive procedure to measure equal loudness. The amount of change in the level of the variable stimulus on each trial is determined by the experimenter and is often reduced as the point of subjective equality is approached. For example, a 5- or 6-dB step size may be used until the second reversal in direction of the level, with a 2-dB step size used thereafter (e.g., Zeng and Turner 1991; Buus and Florentine 2002). The entire series of trials over which the signal level varies according to a single adaptive algorithm is called an "adaptive track" and it results in a single measurement.

The stopping rules for adaptive tracks vary among laboratories and are usually based on a predetermined number of reversals. In general, there is a trade-off between the number of trials and the variability in the data: the more measurements, the less variability. On simple statistical grounds, the standard error of the mean across repeated measurements should be inversely proportional to the square root of the number of observations. But requiring subjects to make large numbers of tedious judgments may produce fatigue, which in turn is likely to increase variability over time. For this reason, it is essential that the psychophysical procedure be efficient and that subjects take breaks from listening to prevent fatigue, especially in long experiments.

Care must be taken to eliminate sources of bias in adaptive procedures that may distort judgments. In addition to the time-order errors mentioned earlier, adjustment biases might affect results obtained with adaptive procedures. Although the control over stimulus levels in adaptive procedures is indirect, the listener may nevertheless become aware of which stimulus is varied and attempt to "adjust" the level by responding in particular ways – for instance, by either perseverating or changing responses. Moreover, responses may be affected if the listener compares the perception of the current stimulus to the memory of stimuli on previous trials. Some of these biases can be minimized by randomizing the order of the test stimulus and comparison on every trial and by interleaving multiple adaptive tracks in which the test stimulus and the comparison are varied (Buus et al. 1998; for a general discussion of possible biases and the use of interleaved tracks, see Cornsweet 1962). Using concurrent tracks with the fixed-level stimulus presented at different levels creates additional, apparently random, variation in overall loudness, which forces the listeners to base their responses only on the loudness judgments presented in a trial. However, caution should be used when roving the stimulus level due to context effects, such as induced loudness reduction, described by Arieh and Marks in Chap. 3.

Comparisons across studies using adaptive procedures show large variability in the resulting equal-loudness matches. In part, the variability is likely due to individual differences; it appears also to result from characteristics of the measurement procedures themselves. In many experiments, the goal is to take measurements at several different levels of intensity (or some other parameter of the acoustic stimulus). Different experimenters may use different experimental designs to determine the sequence of the presentation levels, and the sequence may affect the results. Experimenters may opt to vary stimulus intensity in several ways: increasing level across blocks of trials (Ascending Across Blocks [(AAB)]), decreasing level across blocks of trials (Descending Across Blocks [(DAB)]), randomizing level across blocks of trials (Random Across Blocks [(RAB)]), or randomizing level within blocks (Random Within Blocks [(RWB)]). Most contemporary studies use an RAB paradigm, but all four of the aforementioned designs have been used in one investigation or another. Unfortunately, some studies failed to report the stimulus sequence (for review, see Silva and Florentine 2006).

Researchers have long known that "measurement bias" can affect equal-loudness matching data. For example, Stevens and Greenbaum (1966) found that when listeners adjust the level of stimulus B to match several fixed intensity levels of A, and also adjust A to match several levels of B, the results commonly show a so-called "regression effect": The slope of the function plotting adjusted B against A is flatter than the slope of the function plotting B against adjusted A. This might occur due to the preferences that subjects have for listening to sounds at a comfortable loudness. The implication of the regression effect is that the loudness of the variable sound is "over-estimated" near threshold and "under-estimated" at high levels. Regression-type biases in comparison and matching are ubiquitous. Florentine et al. (1996, 1998) observed a regression effect in an adaptive two-interval, two-alternative forced-choice RAB procedure, originally developed by Jesteadt (1980), when they measured the loudness of two stimuli having different durations. An example of this regression effect is shown in Fig. 2.4.

To examine how different stimulus sequences affect loudness matches measured in an adaptive procedure, Silva and Florentine (2006) compared four different sequences in a study of temporal integration. Specifically, they obtained loudness matches between 1-kHz tones having two durations (5 and 200 ms) in each of six listeners, asking whether different sequences of stimuli might affect the magnitude of temporal integration. Three of the sequences varied the level of the fixed tone either sequentially (AAB, DAB) or randomly (RAB) across blocks of trials. The fourth sequence (RWB) randomized the level within blocks. As shown in Fig. 2.5, when the short-duration tone was fixed, there was a significant difference between the magnitude of temporal integration obtained using the RWB procedure vs. the other three procedures, at moderate levels (50–60-dB SL). When comparing loudness matches obtained over a wide range of levels in different experimental studies, therefore, it is important to consider the sequence of stimulus levels presented within each study.

Methods of measuring equal loudness vary among research laboratories and some of the methods have not been fully evaluated with regard to internal consistency.

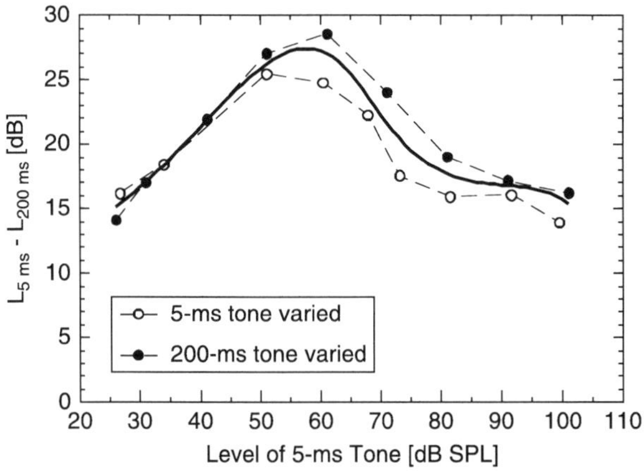

Fig. 2.4 Amount of temporal integration as a function of level. Data of Florentine et al. (1998) for the level difference between equally loud 5- and 200-ms tones at 1 kHz are plotted as a function of the SPL of the 5-ms tone. The filled points show data obtained by a simple up–down method when the 200-ms tones were varied; the unfilled points show the data obtained when the 5-ms tones were varied. Differences between the filled and unfilled points reflect judgment biases, which cause the level of the variable tone to migrate toward a comfortable loudness. The *solid line* shows the difference in level obtained between the 5- and 200-ms loudness functions (the figure, from Buus (2002, Fig. 19), is reproduced with permission. It was published in Tranebjærg L, Christensen-Dalsgaard J, Andersen T, Poulsen T (eds): Genetics and the Function of the Auditory System. Proceedings of the 19th Danavox symposium, Kolding, Denmark. Danavox Jubilee Foundation, ISBN 87-982422-9-6, Copenhagen, 2001)

When a question exists regarding the viability of a particular method, whenever feasible, it is wise to include a check of internal consistency in the experimental design. To be sure, additional testing of consistency can be laborious and time-consuming, but in many circumstances it is critical to ensure that one understands how methodological decisions may affect the results, and therefore the conclusions drawn from them. An example of an experimental design containing a test of consistency can be found in Florentine et al. (1978).

2.3.2 *Loudness Scaling*

Measures such as loudness level serve as a kind of intervening variable, to use the terminology of MacCorquodale and Meehl (1948). Loudness level captures information about a perceptual attribute, indicating, for instance, that any sound having a specific loudness level has the same loudness as any other sound of a specific loudness level. Loudness level also tells us about rank order of loudness, in that loudness level increases as loudness increases. Loudness level indicates nothing more. As an intervening variable, loudness level specifies loudness equivalence – and, by the

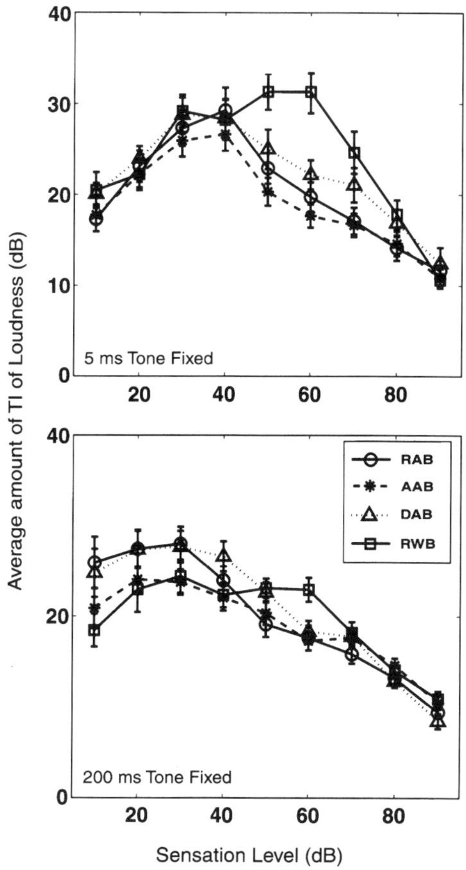

Fig. 2.5 Average difference in level required for equal loudness between 5- and 200-ms tones, plotted as a function of sensation level across six listeners for four adaptive procedures. The random across blocks (RAB), ascending across blocks (AAB), and descending across blocks (DAB) procedures varied the level of the fixed tone in a random, increasing, and decreasing order, respectively, across blocks of trials. The random within blocks (RWB) procedure presented only two blocks of trials, where the level of the fixed tone varied randomly across a range of levels. The *error bars* represent plus and minus one standard error (the figure is a reproduction of Fig. 4 from: Silva and Florentine 2006)

addition of an empirical measure of order, about rank order – but it does not specify to what extent the loudness of one sound exceeds that of another. This is insufficient; we want to know how loudness itself depends on loudness level.

A set of quantitative measures of loudness, per se, can serve as a hypothetical construct – another term used by MacCorquodale and Meehl (1948), this one referring to unobserved variables that go beyond summarizing empirical relations but presumably contain additional information. Quantitative scales of loudness, obtained by methods such as category rating or magnitude estimation, contain all of the information that is available in intervening variables such as loudness level, and more. These scales, when they are free of biases, inform not only about equivalences and rank orders, but also have the potential to inform about quantitative relations, such as how loud one sound is relative to others.

It is convenient to characterize loudness rating scales in terms of two main attributes that help distinguish them: whether the scale is bounded or unbounded, and whether the scale is discrete or continuous. On one end of the spectrum are traditional loudness rating scales (i.e., categorical loudness scales), which are both bounded and discrete. Within a fixed range (bounds), these scales provide the listener with a relatively small number of categorical labels (discrete), such as the integers from 1 through 9, or descriptive labels from "extremely soft" to "extremely loud." On the other end of the spectrum are magnitude-estimation scales (described more fully in the next section), which are unbounded and continuous. In principle, responses on a magnitude-estimation scale may be infinitesimally small or infinitely large.

Hybrid scales are also possible, a common modern adaptation being the so-called visual analog scales. These hybrids are bounded, continuous scales that are often presented as line segments and labeled numerically or adjectivally at their ends (and sometimes at various points between). Hybrid scales allow essentially continuous response along the line segment with continuity being limited only by the precision of the response or its measurement. Visual analog scales are attractive because people generally find it easy to use spatial length or position as a "metaphor" for perceived strength (see Lakoff and Johnson 1980); children as young as 3–4 years readily make graded responses on visual analog scales (e.g., Anderson and Cuneo 1978; Cuneo 1982; Marks et al. 1987). As we shall see, visual analog scales avoid some of the pitfalls of many discrete rating scales, especially those associated with the use of small numbers of discrete labels or categories.

2.3.2.1 Category Scales

The use of categorical rating scales has a long history, going back to the nineteenth century. These scales have been deployed to study not only loudness (and other sensory responses), but just about anything that people are able to judge. Dawes (1972), for example, noted that in the year 1970, about 60% of all of the experimental articles published in the Journal of Personality and Social Psychology reported measures made on these scales. Category Loudness Scales (CLS) are ubiquitous in clinical settings for fitting hearing aids and elsewhere, because they are easy to administer and measurements can be obtained quickly (for clinical usages, see Chap. 9). In this procedure, a listener is presented a series of sounds. After each sound, the listener assigns one of a number of possible categories to its loudness.

To ensure proper use and interpretation of CLS, it is important to understand their limitations and the assumptions that underlie them. It is often assumed that each successive number or adjective on a discrete rating scale marks off, or should mark off, a uniform difference or interval in the quantity being measured. For example, in rating loudness on a nine-point scale, the categories might be the integers from 1 through 9, or they might be nine descriptive labels, such as "extremely soft," "very soft," "soft," "somewhat soft," "medium," "somewhat loud," "loud," "very loud," and "extremely loud." Either way, it is commonly assumed that the step from 1 and 2 on the numerical scale, or from "extremely soft" to "very soft" on the adjectival scale, represents the same difference in loudness as the step from 3 to 4 or from 8 to 9, or from "soft" to "somewhat soft" or from "very loud to "extremely loud." Thus, for computational purposes, successive categories on an adjectival scale are commonly assigned successive integers. The assumption of uniformity implies that the resulting measurements are made on an interval scale.

It is often assumed further that adjectival and numerical labels provide similar quantitative information. Judgments made on rating scales are often relativistic, as subjects tend to use all of the categories equally often, commonly assigning the lowest category to the weakest stimulus presented and the highest category to the strongest (see Chap. 3). The lowest and highest stimulus levels, in turn, may serve as anchors, so responses to the lowest and highest levels often show much less variability than responses to stimulus levels in between – an example of what has been called the "edge resolution effect" (e.g., Berliner and Durlach 1973; Berliner et al. 1977). The tendency toward relativistic judgment and the presence of edge effects have important consequences for any attempt to compare directly category ratings made by different groups of subjects, for example, subjects with normal hearing and subjects with hearing loss. A person who cannot hear very soft sounds (e.g., with "softness imperception" caused by a hearing loss) may label a sound close to threshold as "very soft," not because it is perceived with the same loudness as a person with normal hearing, but because it is the softest sound the person is capable of perceiving.

An explicit method for using descriptive categories in the measurement of loudness was proposed by Heller (1985). In Heller's scheme, the measurement procedure involves two phases: First, in response to a test sound, the listener selects from five broad descriptive categories that cover the range of possible loudness from very soft to very loud. And second, in response to a repetition of the test sound, the listener then selects from ten levels within the initial category. Thus, the overall scale contains 50 possible response categories in all – a sufficiently large number to avoid the biases inherent in the use of small numbers of response alternatives. As a shortcut, one can combine the two steps into one, presenting the subject with all 50 alternatives. Other modifications have also been offered, such as the adaptive method of Brand and Hohmann (2002). A new ISO standard, 16832 (ISO 2006), proposes conditions to help ensure reliability in the use of categorical methods to study loudness. These methods may be useful in applied research (e.g., audiology or environmental noise).

There are different types of CLS and a number of factors to consider when choosing a CLS for a given task, such as stimulus spacing and the number of response categories. Stevens and Galanter's (1957) classic study compared the ratings of loudness on category scales to the corresponding ratings of the same stimuli on unbounded magnitude-estimation scales, and the critical finding – discussed later, in the review of unbounded scales – was the nonlinear relation between judgments on the two scales. For present purposes, however, it is sufficient to note that Stevens and Galanter obtained ratings in several experiments in which the authors compared, albeit somewhat unsystematically, (1) numerical and adjectival labels, (2) different numbers of available response categories, and (3) different sized steps between successive stimulus levels. Presenting numbers vs. adjectives did not have a major effect on the ratings for their normal-hearing subjects, but variations in the number of available response categories and variations in stimulus spacing had more substantial effects. The greater the number of response categories available to the subject, for example, and hence the more nearly continuous the rating scale, the more linear the relation between the resulting responses on the rating scale and responses on a fully continuous magnitude-estimation scale. Heller's (1985) scheme, described in the preceding text, capitalizes on the availability of a relatively large number of response categories.

Research following the study of Stevens and Galanter (1957) has shown systematic effects of both number of categories and spacing of stimulus levels on rating-scale responses, although these studies have largely investigated sensory dimensions other than loudness. Marks (1968) reported results of a systematic study of the perception of brightness of flashes of light, examining the effects of both stimulus spacing (size of the log intensity step between successive stimuli) and number of available numerical categories. The ratings could be described by a power function of stimulus intensity, having the form

$$C - C_0 = cS^a \tag{2.10}$$

where C is the average rating on the category scale and a is the fitted exponent (the additive constant C_0 is necessary in order to adjust for the scale's arbitrary zero-point). The value of the exponent a increased with increasing stimulus spacing and with increasing number of available responses. Results of Stevens and Galanter (1957) suggest that category ratings of loudness would behave similarly.

Given the historic popularity of seven-point category scales, it may be tempting to assume that a relatively small number of categories is sufficient. Incorrect though it is, this temptation may be increased by the evidence, famously reviewed by Miller (1956) that the channel capacity for absolute identification of stimuli on a univariate continuum, such as loudness, is roughly seven items. It is important to keep in mind that the "magical number seven," as Miller dubbed it, does not mean that performance ceases to improve by presenting more than seven stimuli and seven response categories. The channel capacity of seven refers to the level of asymptotic performance when the number of stimuli – and the number of possible responses – is considerably larger than seven. Reanalysis (Marks 1996) of the category ratings reported by Marks (1968) suggests that increasing the number of response categories from 4–20 to 100 increases the amount of information transmitted, a measure of the mutual discriminability among the stimuli (see also Garner 1960). When bounded scales are appropriate or desirable, therefore, it is crucial that the scale contain a sufficient number of categories, on the order of 15 or more. An alternative to a discrete category scale is a visual analog scale, a line scale that permits the subject virtually continuous response between the end points (e.g., Anderson, 1981).

Category rating scales appear to be especially sensitive to the selection of the stimulus and response alternatives. Half a century ago, Stevens (1958a) examined the role of stimulus spacing in categorical judgments of loudness. In particular, Stevens compared uniform decibel spacing to two kinds of nonuniform spacing: spacing with stimuli bunched at the lower end and spacing with stimuli bunched at the upper end. Spacing exerted a substantial effect on judgment: Over the region in which stimulus levels were bunched together, subjects tended to spread out their responses to a greater extent than they did over comparable ranges where stimuli were spaced more sparsely. Because the scale is bounded, if a subject "uses up" more categories where spacing is narrow, the subject must perforce change the relation of the ratings to the stimuli in the remaining region of the stimulus range, where spacing is broader. With unbounded scales, however, this constraint could disappear, or at least diminish. Not surprisingly, in the same study, Stevens found that stimulus spacing exerted a much smaller effect on magnitude estimations of loudness than it did on category ratings.

Effects similar to those of stimulus spacing, just mentioned, arise when the ensemble of stimulus levels remains constant, but the frequency of presentation of the various stimuli changes across conditions. When fixing the stimulus levels, one might present the lower level twice as often as the higher ones, or the higher ones twice as often as the lower ones. From the perspective of the subject, the effect is much like bunching stimulus levels at the low and high ends, respectively, and the resulting patterns of response are similar. From results of this sort, Parducci (1965, 1974) developed a range-frequency model of categorical judgment, using as one of his primary principles the notion that, with discrete scales, subjects tend to use all of the available responses equally often. This tendency underlies the effects of stimulus spacing and frequency of presentation. Given this tendency, the "ideal" or "unbiased" function would be one that spaced the stimuli uniformly with regard to loudness, so that successive categories marked off uniform changes in loudness. Pollack (1965a, b) has shown how one can use an iterative experimental method to reduce equal-response tendencies – capitalizing on evidence that these tendencies do not wholly determine the results. In Pollack's method, one uses the results of an initial experiment to adjust the spacing so as to try to increase the uniformity of the subjects' responses in a subsequent experiment, and the procedure is repeated until a uniform scale is achieved.

One concern about using categorical and other rating scales is that they may not provide adequate measures of internal consistency. Relatively few studies have addressed the question whether rating scales provide results consistent with the principle of equality. One pair of studies did address the question, albeit indirectly, by examining binaural summation of loudness using several methods, including loudness matching, loudness scaling on a visual analog scale, and magnitude estimation, which is described in the next section (Marks 1978, 1979). All three methods gave comparable measures of summation, quantified in terms of matching the loudness of monaural and binaural sounds. It is possible, however, that some rating scales may fail this "test of internal consistency." Support for this contention comes from two small studies in vision that obtained category ratings (on 9-point and 11-point scales, respectively) in order to determine how brightness depends on both the duration and the luminance of flashes of light (Raab et al. 1961; Lewis 1965). Results of both studies failed to show the presence of peaks in brightness as a function of duration (Broca-Sulzer effect); these peaks are readily shown with other methods, including direct matching (Aiba and Stevens 1964) and magnitude estimation (Raab 1962; Stevens and Hall 1966). Lewis's study also failed to show the level-dependent change in critical duration for integration, another visual phenomenon revealed in both matches and magnitude estimations. These findings suggest that category scales, especially ones using relatively small numbers of possible responses, may give results incompatible with the principle of equality.

Another potential problem with using CLS is that rating scale responses are typically averaged and treated as if they provide reasonably uniform (interval scale) measures of the underlying perceptual representations. Yet, as discussed in detail by Arieh and Marks in Chap. 3, the pervasive effects of stimulus spacing, presentation frequency, and number of available responses suggest that decisional processes play a substantial role in determining categorical judgments, hence in determining the relation of mean category judgment to stimulus level.

2.3.2.2 Magnitude Estimation Scales of Loudness

Magnitude estimation is a type of unbounded, continuous scaling procedure. In the method of magnitude estimation, a listener is presented a series of stimulus levels in random order. After each stimulus presentation the listener is asked to respond with a number that matches its loudness. Any positive number that seems appropriate to the listener may be used. Stevens (1956, 1975), Hellman (1991), and others (e.g., Zwislocki 1983) have argued that this type of unbounded response scale is most effective in producing responses that are approximately proportional to loudness.

Magnitude estimation comes in several varieties. In Stevens's earliest version of the method, subjects were presented at the start of a session with a standard stimulus of fixed sound level, together with a numerical modulus assigned to represent each stimulus. The standard typically came from somewhere in the middle range of levels, and the modulus commonly had a value of "10," a numeral deemed neither "too large" nor "too small." Subjects were instructed to assign numbers to the loudness of other sounds in proportion – that is, to maintain the appropriate ratio between numbers and sounds. If another sound was twice as loud as the standard, it should receive a response of "20." If it was one fifth as loud, it should receive a response of "2." Some investigators may omit the standard stimulus, but continue to emphasize the relative, ratio relations of responses, by explicitly asking subjects to judge the loudness of the current stimulus in terms of the loudness of the previous stimulus; if the previous stimulus was assigned the numeral "5" and the current stimulus appears three times as loud, the subject should assign it the numeral "15" (e.g., Luce and Green 1974). Luce and Green dubbed this method "ratio magnitude estimation." Ratio magnitude estimation is likely to enhance sequential (contextual) effects, that is, the way that stimuli and responses on trial n affect responses on trial $n + 1$ (for a more thorough discussion of sequential effects, see Arieh and Marks, Chap. 3). Although useful for studying decisional processes, ratio magnitude estimation is probably not a method of choice when the goal is to measure loudness in ways that minimize such sequential effects.

Eventually, Stevens (e.g., 1956) abandoned the use of both a standard stimulus and numerical modulus. Following many earlier studies on the topic, Hellman and Zwicklocki (1961) found that the values of both standard and modulus affected the numerical responses, and, in particular, affected the observed exponent of the power function relating judgments of loudness to sound intensity. Most notably, the exponent remained constant if both standard and modulus increased or decreased in tandem, but not if either standard or modulus changed while the other remained constant. This pattern of results suggested the possible existence of a "natural" connection between sensation magnitude and numerical response, and hence the possibility that the experimenter's arbitrary choice of standard and modulus may induce biases in responses (Hellman and Zwislocki 1963, 1964; Hellman 1991). This eventually led the way to the development of a method known as "absolute magnitude estimation," in which instructions avoid any reference to ratio relations, but instead encourage subjects to assign numerals to stimuli such that the "perceived magnitude of the numbers match the perceived magnitudes of the sensations." Subjects may be allowed to hear a stimulus as often as desired before rendering a judgment (Cross 1973; Hellman 1976).

An example of instructions using absolute magnitude estimation follows:

> You are going to hear a series of sounds. Your task is to specify how loud each sound is by assigning numbers. Louder sounds should be assigned larger numbers. You are free to use any positive numbers that seem appropriate–whole numbers, decimals, or fractions. Do not worry about running out of numbers; there will always be a smaller number than the smallest you use and a larger number than the largest you use. If you do not hear a sound, please assign it zero, otherwise all numbers should be larger than zero. Do not worry about the number you assigned to previous sounds, simply try to match the appropriate number to each sound regardless of what number you may have assigned the pervious sound.

Although there appear to be conditions in which the method of absolute magnitude estimation encourages subjects to map their numerical responses to sensations in a way that, per the method's label, is "absolute" (Zwislocki and Goodman 1980; Zwislocki 1983), absolute magnitude estimation shows at least some of the contextual effects in the judgment of loudness that are shown by other methods, such as category rating and ratio magnitude estimation (Ward 1987).

The range and spacing of the stimuli presented to the subject also influence magnitude estimates of loudness. Although Arieh and Marks (Chap. 3) discuss the role of contextual effects on loudness, it is important to consider here the role of stimulus range and stimulus distribution. Several investigations have shown that the form of the loudness function, and in particular the exponent of the power function, can vary systematically with the range of test levels: the larger the range, the smaller the exponent (Poulton 1968, 1989; Teghtsoonian 1973). Keep in mind, however, that the effect of stimulus range is typically fairly modest, appearing only when the range of levels becomes very small (smaller than about 20 dB). Over larger ranges, the exponent is more or less independent of the range (Teghtsoonian 1971, 1973). This effect can be explained, at least in part, by the fact that the slope of the loudness function is shallower at moderate levels than at low and high levels.

Although stimulus range exerts a relatively small effect on power–function exponents, the effect is systematic, and the very presence of the range effect points to the importance of distinguishing between the underlying perception of loudness and the overt responses that listeners give to a particular set of stimuli, in a particular contextual setting, under a particular set of instructions. Overt responses, such as magnitude estimations, represent the end product of at least two sets of processes. The first is the set of sensory processes by which patterns of stimulus energy are transformed into internal representations of sounds, including their loudness. The second is the set of decisional and judgmental processes by which the internal representations of loudness map into the numerical responses (see Gescheider 1997; Marks and Algom 1998; Marks and Gescheider 2002). To explain the effect of stimulus range on the exponent of the loudness function, therefore, one would hypothesize an initial sensory, power–function transformation of sound pressure or energy to loudness, followed by a subsequent decisional, power–function transformation of loudness to numerical response. Only when the exponent of the decisional power function is 1.0 – that is, when the function is linear – would the numerical responses provide "valid" measures of loudness.

The modest size of the range effect contravenes the hypothesis (e.g., Poulton 1968, 1989) that exponents are simply accidental byproducts of the choice of stimuli presented by the experimenter, along with the predilections for particular numerical responses on the part of subjects. Were this so, then the subjects would presumably give the same range of numerical responses regardless of the stimuli presented. This does not occur. Instead, as stimulus range increases, so does the range of numerical responses, implying that stimulus range has only a modest effect on the exponent of the decisional power function (Teghtsoonian 1971). Nevertheless, to help circumvent effects of stimulus range, and other factors that influence decisional processes, one might choose to "calibrate" the subjects in advance of testing, by teaching them a particular stimulus-response function, as suggested by West et al. (2000; see also Marks et al. 1995).

One should be cautious, however, about making the implicit assumption that stimulus range affects only the decisional and judgmental processes that intervene between loudness and overt responses. Algom and Marks (1990) have provided some evidence that stimulus range may have two effects: As already discussed, changing the stimulus range can influence the decisional function relating numerical responses to the underlying values of loudness. But changing range may also affect the sensory function relating the underlying values of loudness to stimulus level. Algom and Marks drew this conclusion from the observation that stimulus range affected the implicit loudness matches between tones heard monaurally and binaurally.

Loudness functions can vary not only with the overall dynamic range of stimuli but also with their spacing and distribution. For example, if the sound levels are spaced unevenly, with smaller steps between successive levels in one region of the overall range compared to others (or if a subset of levels is presented more frequently than others), the exponent of the power function will tend not to be uniform over the entire stimulus range, but instead will be greater over the local region in which the stimulus levels are bunched (Stevens 1958a).

2.3.2.3 Magnitude Production and Cross-Modality Matching

In some methods, the subject controls the stimulus and sets it to a target loudness, which can be specified in several ways. In magnitude production, the subject hears only the variable stimulus and is instructed on each trial to adjust its loudness to match the number assigned on that trial. If the perceived magnitude of numbers is considered a separate modality, then magnitude production and magnitude estimation become special cases of cross-modality matching (e.g., Stevens 1959; Reynolds and Stevens 1960; Hellman and Zwislocki 1961, 1963; Hellman 1991).

Some investigators have suggested that the results of magnitude estimation and magnitude production be averaged in order to compensate for biases in each method (Hellman and Meiselman 1993). The combination of these two methods is sometimes called "numerical magnitude balance" (Hellman and Zwislocki 1963; Hellman 1976). The recommendation to average results obtained by estimation and production assumes that the biases in the two methods are equal and opposite. In this regard, Hellman and Zwislocki (1961) reported excellent agreement between results obtained by directly matching tones in the absence and presence of masking noise and results obtained by magnitude production alone.

In methods involving "ratio determinations," the subject is presented a fixed stimulus alternating with the variable stimulus and is instructed to adjust the loudness of the variable to some given ratio (or fraction) of the fixed stimulus's loudness. Often the subject is asked to halve or double the loudness, but other ratios have also been used. There are undoubtedly biases in these procedures, in that, for example, doubling loudness twice is not the same as quadrupling loudness (see Ellermeier and Faulhammer 2000; Zimmer 2005). In the method of "bisection" the subject is presented two reference stimuli differing in loudness and is instructed to adjust the variable to be midway between them. Although the method of bisection too has its biases (e.g., Gage 1934), carefully measured bisections of loudness (Garner 1954; Carterette and Anderson 1979) produce scales that, like those produced by magnitude estimation and production, can be described as power functions of sound level. The scales obtained by bisection – and by other methods in which subjects judge or compare intervals of loudness (Parker and Schneider 1974; Schneider et al. 1974) – generally have much smaller exponents than do scales obtained by magnitude estimation and production (see Marks 1974a). Because the subject controls the level of the variable, all of these methods are likely to be affected by the adjustment biases described earlier. In fact, Stevens and Poulton (1956) found that results obtained in these adjustment procedures depend on the attenuation characteristics of the device and advocate the use of a "sone potentiometer," which is designed to make loudness in sones an approximately linear function of angular position of the unmarked, smoothly rotating knob.

In the method of cross-modality matching, the loudness of a sound is matched to the magnitude of a percept in another modality, such as line length (or string length), brightness, tactile vibration, or the magnitude of the other percept is matched to the loudness. Cross-modality matching between loudness and line length is most common (Teghtsoonian and Teghtsoonian 1983). Cross-modality matches are consistent with results obtained by magnitude estimation in subjects with normal hearing and hearing losses (Hellman 1991). Results obtained in individual subjects are more consistent in cross-modality matching with line length or string length than in magnitude estimation, especially for short-duration sounds (Green and Luce 1974; Hellman and Meiselman 1988; Epstein and Florentine 2005, 2006a).

2.3.2.4 Magnitude Estimation, Magnitude Production, Cross-Modality Matching, and the Principle of Equality

Over the past half-century, the methods of magnitude estimation, magnitude production, and cross-modality matching have shown themselves to be especially versatile, readily applied to study loudness perception of groups of listeners in a variety of settings and under a variety of conditions; the methods have not been tested nearly so thoroughly, however, in individual listeners. To give just a few examples, the method of magnitude estimation in particular has been used to study how loudness is affected by factors such as stimulus duration (Stevens and Hall 1966; Epstein and Florentine 2006a), the presence of masking noise (Hellman and Zwislocki 1964), delivery to one ear or two (Hellman and Zwislocki 1963; Scharf and Fishken 1970; Marks 1978; Epstein and Florentine 2009), and normal hearing vs. hearing loss (e.g., Hellman and Meiselman 1991, 1993; Marozeau and Florentine 2009).

Results obtained with scaling methods potentially provide two kinds of information, information about relative magnitude and information about equality – assuming in each case that one can minimize or take account of the pertinent sources of potential bias. The use of the qualifier "pertinent" is intended to indicate the possibility that certain biases may selectively affect one kind of information but not the other. For example, the so-called regression effect (Stevens and Greenbaum 1966) points to nonlinear relations between numerical judgments, such as magnitude estimations, and stimulus level. According to Stevens and Greenbaum, subjects tend to compress the range of whatever response variable is under their control, compressing the range of numerical responses in magnitude estimation and compressing the range of stimulus levels in magnitude production. For magnitude estimations to be unbiased, the numerical responses must be directly proportional to loudness: Quadruple the underlying loudness, and the subject should give a number four times as great. With a tendency to compress the range of numbers, subjects might only double their numerical responses when loudness quadruples. In this case, the exponent obtained in magnitude estimation would be half the size of the exponent that governs the underlying perceptions of loudness.

Although regression and similar biases affect the quantitative properties of the results, they need not necessarily affect the underlying loudness equalities. Consider the situation in which several sounds have underlying loudness values of X, while other sounds have underlying values of loudness 4X. Then as long as all of the sounds with loudness X receive the same average judgment of loudness and all of the sounds with loudness 4X receive the same average judgment of loudness (whether four or only two times as great), the resulting numerical judgments will preserve the loudness equalities. Simply put, as long as the loudness of every sound is mapped to a single, uniform numerical scale, the loudness judgments will conform to the principle of equality. It is possible, of course, that subjects may use different numerical scales to judge different sounds. Consequently, it is often helpful to obtain converging information about loudness equalities with other methods, such as loudness matching. A few studies have asked to what extent results obtained by scaling methods such as magnitude estimation and magnitude production agree with equal-loudness matches. Hellman and Zwislocki (1964) found excellent agreement between measures of masking of a 1,000-Hz tone by noise as determined by magnitude production and by loudness matching, and Marks (1978) reported good agreement between measures of binaural addition predicted from magnitude estimations and determined directly by loudness matches between tones of equal and unequal SPL to the two ears. Epstein and Florentine (2006a) compared loudness measures for 5- and 200-ms tones, obtaining magnitude estimations and equal-loudness matches from the same subjects. Results indicated that both procedures provide rapid and accurate assessments of group loudness functions for brief tones, although the assessments may not be reliable enough to reveal specific characteristics of loudness in individual subjects. Comparisons of scaling data and direct matches are especially important in studies of individual differences, where magnitude estimations and loudness matches may not give equivalent measures.

Almost all of the studies discussed thus far presented listeners with static, steady-state sounds, that is, with stimuli whose levels remained constant over a single trial (except for initial rise and final decay). Most sounds encountered in the world, however, are dynamic. The levels of speech, music, and environmental noises commonly rise and fall over time, either because the levels emitted from the sources themselves change, or because the sound source, the listener, or both change their spatial locations over time. Assessing the loudness of dynamic sounds poses special questions: Can listeners judge momentary loudness? Overall or average loudness? In judging overall or average loudness, how might the listener weight the loudness experienced at different points in time?

Several experiments have studied what has been called "decruitment" in loudness: the marked decrease in loudness when sound level decreases over time, compared to comparable increases over time (Canévet and Scharf 1990; Teghtsoonian et al. 2000). When a sound decreases steadily from a high level to a low one, at the low-level the sound appears softer than it does when it is presented discretely (statically), following the same high-level sound at a comparable point in time. Testing sounds that increased steadily in their level, Marks and Slawson (1966) asked a rather different kind of question about the perception of dynamic sounds: how linear do listeners perceive the change in loudness to be when sound level increases as a power function of time? Marks and Slawson tested a wide range of different exponents and found that subjects judged the increase in loudness to be most linear when sound intensity increased as 3.3 power of time: $I = t^{3.3}$. Given that loudness in sones equals (given appropriate units) the 0.3 power of intensity, $LS = I^{0.3}$, this outcome means that the increase in loudness was judged most linear when loudness increased linearly in sones: $LS = (t^{3.3})^{0.3} = t$.

2.3.3 Measuring Loudness of Long-Duration Sounds

Equal-loudness matching and loudness scaling are especially useful in measuring the loudness of steady state and non-steady-state sounds of relatively short duration – usually no more than a few seconds. The methods that are used in laboratories to measure the loudness of short-duration sounds are not generally useful for measuring the subjective impressions of sounds along a sound stream that varies over time and can last for long durations, such as those in daily environments (see Teghtsoonian et al. 2005). Although attempts have been made to adapt category scaling and magnitude estimation to the assessment of long-duration sounds, these methods have not been carefully evaluated. There is a need to develop and rigorously test methods for measuring the loudness of relatively long dynamic stimuli.

Two methods that may be useful are "the method of continuous judgment by category" (see Chap. 6) and the "acoustic menu" (Molino et al. 1979). The original method of continuous judgment by category uses a modified category scale to record subjective judgments over time, but the method has a number of variations. For example, continuous judgments may be made using cross-modality matching of muscular effort (Susini et al. 2002) or line-length (Kuwano and Namba 1990). The acoustic menu method uses an avoidance paradigm to measure the unpleasantness of loud sounds. Kuwano and Namba describe these methods in detail in Chap. 6.

2.4 Evaluative Summary

Most researchers studying loudness are primarily interested in obtaining measurements of loudness and are less interested in details of the methods *per se*. This is understandable given time constraints in research settings, but it is unwise to choose a method without understanding its limitations as well as its strengths. Errors made in the acquisition, treatment, and interpretation of the data can arise from ignorance of basic concepts regarding methods of measuring loudness. Every method, technique, and paradigm designed to measure loudness (or probably anything else) rests tacitly or explicitly on a set of underlying assumptions, hypotheses, or theoretical principles. For example, it had long been assumed that loudness at threshold is zero. This assumption influenced models of loudness in people with normal hearing and hearing losses. When Buus et al. (1998) actually measured loudness at threshold, the data showed a small but positive value. Models of loudness (see Chap. 10) and standards (e.g., ANSI S3.4-2007) are now being revised in light of this new finding. The collapse of this old assumption about loudness at threshold has opened the door to question other assumptions (see Chap. 1).

To measure loudness means, *ipso facto*, to be able to determine how loudness depends, quantitatively, on all of the variables that affect it: not only on level of an acoustical signal, but also on its other stimulus variables (such as frequency, spectral content, duration, presence of background sounds, etc.). The fact that loudness depends on a multiplicity of physical, psychological, and physiological factors – that an enormous number of different stimuli and conditions can produce the same loudness – sets a minimal empirical requirement for any method to measure loudness adequately. To measure loudness adequately, the method must provide measures that are internally consistent. That is, the method must assign the same value to all of the different conditions of stimulation that produce a given level of loudness. In other words, acoustical signals that have the same loudness should have the same loudness level. In addition, loudness equalities must be transitive: If acoustic signal A is as loud as signal B, and B is as loud as C, then A must be as loud as C.

In theory, once an adequate system for loudness measurement is established, the system itself will be able to provide information about loudness and known sources of bias can be taken into account. Unfortunately, psychophysical methods in their many forms have not been tested for all potential sources of bias, and the design of very few experiments permits the ready assessment of internal consistency in the data. In such cases, it is wise to ensure that the experimental designs include checks of internal consistency.

Ignorance of the limitations of a measurement method is only one of a number of pitfalls that an experimenter must avoid. Measurements are determined not only by the experimental method, but also by the way the data are treated. A review of all the possible errors is not possible, given the many potential pitfalls in data analysis, so an example will have to suffice.

It is well known that the distributions of magnitude estimations typically are highly skewed and often log normal, leading many investigators, appropriately, to use geometric averages. This approach becomes problematic, however, if a few subjects occasionally give judgments of "zero," because the geometric mean of a distribution containing a value of zero will be zero. An investigator may be tempted to try to circumvent the problem by adding a positive constant to all of the magnitude estimations, calculate geometric averages, then subtract out the constant. This may be satisfactory if the data have appropriate statistical properties, but these properties must be ascertained. Other, simpler, solutions include calculating medians.

The final pitfall discussed in this chapter lies in errors in the interpretation of the data. For example, whereas loudness level provides a useful scale, it informs only how the loudness of a given sound compares to that of a 1-kHz tone. That is, loudness level provides information only about loudness equalities and rank order. Importantly, loudness level does not correspond directly to the subjective magnitude of the perception. Loudness level is not the same as loudness. For example, a sound with the loudness level of 100 phons is much more than twice as loud as a sound with the loudness level of 50 phons.

The implementation of every psychophysical method is based on a set of underlying assumptions. So too is every analytical and statistical treatment of the data and so is every interpretation of the results. Progress in every discipline of science, including psychoacoustics, comes with advances in technology, methodology, and conceptualizations. But progress also requires a firm understanding of the assumptions that underlie interpretation, analyses, and, notably, the methods.

References

Aiba RS, Stevens SS (1964) Relation of brightness to duration and luminance under light- and dark-adaptation. Vision Res 4:391–401.

Algom D, Marks LE (1984) Individual differences in loudness processing and loudness scales. J Exp Psychol Gen 113:571–593.

Algom D, Marks LE (1990) Range and regression, loudness processing and loudness scales: Toward a context-bound psychophysics. J Exp Psychol Hum Percept Perform 16:706–727.

Anderson NH (1970) Functional measurement and psychophysical judgment. Psychol Rev 77:153–170.

Anderson NH (1981) Foundations of Information Integration Theory. New York: Academic Press.

Anderson NH, Cuneo DO (1978) The height + width rule in children's judgments of quantity. J Exp Psychol Gen 107:335–378.

ANSI-S3.4 (2007) American National Standard Procedure for the Computation of Loudness of Steady Sounds. New York: American National Standards Institute.

Berliner MH, Durlach NI (1973) Intensity perception. IV. Resolution in roving-level discrimination. J Acoust Soc Am 53:1270–1287.

Berliner JE, Durlach NI, Braida LD (1977) Intensity perception. VII. Further data on roving-level discrimination and the resolution and bias edge effects. J Acoust Soc Am 61:1577–1585.

Brand T, Hohmann F (2002) An adaptive procedure for categorical loudness scaling. J Acoust Soc Am 112:1597–1604.

Buus S (2002) Psychophysical methods and other factors that affect the outcome of psychoacoustic measurements. In: Tranebjærg L, Christensen-Dalsgaard J, Andersen T, Poulsen T (eds), Genetics and the Function of the Auditory System: Proceedings of the 19th Danavox Symposium. Copenhagen, Denmark: Holmens Trykkeri, pp. 183–225.

Buus S, Florentine M (2001) Modifications to the power function for loudness. In: Summerfield E, Kompass R, Lachmann T (eds), Fechner Day 2001. Proceedings of the 17th Annual Meeting of the International Society for Psychophysics. Berlin: Pabst, pp. 236–241.

Buus S, Florentine M (2002) Growth of loudness in listeners with cochlear hearing losses: Recruitment reconsidered. J Assoc Res Otolaryngol 3:120–139.

Buus S, Florentine M, Poulsen T (1997) Temporal integration of loudness, loudness discrimination, and the form of the loudness function. J Acoust Soc Am 101:669–680.

Buus S, Müsch H, Florentine M (1998) On loudness at threshold. J Acoust Soc Am 104:399–410.

Canévet G, Scharf B (1990) The loudness of sounds that increase and decrease continuously in level. J Acoust Soc Am 88:2136–2142.

Carterette EC, Anderson NH (1979) Bisection of loudness. Percept Psychophys 26:265–280.

Cattell JMcK (1886) The influence of the intensity of the stimulus on the length of the reaction time. Brain 8:510–515.

Chocholle R, Greenbaum HB (1966) La sonie de sons purs partiallement masqués: Étude comparative par une méthode d'égalisation et par la méthode des temps de réaction [Loudness of partially masked pure tones: Comparative study by an equalization method and by the reaction time method]. J Psychol Norm Pathol 63:387–414.
Churcher BG (1935) A loudness scale for industrial noise measurements. J Acoust Soc Am 6:216–225.
Clark A (1993) Sensory Qualities. Oxford: Oxford University Press.
Cornsweet TN (1962) The staircase-method in psychophysics. Am J Psychol 75:485–491.
Cross DV (1973) Sequential dependencies and regression in psychophysical judgments. Percept Psychophys 14:547–552.
Cuneo DO (1982) Children's judgments of numerical quantity: A new view of early quantification. Cogn Psychol 14:13–44.
Dawes RM (1972) Fundamentals of Attitude Measurement. New York: Wiley.
Delboeuf JR (1873) Étude psychophysique: Recherches théoretiques et expérimentales sur la mesure des sensations, et spécialement des sensations de lumière et de fatigue [Psychophysical study: Theoretical and experimental research on the measurement of sensations, especially sensations of light and fatigue]. Mémoires de l'Académie Royale de Belgique 23:3–115.
Dzhafarov EN, Colonius H (2005) Psychophysics without physics: A purely psychological theory of Fechnerian scaling in continuous stimulus spaces. J Math Psychol 49:1–50.
Ellermeier W, Faulhammer G (2000) Empirical evaluation of axioms fundamental to Stevens's ratio-scaling approach: I. Loudness production. Percept Psychophys 62:1505–1511.
Epstein M, Florentine M (2005) A test of the equal-loudness-ratio hypothesis using cross-modality matching functions. J Acoust Soc Am 118:907–913.
Epstein M, Florentine M (2006a) Loudness of brief tones measured by magnitude estimation and loudness matching. J Acoust Soc Am 119:1943–1945.
Epstein M, Florentine M (2006b) Reaction time to 1- and 4-kHz tones as a function of sensation level in listeners with normal hearing. Ear Hear 27:424–429.
Epstein M, Florentine M (2009) Binaural loudness summation for speech and tones presented via earphones and loudspeakers. Ear Hear 30:234–237.
Falmagne JC (1985) Elements of Psychophysical Theory. Oxford: Oxford University Press.
Fechner GT (1860/1966) Elemente der Psychophysik. Leipzig, Germany: Breitkopf und Härtel. [Elements of Psychophysics, Adler HE (trans), Howes DH, Boring EG (eds). New York: Holt, Rinehart, and Winston].
Fletcher H, Munson WA (1933) Loudness, its definition, measurement, and calculation. J Acoust Soc Am 5:82–108.
Florentine M, Epstein M (2006) To honor Stevens and repeal his law (for the auditory system) In: Kornbrot DE, Msetfi RM, MacRae AW (eds), Fechner Day 2006. Proceedings of the 22nd Annual Meeting of the International Society for Psychophysics. St. Albans, England: ISP, pp. 37–42.
Florentine M, Buus S, Bonding P (1978) Loudness of complex sounds as a function of the standard stimulus and the number of components. J Acoust Soc Am 64:1036–1040.
Florentine M, Reed C, Durlach NI, Braida LD (1979) Intensity discrimination and loudness matches in subjects with sensorineural hearing loss. In: Wolf JJ, Klatt DH (eds), Speech Communication Papers. New York: Acoustical Society of America. J Acoust Soc Am Speech Commun Papers, pp. 575–578.
Florentine M, Buus S, Mason CR (1987) Level discrimination as a function of level for tones from 0.25 to 16 kHz. J Acoust Soc Am 81:1528–1541.
Florentine M, Buus S, Poulsen T (1996) Temporal integration of loudness as a function of level. J Acoust Soc Am 99:1633–1644.
Florentine M, Buus S, Robinson M (1998) Temporal integration of loudness under partial masking. J Acoust Soc Am 104:999–1007.
Freeman K (1948) Ancilla to the Pre-Socratic Philosophers: A Complete Translation of the Fragments in Diels Fragmente der Vorsokratiker. Oxford: Blackwell.

Gage FH (1934) An experimental investigation of the measurability of auditory sensation. Proc R Soc Lond 116B:103–122.
Garner WR (1954) Context effects and the validity of loudness scales. J Exp Psychol 48:218–224.
Garner WR (1960) Rating scales, discriminability, and information transmission. Psychol Rev 67:343–352.
Geiger PH, Firestone FA (1933) The estimation of fractional loudness. J Acoust Soc Am 5:25–30.
Gelfand SA (2004). Hearing – An Introduction to Psychological and Physiological Acoustics (4th Ed). New York: Marcel Dekker.
Gescheider G (1997) Psychophysics: The Fundamentals. Mahwah, NJ: Lawrence Erlbaum.
Gigerenzer G, Strube G (1983) Are there limits to binaural additivity of loudness? J Exp Psychol 9:126–136.
Green DM, Luce RD (1974) Variability of magnitude estimates: A timing theory analysis. Percept Psychophys 15:291–300.
Guilford JP (1954) Psychometric Methods (2nd Ed). New York: McGraw-Hill.
Gulick WL, Gescheider GA, Frisina RD (1989) Hearing: Physiological Acoustics, Neural Coding, and Psychoacoustics. New York: Oxford University Press.
Hall JL (1981) Hybrid adaptive procedure for estimation of psychometric functions. J Acoust Soc Am 69:1763–1769.
Ham LB, Parkinson JS (1932) Loudness and intensity relations. J Acoust Soc Am 3:511–534.
Heller O (1985) Hörfeldaudiometrie mit dem Verfahren der Kategorienunterteilung (KU) [Listening field audiometry by the process of categorical subdivision (KU)]. Psychologische Beiträge 27:478–493.
Hellman RP (1976) Growth of loudness at 1000 and 3000 Hz. J Acoust Soc Am 60:672–679.
Hellman RP (1991) Loudness measurement by magnitude scaling: Implications for intensity coding. In: Bolanowski SJ Jr, Gescheider GA (eds), Ratio Scaling of Psychological Magnitude. Hillsdale, NJ: Lawrence Erlbaum, pp. 215–228.
Hellman RP, Meiselman CH (1988) Prediction of individual loudness exponents from cross-modality matching. J Speech Hear Res 31:605–615.
Hellman RP, Meiselman CH (1991) Loudness relations for individuals and groups in normal and impaired hearing. J Acoust Soc Am 86:2596–2606.
Hellman RP, Meiselman CH (1993) Rate of loudness growth for pure tones in normal and impaired hearing. J Acoust Soc Am 93:966–975.
Hellman RP, Zwislocki JJ (1961) Some factors affecting the estimation of loudness. J Acoust Soc Am 33:687–694.
Hellman RP, Zwislocki JJ (1963) Monaural loudness function at 1000 cps and interaural summation. J Acoust Soc Am 35:856–865.
Hellman RP, Zwislocki JJ (1964) Loudness function of a 1000–cps tone in the presence of a masking noise. J Acoust Soc Am 36:1618–1627.
Hellman RP, Scharf B, Teghtsoonian M, Teghtsoonian R (1987) On the relation between growth of loudness and the discrimination of intensity of pure tones. J Acoust Soc Am 82:448–453.
Hellström A (1979) Time errors and differential sensation weighting. J Exp Psychol Hum Percept Perform 5:460–477.
Houtsma AJM, Durlach NI, Braida LD (1980) Intensity perception. XI. Experimental results on the relation of intensity resolution to loudness matching. J Acoust Soc Am 68:807–813.
Hübner R, Ellermeier W (1993) Additivity of loudness across critical bands: A critical test. Percept Psychophys 54:185–189.
International Organization for Standardization (1959) ISO/R 131:1959 Acoustics. Expression of the Physical and Subjective Magnitudes of Sound. Geneva: International Organization for Standardization.
International Organization for Standardization (2003) ISO 226:2003 Acoustics. Normal Equal-Loudness Contours. Geneva: International Organization for Standardization.
International Organization for Standardization (2006) ISO 16832 Acoustics. Loudness Scaling by Means of Categories. Geneva: International Organization for Standardization.

Jesteadt W (1980) An adaptive procedure for subjective judgments. Percept Psychophys 28:85–88.
Jesteadt W, Luce RD, Green DM (1977) Sequential effects in judgments of loudness. J Exp Psychol Hum Percept Perform 3:92–104.
Johnson JH, Turner CW, Zwislocki JJ, Margolis RH (1993) Just noticeable differences for intensity and their relation to loudness. J Acoust Soc Am 93:983–991.
Kohfeld DL (1971) Simple reaction time as a function of stimulus intensity in decibels of light and sound. J Exp Psychol 88:251–257.
Kohfeld DL, Santee JL, Wallace ND (1981a) Loudness and reaction time: I. Percept Psychophys 29:535–549.
Kohfeld DL, Santee JL, Wallace ND (1981b) Loudness and reaction time: II. Identification of detection components at different intensities and frequencies. Percept Psychophys 29:550–562.
Krüger JG (1743) Naturlehre [Lectures on nature]. Halle-Magdeburg: Hemmerde.
Kuwano S, Namba S (1990) Continuous judgment of loudness and annoyance. In: Müller F (ed), Fechner Day 90. Proceedings of the 6th Annual Meeting of the International Society for Psychophysics. Würzburg, Germany: ISP, pp. 129–134.
Laird DA, Taylor E, Wille HH Jr (1932) The apparent reduction of loudness. J Acoust Soc Am 3: 393–401.
Lakoff G, Johnson M (1980) Metaphors We Live By. Chicago, IL: University of Chicago Press.
Laming DRJ (1997) The Measurement of Sensation. Oxford: Oxford University Press.
Leek MR (2001) Adaptive procedures in psychophysical research. Percept Psychophys 63:1279–1292.
Levelt WJM, Riemersma JB, Bunt AA (1972) Binaural additivity of loudness. Br J Math Statist Psychol 25:51–68.
Levitt H (1971) Transformed up-down methods in psychoacoustics. J Acoust Soc Am 49:467–477.
Lewis MF (1965) Category judgments as functions of flash luminance and duration. J Opt Soc Am 55:1555–1560.
Lim LS, Rabinowitz WM, Braida LD, Durlach NI (1977) Intensity perception. VIII. Loudness comparisons between different types of stimuli. J Acoust Soc Am 62:1256–1267.
Link SW (1992) The Wave Theory of Difference and Similarity. Hillsdale, NJ: Lawrence Erlbaum.
Locke J (1690) An Essay Concerning Humane Understanding. London: Basset.
Luce RD (1994) Thurstone and sensory scaling: Then and now. Psychol Rev 101:271–277.
Luce RD, Edwards W (1958) The derivation of subjective scales from just noticeable differences. Psychol Rev 65:222–237.
Luce RD, Green DM (1972) A neural timing theory for response times and the psychophysics of intensity. Psychol Rev 79:14–57.
Luce RD, Green DM (1974) The response ratio hypothesis for magnitude estimation. J Math Psychol 11:1–14.
Luce RD, Krumhansl CL (1988) Measurement, scaling, and psychophysics. In: Atkinson RC, Herrnstein RJ, Lindzay G, Luce RD (eds), Stevens' Handbook of Experimental Psychology (2nd Ed), Vol. 1. New York: Wiley, pp. 3–74.
Luce RD, Tukey JW (1964) Simultaneous conjoint measurement: A new type of fundamental measurement. J Math Psychol 1:1–27.
MacCorquodale K, Meehl PE (1948) On a distinction between hypothetical constructs and intervening variables. Psychol Rev 55:95–107.
Marks LE (1968) Stimulus-range, number of categories, and form of the category-scale. Am J Psychol 81:467–479
Marks LE (1974a) On scales of sensation: Prolegomena to any future psychophysics that will be able to come forth as science. Percept Psychophys 16:358–375.
Marks LE (1974b) Sensory Processes: The New Psychophysics. New York: Academic Press.
Marks LE (1978) Binaural summation of the loudness of pure tones. J Acoust Soc Am 64:107–113.

Marks LE (1979) A theory of loudness and loudness judgments. Psychol Rev 86:256–285.
Marks LE (1980) Binaural summation of loudness: Noise and two-tone complexes. Percept Psychophys 27:489–498.
Marks LE (1996) Psychophysics in the scientific market-place: Peer review of grant applications. In: Masin S (ed), Fechner Day 96. Proceedings of the12th Annual Meeting of the International Society for Psychophysics. Padua, Italy: ISP, pp. 329–334.
Marks LE, Algom D (1998) Psychophysical scaling. In: Birnbaum MH (ed), Measurement, Judgment, and Decision Making. San Diego, CA: Academic Press, pp. 81–178.
Marks LE, Gescheider GA (2002) Psychophysical scaling. In: Wixted J, Pashler H (eds), Stevens's Handbook of Experimental Psychology (3rd Ed). Vol. 4. Methodology. New York: Wiley, pp. 91–138.
Marks LE, Slawson AW (1966) Direct test of the power function for loudness. Science 154: 1036–1037.
Marks LE, Hammeal RJ, Bornstein MH (1987) Perceiving similarity and comprehending metaphor. Monogr Soc Res Child Dev 42:1–91.
Marks LE, Galanter E, Baird JC (1995) Binaural summation after learning psychophysical functions for loudness. Percept Psychophys 57:1209–1216.
Marozeau J, Florentine M (2009) Testing the binaural equal-loudness-ratio hypothesis with hearing-impaired listeners. J Acoust Soc Am 126:310–317.
McGill W J (1961) Loudness and reaction time: A guided tour of the listener's private world. Acta Psychologica 19:193–199.
McGill WJ, Goldberg JP (1968) Pure-tone intensity discrimination and energy detection. J Acoust Soc Am 44:576–581.
Merkel J (1888) Die Abhängigkeit zwischen Reiz und Empfingung [The relation between stimulus and sensation]. Philosophische Studien 4:541–594.
Miller GA (1956) The magical number seven, plus or minus two: Some limits on our capacity for processing information. Psychol Rev 63:81–97.
Molino JA, Zerdy GA, Lerner ND, Harwood DL (1979) Use of the "acoustic menu" in assessing human response to audible (corona) noise from electric transmission lines. J Acoust Soc Am 66:1435–1445.
Newman E B (1933) The validity of the just noticeable difference as a unit of psychological magnitude. Trans Kans Acad Sci 36:172–175.
Ozimek E, Zwislocki JJ (1996) Relationships of intensity discrimination to sensation and loudness levels: Dependence on sound frequency. J Acoust Soc Am 100:3304–3320.
Parducci A (1965) Category judgment: A range-frequency model. Psychol Rev 72:407–418.
Parducci A (1974) Contextual effects: A range-frequency analysis. In: Carterette EC, Friedman MP (eds), Handbook of Perception, Vol. 2. Psychophysical Judgment and Measurement. New York: Academic Press, pp. 127–141.
Parker S, Schneider B (1974) Non-metric scaling of loudness and pitch using similarity and difference estimates. Percept Psychophys 15: 238–242.
Parker S, Schneider B (1980) Loudness and loudness discrimination. Percept Psychophys 28:398–406.
Piéron H (1914) Recherches sur les lois de variation des temps de latence sensorielle en fonction des intensities excitatrices [Research on the laws of variation of sensory latency as a function of excitatory intensity]. L'Année Psychologique 20:2–96.
Piéron H (1952) The Sensations: Their Functions, Processes and Mechanisms. New Haven: Yale University Press.
Pollack I (1965a) Iterative techniques for unbiased rating scales. Q J Exp Psychol 17:139–148.
Pollack I (1965b) Neutralization of stimulus bias in the rating of grays. J Exp Psychol 69:564–578.
Poulton EC (1968) The new psychophysics: Six models for magnitude estimation. Psychol Bull 69:1–19.
Poulton EC (1989) Bias in Quantifying Judgments. Hove, England: Lawrence Erlbaum.

Raab DH (1962) Magnitude estimation of the brightness of brief foveal stimuli. Science 135:42–44.

Raab D, Fehrer E, Hershenson M (1961) Visual reaction time and the Broca-Sulzer phenomenon. J Exp Psychol 61:193–199.

Reynolds GS, Stevens SS (1960) Binaural summation of loudness. J Acoust Soc Am 32: 1337–1344.

Richardson LF, Ross JS (1930) Loudness and telephone current. J Gen Psychol 3:288–306.

Riesz RR (1933) The relationship between loudness and the minimum perceptible increment of intensity. J Acoust Soc Am 5:211–216.

Robinson DW, Dadson RS (1956) A re-determination of the equal-loudness relations for pure tones. Brit J Appl Phys 7:166–181.

Savage CW (1970) The Measurement of Sensation: A Critique of Perceptual Psychophysics. Berkeley: University of California Press.

Scharf B (1959) Loudness of complex sounds as a function of the number of components. J Acoust Soc Am 31:783–785.

Scharf B (1961) Loudness summation under masking. J Acoust Soc Am 33:503–511.

Scharf B, Fishken D (1970) Binaural summation of loudness: Reconsidered. J Exp Psychol 86:374–379.

Schneider B, Parker S (1987) Intensity discrimination and loudness for tones in notched noise. Percept Psychophys 41:253–261.

Schneider B, Parker S, Stein D (1974) The measurement of loudness using direct comparisons of sensory intervals. J Math Psychol 11:259–273.

Silva I, Florentine M (2006) Effect of adaptive psychophysical procedure on loudness matches. J Acoust Soc Am 120:2124–2131.

Stevens JC (1958a) Stimulus spacing and the judgment of loudness. J Exp Psychol 56:246–250.

Stevens JC, Hall JW (1966) Brightness and loudness as a function of stimulus duration. Percept Psychophys 1:319–327.

Stevens SS (1936) A scale for the measurement of a psychological magnitude: Loudness. Psychol Rev 43:405–416.

Stevens SS (1946) On the theory of scales of measurement. Science 103:677–680.

Stevens SS (1955) The measurement of loudness. J Acoust Soc Am 27:815–829.

Stevens SS (1956) The direct estimation of sensory magnitudes – loudness. Am J Psychol 69:1–25.

Stevens SS (1958b) Problems and methods of psychophysics. Psychol Bull 55:177–196.

Stevens SS (1959) Cross-modality validation of subjective scales for loudness, vibration, and electric shock. J Exp Psychol 57:201–209.

Stevens SS (1975) Psychophysics: Introduction to Its Perceptual, Neural and Social Prospects. New York: Wiley.

Stevens SS, Galanter EH (1957) Ratio scales and category scales for a dozen perceptual continua. J Exp Psychol 54:377–411.

Stevens SS, Greenbaum HB (1966) Regression effect in psychophysical judgment. Percept Psychophys 1:439–446.

Stevens SS, Poulton EC (1956) The estimation of loudness by unpracticed observers. J Exp Psychol 51:71–78.

Stillman JA, Zwislocki JJ, Zhang M, Cefaratti LK (1993) Intensity just-noticeable differences at equal-loudness levels in normal and pathological ears. J Acoust Soc Am 93:425–434.

Susini P, McAdams S, Smith Benett K (2002) Global and continuous estimation of sounds with time-varying intensity. Acta Acoustica united with Acustica 88:536–548.

Takeshima H, Suzuki Y, Fujii H, Kumagai M, Ashihara K, Fujimori T, Sone T (2001) Equal-loudness contours measured by the randomized maximum likelihood sequential procedure. Acta Acoustica united with Acustica 87:389–399.

Teghtsoonian M, Teghtsoonian R (1983) Consistency of individual exponents in cross-modal matching. Percept Psychophys 33:203–214.

Teghtsoonian R. (1971). On the exponents in Stevens' law and the constants in Ekman's law. Psychol Rev 78:71–80.
Teghtsoonian R (1973) Range effects of psychophysical scaling and a revision of Stevens' law. Am J Psychol 86:3–27.
Teghtsoonian R, Teghtsoonian M, Canévet G (2000) The perception of waning signals: Decruitment in loudness and perceived size. Percept Psychophys 62:637–646.
Thurstone LL (1927) A law of comparative judgment. Psychol Rev 34:273–286.
Wagner E, Florentine M, Buus S, McCormack J (2004) Spectral loudness summation and simple reaction time. J Acoust Soc Am 116:1681–1686.
Ward LM (1987) Remembrance of sounds past: Memory and psychophysical scaling. J Exp Psychol Hum Percept Perform 13:216–227.
West R, Ward M, Khosla R (2000) Beyond magnitude estimation: Constrained scaling and the elimination of idiosyncratic response bias. Percept Psychophys 62:137–151.
Zeng F-G, Turner CW (1991) Binaural loudness matches in unilaterally impaired listeners. Quart J Exp Psychol 43A; 565–583.
Zimmer K (2005) Examining the validity of numerical ratios in loudness fractionation. Percept Psychophys 67:569–579.
Zwicker E (1958) Über psychologische und methodische Grundlagen der Lautheit [On psychological and methodological bases of loudness]. Acustica 8:237–258.
Zwicker E, Flottorp G, Stevens SS (1957) Critical band width in loudness summation. J Acoust Soc Am 29:548–557.
Zwislocki JJ (1965) Analysis of some auditory characteristics. In: Luce RD, Bush RR, Galanter E (eds), Handbook of Mathematical Psychology, Vol. 3. New York: Wiley, pp. 1–97.
Zwislocki JJ (1983) Group and individual relations between sensation magnitudes and their numerical estimates. Percept Psychophys 33:460–468.
Zwislocki JJ, Goodman DA (1980) Absolute scaling of sensory magnitudes: A validation. Percept Psychophys 28:28–38.
Zwislocki JJ, Jordan HN (1986) On the relations of intensity jnd's to loudness and neural noise. J Acoust Soc Am 79:772–780.

Chapter 3
Measurement of Loudness, Part II: Context Effects

Yoav Arieh and Lawrence E. Marks

3.1 Introduction

The acoustic environment is typically in a constant flux. Not only do sounds often change over time in their intensity and spectral composition, but they also commonly impinge on our ears in the company of other sounds. The dynamic ensemble of acoustic energies constitutes a Heraclitean context for the perception of auditory intensity, or loudness. Depending on how listeners direct their attention, they may focus on the loudness of an individual component discernible within the complex, on a set of components, or on the entire auditory experience – the *Gesamtempfindung*. Indeed, researchers have long recognized that the loudness of a sound heard at any moment reflects not only the acoustical energy of that particular sound, but also other sounds heard at the same time as well as the history of acoustical stimulation to which the listeners have been exposed. In other words, a sound of fixed physical properties may be judged as louder or softer depending on the context in which it is perceived. The purpose of this chapter is twofold: to review and summarize current understanding of the ways that context affects loudness and loudness judgments and to put forth a general, sequential, information-processing framework in which to describe and explain the possible sources of these effects.

A long-standing tradition in psychophysics holds that most contextual effects, especially those that arise from the stimulus set, that is, from the set of possible stimuli presented to a listener, reflect relatively late processes of judgment rather than changes in the internal perceptual representation of sound intensity (Anderson 1975; Stevens 1958b). Consider, for example, the so-called range effect in magnitude estimation: According to the "normative" loudness function, loudness judgments, such as magnitude estimations, follow a power function with an exponent of 0.6 relative to sound pressure or 0.3 relative to sound intensity. This means that the loudness judgments should double with every 10-dB increase in sound pressure. Thus, a relatively small, 10-dB stimulus range should elicit a correspondingly small

Y. Arieh (✉)
Department of Psychology, Montclair State University, Montclair, NJ 07043, USA
e-mail: ariehy@mail.montclair.edu

M. Florentine et al. (eds.), *Loudness*, Springer Handbook of Auditory Research 37,
DOI 10.1007/978-1-4419-6712-1_3, © Springer Science+Business Media, LLC 2011

response range of 2:1, and a 30-dB range should elicit a correspondingly larger response range of 8:1. But often they do not. Larger stimulus ranges commonly elicit smaller power-function exponents, meaning that larger stimulus ranges elicit smaller-than-predicted ranges of response. This is the range effect (Poulton 1968; Teghtsoonian 1973).

Most accounts of range effects attribute them to a bias in response, that is, to a shift in the rule by which subjects assign numbers to underlying percepts (Treisman 1984). Such an account, when applied to the exclusion of other possible explanations, relegates context effects to the status of unavoidable nuisances that may never be wholly eliminated, but should be minimized or neutralized as far as possible in order to reveal underlying principles of sensory functioning (Stevens 1975).

To be sure, decisional processes and response variables undoubtedly play an important role in some contextual effects. Trial-to-trial (sequential) effects provide a good example: The judgment given on trial n often depends on the stimulus and/or response given on the previous trial, $n - 1$. In particular, ratings of loudness often show sequential assimilation, a positive correlation between ratings on successive trials. The rating of loudness on trial n is typically greater if the rating on trial $n - 1$ was relatively great rather than small (Ward and Lockhead 1971; Jesteadt et al. 1977). Assimilation occurs even if the stimuli on trial n and trial $n - 1$ have different sound frequencies or are delivered to different modalities, such as vision and hearing (Ward 1985). Assimilation apparently depends on prior responses alone, and therefore presumably represents the result of decisional processes in judgment. We consider assimilation effects in detail in Sect. 3.6.

Although contextual effects may operate through decisional mechanisms, not all do. Consequently, the present chapter reviews relatively recent research on contextual effects that strongly implicates the presence of nondecisional changes, that is, changes in the underlying representation of loudness per se. For example, there is evidence that changing the range of the stimuli can modify the representation of sound intensity (Schneider and Parker 1990) and may lead to concomitant changes in loudness summation (Algom and Marks 1990). These new insights not only have important theoretical implications for understanding basic sensory processes, but also have practical, methodological implications for the design of psychophysical studies of sensory processes. In designing experiments, it is critical to be aware of the ways that context might affect the underlying sensory as well as decisional processes.

To say that no sound, indeed no stimulus, is perceived without a context is a truism, suggesting that the domain of "context" is virtually boundless and making a limited review on context effects in loudness difficult. To be sure, every aspect of the experimental situation can be broadly thought of as a context. The perception of loudness and the psychophysical functions that relate loudness to sound intensity depend on variables such as the method of measurement (e.g., category rating vs. magnitude estimation: Stevens and Galanter 1957; see Chap. 2), the inclusion of a reference sound and its value (Hellman and Zwislocki 1961; Robinson 1976), and the physical environment in which the study takes place (Marks and Aylor 1976). Indeed, when sounds are delivered from loudspeakers located behind barriers that differ in their visually perceived solidity, the judgments of loudness are directly related to the solidity of the barrier – the more solid the barrier, and hence the less visible the sound source, the louder the sounds are judged (Marks and Aylor 1976). These types of context effects are reviewed by Fastl and Florentine, Chap. 8.

The present chapter emphasizes those context effects that are contingent on the set of all possible stimuli presented in an experimental session. We review the effects on loudness judgments of sequential presentation of discrete suprathreshold sounds, usually of relatively short duration, focusing in particular on stimulus distribution effects, differential context effects, induced loudness reduction, loudness enhancement, and assimilation and contrast effects.

Note that context is defined in terms of the set of possible stimuli and not just the set of possible auditory stimuli. Contextual effects in the judgment of loudness can arise from stimuli presented in other modalities, such as vision and touch. There is evidence, for example, that sounds may be perceived as louder when accompanied by irrelevant flashes of light (Odgaard et al. 2004). The chapter concludes by reviewing recent developments in the study of cross-modal interactions in loudness judgment, where the presence of nonauditory stimuli that provide no information about the acoustic stimulus can nevertheless modify the judgment and perhaps the perception of loudness.

3.2 Hypothetical Stages in Processing Auditory Intensity

Before we explore the empirical results, we introduce a simple information-processing scheme that proposes a series of stages that lead to the perception of sound intensity. The subjective experience of loudness begins when the auditory apparatus transduces stimulus energy into nerve impulses. These impulses are propagated through several anatomical sites or stages at which auditory intensity is presumably represented by the level of neural excitation – in particular, by the discharge rate of individual neurons and the number of active neurons. Processes that determine loudness include nonlinear transduction by hair cells in the inner ear and binaural interactions in the central nervous system. Note that the distinction between peripheral and central processes is somewhat arbitrary, although many researchers consider the locus at which information from both ears combines as the demarcation between peripheral and central. Anatomically, this locus might be the superior olivary complex, where information is first extracted about interaural differences in stimulus level and time.

It is important to emphasize that the flow of information in the nervous system is bidirectional. As stimulus information is propagated from the sense organs to the central nervous system, neural signals are also delivered back to the periphery. This feedback can modify the responsiveness of the receptive organs and thus might serve as a mechanism to extend the effects of context to relatively early stages of auditory processing. Parker and Schneider (1994) have suggested, for example, that loud sounds activate a top-down gain-control mechanism that attenuates the responsiveness of the system to subsequent sounds.

Presumably, organisms have phenomenal access to loudness information only at a relatively late stage of central processing, after a sensory/perceptual representation has been formed, most likely in a network that includes the auditory cortex. This representation is considered the "pure" experience of sound intensity and traditionally is a prime target of psychophysical research seeking to elucidate the lawful relationships between the experience of loudness and parameters of the physical stimulus. Access to this subjective experience is usually gained only indirectly, however, by asking subjects to report on the magnitude of their sensations, using procedures that invariably involve cognitive processes of decision and judgment. In the magnitude-estimation procedure, for example, subjects are asked to use numbers to represent the magnitudes of sound sensations, comparing each sound either to a standard sound provided by the experimenter or to the sound heard in the previous trial. This procedure requires the subjects, at least implicitly, to decide on a range within which to choose the first response, to decide how often to use each numerical response, to decide how much the perceived intensity must change before changing the numerical response, and to decide whether and how to keep the response to a given level of loudness invariant in the face of contextual changes in the stimulus set.

The last point reflects the traditional view that context affects psychophysical judgments mainly, if not exclusively, at a late decisional stage (Stevens 1958b; Anderson 1975). According to this view, loudness is invariant as long as the sensory system, and hence the level of excitation that underlies loudness, is not perturbed (as it might be by masking noise, adaptation, or fatigue). One aim of this review is to highlight the changing conception of the role of context in loudness perception and judgment. Evidence from the present authors' own investigations and from those of other laboratories imply that manipulating the psychophysical context – especially, changing the stimulus levels presented – may affect not only decisional processes but also the sensory/perceptual representations of loudness itself.

3.3 Effects of Stimulus Distribution

3.3.1 Effects of Stimulus Distribution I: Changing Stimulus Range

How does the loudness judgment of a given sound depend on the levels of other sounds that were recently presented? In other words, how does the distribution of the stimuli affect loudness? We start by considering the effect of the intensity range on loudness judgments, a topic that stands among the best-established contextual effects. A set of stimulus levels (say, dB values) can be described by two measures. One is a measure of the central tendency (typically, the mean) and the other is a measure of the dispersion (typically, the dB difference between maximum and minimum). Figure 3.1 shows two ways in which the mean and the dispersion may be manipulated: (1) On the left side, the initial 50–80-dB range has been stretched,

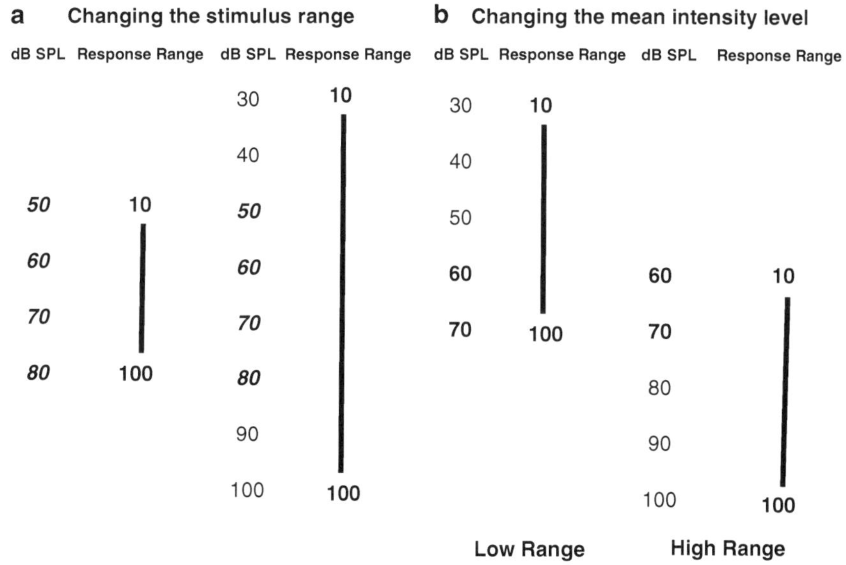

Fig. 3.1 Two examples of manipulating the distribution of stimulus intensities. In (**a**), the stimulus range is expanded symmetrically, and in (**b**), it shifted from relatively low values to relatively high values, thereby changing the mean intensity level. Also shown is a hypothetical range of numerical responses between 10 and 100, which remains constant as the stimulus range changes

in this case symmetrically, to 30–100 dB, leaving the mean unchanged at 65 dB. We dub this manipulation a change in stimulus range. (2) On the right side, the entire set of stimuli is increased uniformly in level, from 30–70 dB to 60–100 dB, increasing the mean by 30 dB, but leaving the dispersion unchanged. We dub this manipulation a change in mean level.

When magnitude estimation (ME) is used to assess loudness, the numerical judgments of loudness can typically be described to a first approximation by a psychophysical power function of physical intensity, where the exponent of the function depends systematically on the range of the stimuli. The larger the stimulus range, the smaller the exponent (Engen and Levy 1958; Poulton 1968, 1989; Teghtsoonian 1973). The exponent of the power function describes the rate at which loudness grows with intensity, and the exponent's value, in principle, should be independent of measurement method, instructions or stimulus range. Yet loudness often seems to grow more slowly with physical intensity when the stimulus range is larger, implying a change in the dynamics of loudness perception.

How does range influence loudness judgments? A simple explanation is that the changes are due to a response bias in the judgment phase and not to changes in the sensory representation of loudness. According to a common version of this explanation, subjects tend to use a constant range of numerical responses independent of the range of physical stimuli and therefore independent of the range of underlying perceptual events. Figure 3.1a shows an example in which the same response range is applied to the small range and to the large range. In this extreme relativistic scenario, there would be a perfect inverse relation between log stimulus range and exponent. In fact, a perfect inverse relation is not found with intramodal manipulations of stimulus range, because subjects show only a modest tendency to keep response range constant when stimulus range changes (Teghtsoonian 1973). This finding suggests in turn that the subjects' responses reflect a compromise between wholly absolute judgment and wholly relative judgment. In any case, this explanation places the effect of stimulus range on loudness judgment firmly in the decision and judgment, and, more explicitly, in the way that subjects use numbers to judge perceived intensity.

There is an analogous effect observed in results obtained with the method of magnitude production (MP), which inverts the magnitude-estimation task. In MP, subjects are asked to set the sound pressure levels (SPLs) of sounds so their loudness matches numbers given by the experimenter (see Chap. 2). The exponent of the loudness function obtained with MP changes in the opposite direction: increasing the range of stimuli (the numbers given to the subjects) leads to larger exponents (see Teghtsoonian, 1973; Teghtsoonian and Teghtsoonian, 1978). This happens because the subjects tend to compress the range of the response variable when the input range increases, truncating the range of numerical responses in ME and truncating the range of dB settings in MP. Truncating the range of numerical responses leads to a decrease in exponent, whereas truncating the range of dB leads to an increase in exponent. The general implication, in turn, is that the range bias is not specifically numerical, but is a more general property of judgment (e.g., Stevens and Greenbaum 1966; Stevens 1975).

On the other side of the coin, there is evidence that changing the stimulus range may also modify the underlying sensory representations themselves. For example, changes in stimulus range have been shown to affect measures of binaural loudness summation, in particular, the decibel difference between implicitly matching SPLs of monaural and binaural tones (Algom and Marks 1990). By implication, either the sensory representations of sound intensity at the two ears or the rule of binaural summation changed when context (stimulus range) changed. Importantly, analogous range effects have been reported in tasks that did not require subjects to give numerical responses. Schneider and Parker (1990) asked subjects on each trial to choose which of two pairs of tones defined the larger loudness interval. The exponent of the loudness function extracted from the judgments of intervals decreased with increasing stimulus range.

3.3.2 *Effects of Stimulus Distribution II: Changing Mean Intensity Level*

Figure 3.1b shows a second manipulation of the stimulus distribution, where the difference between the minimum and maximum intensities remains constant but the mean SPL shifts by 30 dB. The data points of interest in this design are the loudness judgments given to the subset of stimuli common to the high and low ranges – the

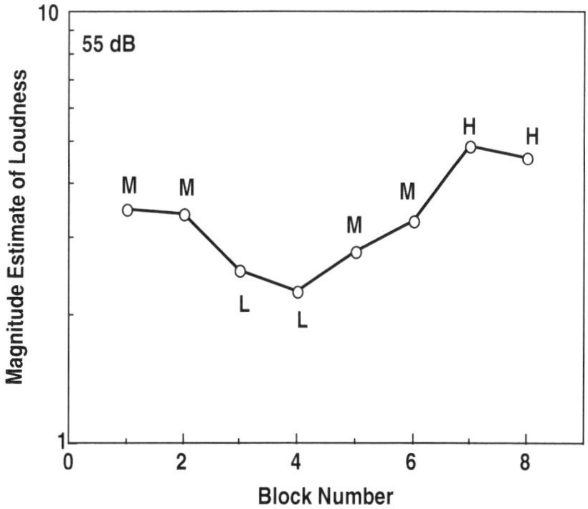

Fig. 3.2 Magnitude estimates of a 55-dB 500-Hz tone as the mean level of the stimulus in which the tone was embedded changes from medium (M) to low (L) to medium to high (H). The positive relation between averaged response and range values is an example of response assimilation (data from Marks 1993)

sounds at 60 and 70 dB (Marks 1988, 1993, for similar designs in other domains, see also Hollingworth 1910; Parducci et al. 1976; Mellers and Birnbaum 1982). The upshot of these studies is that an identical stimulus can receive different judgments when it appears in different stimulus sets. For example, using a magnitude-estimation procedure, Marks (1993) presented subjects with three sets of intensities of a 500-Hz tone: low 25–55 dB, medium 40–70 dB and high 55–85 dB SPL, all in steps of 5 dB. The order of presentations cycled through the different ranges, alternating the medium (M) set with either the high (H) or low (L) set, such that a possible block might look like: M-M-L-L-M-M-H-H. As can be seen in Fig. 3.2, the loudness judgments of the 55-dB stimulus, which was common to all three sets, shifted systematically with the stimulus range, being judged softest in the L set and loudest in the H set. Overall, mean loudness judgments given to the 55-dB stimulus varied by a factor of two across the three sets of stimuli.

This robust assimilation is not affected when a sequence of nonjudged tones or a 60-s silent interval is inserted at each point in time where the contexts (sets) switch. How can we account for such dramatic shifts in loudness judgments? Certainly, assuming that the subjects use a constant response range across shifting contexts cannot do the trick. As we show in Fig. 3.1, this hypothesis predicts the opposite result, namely, contrast. If the subjects applied a constant response range to all stimulus ranges, the common stimulus would be judged louder in the L set and softer in the H set. Helson's (1964) adaptation-level model makes the same prediction. According to Helson's model, the intensity judgment given to each stimulus depends on its position relative to an internal adaptation level (AL), which is itself determined by contextual factors such as the distribution of the stimulus levels in the set. Thus, a given stimulus will be judged softer when it falls below the

AL than when it falls above the AL. Unfortunately, the adaptation-level model, like the constant-response range hypothesis, predicts contrast rather than assimilation.

The results shown in Fig. 3.2 are consistent with the well-documented tendency for psychophysical judgments to show sequential response assimilation. That is, the judgment made to a stimulus on trial *n* correlates positively with, that is, is similar to, the response made to the stimulus on trial $n - 1$ (the correlation might extend in diminishing magnitude up to $n - 5$: Ward 1973a, 1979, 1990; Jesteadt et al. 1977; Staddon et al. 1980). With a shifting-level paradigm, as in Marks's study, the mean response to all of the stimuli within a given set increases or decreases as the mean level of the stimuli increases or decreases. Consequently, the 55-dB stimulus, common to all sets, is more likely to follow a large response in the H set than in the L set. Given assimilation of responses, the net result will be higher judgments in the H set. We elaborate on the nature of sequential assimilation (and contrast) in Sect. 3.6.

Shifting the range of stimulus intensities also affects non-numerical judgments or comparisons of loudness. Garner (1954) presented subjects with a 90-dB standard sound and asked them to decide whether various target sounds were more or less than half as loud as the standard. The target sounds could be drawn from three different ranges: 55–65, 65–75, or 75–85 dB. The values of half-loudness judgments extracted from the cumulative response frequencies increased with the intensity range, equaling 60.8, 70.1, and 80.2 dB, respectively. These values fall at the midpoint of each stimulus range, as if the subjects' responses were completely determined by the context provided by the range. It is not clear, however, whether these results are connected to the assimilation effects observed in numerical responses or whether they reflect another, more general kind of experimental bias, perhaps a demand characteristic in which subjects try to infer the "correct" behavior. In Garner's study, the subjects could have reasoned, erroneously of course, that the "correct" response for the half loudness judgments is approximately the middle of the intensity range presented, regardless of the absolute levels. Whether an example of assimilation or some other kind of effect, it is clear that stimulus context exerted a powerful influence on these loudness judgments.

3.3.3 Effects of Stimulus Distribution III: Changing Presentation Frequency and Stimulus Spacing

The preceding sections make it evident that the distribution of auditory intensities, as defined by the end points of the stimulus range and the mean level, plays a substantial role in determining subjects' responses to individual sounds. But other properties of the stimulus distribution also affect judgments. Many studies, especially those conducted from the vantage point of Parducci's range-frequency model, hold constant the end points of the stimulus range but manipulate either the relative frequency with which each stimuli is presented (leaving all of the stimulus intensities intact) or the spacing of the stimuli between the end points (Parducci 1965; Parducci and Perrett 1971; Mellers and Birnbaum 1982). These studies have mainly used a method, category rating, which requires the subjects to subdivide the stimulus continuum into a finite number of equally spaced categories (Stevens and Galanter 1957). For example, subjects may be asked to judge the loudness of sounds using six categories, characterized by the labels "very soft," "soft," "slightly softer than average," "slightly louder than average," "loud," and "very loud" (Parducci et al. 1968). In quantifying the results of experiments that use adjectival labels, successive labels are usually assigned successive integers, such as 1 through 6.

The range-frequency model proposes that the response given to a stimulus in a category-rating task is a compromise between the categories predicted by a range principle and by a frequency principle. According to the range principle, subjects tend to divide the stimulus range into equal subranges, corresponding to the available categories for judgment. According to the frequency principle, subjects tend to use the resulting categories equally often. These principles imply that the spacing of the stimuli along their physical dimension and their relative frequency of presentation within an experimental session should affect or even determine the categorical responses given to each stimulus.

In Fig. 3.3, we present three possible spacings of dB levels and use the range-frequency principle to predict how stimulus spacing will affect the psychophysical function relating category ratings to intensity level. We assume that each stimulus

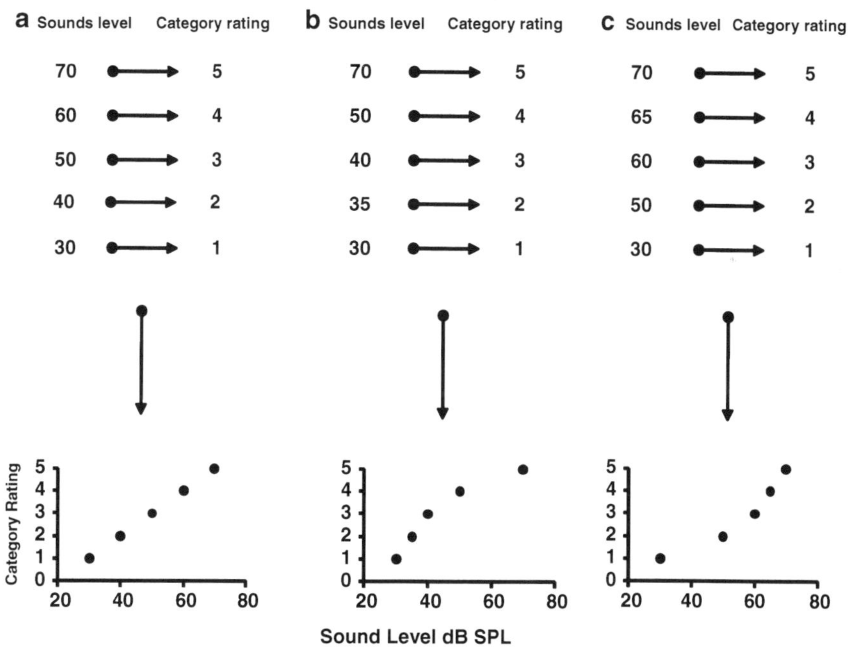

Fig. 3.3 Hypothetical example showing the effect of changing the stimulus intensity spacing on psychophysical functions for loudness, assuming that subjects tend to use response categories equally often. Equal spacing in (**a**) leads to a linear function, whereas unequal spacing in (**b**) and (**c**) leads to a steeper slope in the regions where the stimuli are bunched up

is judged once on a five-point category scale (1 being the softest sound and 5 the loudest). In case A, the five sounds are equally spaced in dB (of course, the dB scale is actually a logarithmic function of sound pressure, and of sound energy, so the stimuli are not spaced linearly in terms of either). According to the frequency principle, this spacing will lead to a linear function between average rating and stimulus intensity in dB. In case B, the stimuli are bunched at the lower end of the scale. Because subjects tend to use all five categories equally often, in case B the unequal spacing will lead to a steeper slope in the dense region of the stimulus scale – at the bottom of the function. In case C, the stimuli are bunched at the upper end of the scale, and consequently, the function's slope becomes steeper at the top.

It is not surprising that many experiments give results such as those sketched in Fig. 3.3. In category-rating experiments in particular, the instructions often encourage the subjects to use the entire range of categorical responses. Parducci's range principle applies directly to category rating, where the response scale has a fixed range and contains discrete categories, but it can also apply to magnitude estimation, where the response scale is unbounded and continuous. Not so with Parducci's frequency principle, however, which applies directly only to scales containing a limited number of discrete categories (although the frequency principle might extend to implicit sub-ranges of response in magnitude estimation). Consequently, we might expect changes in stimulus context to exert relatively greater effects on category ratings than on magnitude estimations. This is in fact what Stevens (1958a) found. Bunching sound intensities at the upper or lower end of the range had a greater effect on category ratings of loudness than on magnitude estimations.

Context-induced changes in the form of the psychophysical function imply relatively smaller or greater judgments of loudness given to individual stimuli. Note that the 50-dB tone is judged louder in case B than in case A, but softer in case C. Similar results were obtained by Stevens and Galanter (1957), who varied stimulus spacing in a classic series of category-scaling studies of loudness and many other perceptual continua. When the stimuli were spaced logarithmically (e.g., in equal dB steps), the psychophysical function relating category ratings to stimulus intensity was strongly curvilinear, being virtually a logarithmic function of stimulus intensity. As the spacing between successive stimuli became less logarithmic and more linear, the psychophysical function became less curvilinear.

A method analogous to bunching the stimuli in different regions of the range is presenting some stimuli more often than others. At the top of Fig. 3.4 appear three possible distributions of stimulus intensity of 20 sounds: (1) a rectangular distribution, where all five intensity levels are presented equally often (4 times each); (2) a positively skewed distribution, where weaker levels are presented more often than stronger ones; and (3) a negatively skewed distribution, where the stronger levels dominate. Again, we assume that each stimulus is judged on a five-point category scale (1 being the softest and 5 the loudest). According to the range principle, with stimulus intensities spaced uniformly, the boundaries around each successive category, from 1 to 5, will symmetrically bracket each successive stimulus, from lowest to highest. According to the frequency principle, subjects tend to use the categories

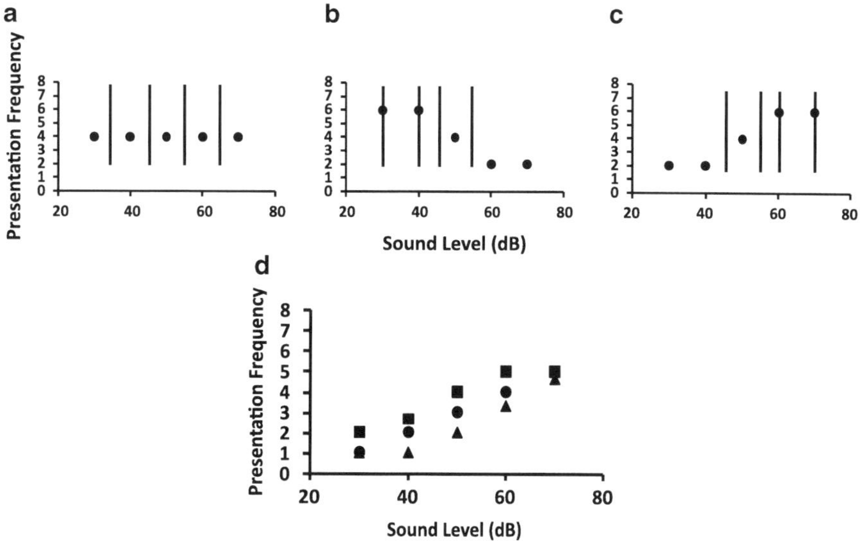

Fig. 3.4 Hypothetical example showing the effect of three different stimulus frequency distributions on psychophysical functions for loudness, assuming that subjects tend to use response categories equally often. The flat distribution shown in (**a**) leads to a linear function (*filled circles in panel* **d**), whereas the negatively (**b**, *filled triangles*) and positively (**c**, *filled squares*) skewed distributions are displaced below and above it, respectively

equally often and because there are 20 presentations in all, each of the five categories will be used 4 times. For the set of present examples, we assume that the frequency principle dominates.

With a rectangular distribution of stimuli, the frequency principle and range principle are in agreement. Each stimulus is presented four times, so each pair of category boundaries will bracket a single stimulus, and as a result each successive category is assigned four times to each successive stimulus. With a positively skewed distribution, the category boundaries are compressed at the bottom of the scale and spread out at the top. As a result, category 5 is assigned to two stimulus levels (60 dB and 70 dB, each being presented twice, making four presentations in all). On the other hand, both category 2 and category 3 are assigned to a single level (40 dB, presented six times and therefore assigned twice to category 2, and four times to category 3). Finally, with a negatively skewed distribution, the category boundaries will be a mirror image of those of the positively skewed distribution. If we average the categorical responses made to each stimulus in the three distributions and plot the averages against sound intensity, the result is the graph at the bottom of Fig. 3.4. The effect of manipulating the frequency of presentation is evident in the displacement of the three psychophysical functions. Compared to the rectangular distribution, sounds are judged louder in the positively skewed distribution (akin to what happens when bunched at the bottom of the scale) and softer in the negatively skewed distribution (akin to what happens when bunched at the top).

a Condition		b Condition	
500 Hz	2500 Hz	500 Hz	2500 Hz
35			30
40			35
45			40
50			45
55	50	**55**	50
60	55	**60**	55
65	60	**65**	60
70	65	**70**	65
	70	75	
	75	80	
	80	85	
	85	90	

Fig. 3.5 Shifting stimulus range across two frequencies. The ensembles of stimuli used to produce and measure differential context effects in loudness. Stimuli that are common to the two contextual conditions in dB SPL are underlined for the 2,500-Hz tone and marked in *bold* for the 500-Hz tone

Comparable results were reported by Parducci et al. (1968). They asked subjects to judge the combined loudness of four bursts of noise. The combined dB sum of these sounds varied between 260 and 320 dB in 5-dB steps, resulting in 13 different quadruplets. These sets of four were then presented in three frequency distributions, similar to the ones sketched in Fig. 3.5 (except that the rectangular distribution was replaced by an equivalently symmetrical normal distribution). The results were clear: the same sets of noise bursts were judged louder in the positively skewed distribution and softer in the negatively skewed distribution.

The effects on category judgments of manipulating spacing and frequency are quite general and have been observed in many domains, such as judgments of square size, social comparison, salary and grade fairness and sweetness (for a review, see Wedell and Parducci 2000). It is reasonable to assume that these effects often (but perhaps not always) originate relatively late in the processing of stimulus information, that is, in the judgment/decision stage. To account for the effects of stimulus spacing, for example, one might posit that subjects keep track of stimulus values over time, store the values in short-term memory, set category boundaries and apply the semantic labels of the categories to stimuli that fall within a category – all processes that presumably rely on relatively complex cognitive abilities. Changing stimulus spacing and frequency of stimulus presentation, like changing stimulus range, can affect the way that subjects use their response scales – in both magnitude-estimation and category-rating tasks. In each case, processes of decision and judgment appear largely responsible for the changing patterns of judgment. Nevertheless, it is possible that underlying sensory representations of loudness also change. This seems most likely to happen when highest SPLs approach or exceed about 80 dB, a level that can lead to induced loudness reduction, as described in Sects. 3.4 and 3.5.

3.3.4 Durlach and Braida's Theory of Intensity Perception

In an extensive series of articles, Durlach, Braida, and their associates developed and tested a quantitative theory of intensity perception that emphasized the role of contextual processes in auditory identification and discrimination. Although most of the studies focused on the discrimination and identification of auditory intensity, and not on judgments of loudness per se, a few of the studies did consider the relation between intensity discrimination and loudness (Lim et al. 1977; Houtsma et al. 1980; see Chap. 2, in this volume for review). Consequently, the theory and findings on the role of contextual processes in auditory discrimination and identification may be pertinent also to loudness.

In the initial version of the theory, Durlach and Braida (1969) proposed that when subjects compare and judge sounds differing in intensity, they operate in either a sensory mode (comparing and judging directly the sensory representations) or in a context-coding mode (comparing and judging the current sensory representation in relation to the stimulus context). The mode depends largely on the task. In most tasks that present sounds one at a time, subjects presumably operate in the contextual mode, where the judgment of the current stimulus is made relative the context of other stimuli previously encountered (Braida and Durlach 1972). Elaboration of the theory enabled it to account for a diverse number of phenomena, including the ways that stimulus range (Pynn et al. 1972), stimulus distribution (Chase et al. 1983), and stimulus sequence (Purks et al. 1980) influence identification and discrimination.

3.4 Differential Context Effects

The contextual effects reviewed so far all appeared in paradigms in which the contextual changes themselves were unidimensional. The effects arose when the stimuli within the set changed in just one dimension, namely, intensity. That is, subjects were presented sound intensities that varied in their range, mean, spacing, or distribution, while all other characteristics of the sounds, such as their acoustic frequency or timbre, remained constant. What happens, however, when contextual changes are made in more than one stimulus dimension? One line of research on multidimensional contextual shifts grew out of the framework of speeded classification, pioneered by Garner (1974). Those studies sought to uncover how variations along an irrelevant dimension influence how quickly and accurately people classify stimuli that differ along another, relevant dimension. Is it harder for people to classify sounds as soft or loud, for example, when frequency (pitch) varies randomly from trial to trial, rather than remaining constant over trials? Can people attend selectively to loudness in the face of concomitant variation in pitch? Research has revealed several kinds of interaction involving loudness – for instance, how variations in pitch interfere with classification of tones differing in loudness (Grau and Kemler Nelson 1988). It is not clear, however, whether these failures in selective attention involve changes in loudness per se.

Consider the following question: what happens to the loudness of sounds if the mean intensity levels vary across two frequencies instead of just one? In particular, what happens when the overall intensity range remains roughly constant across contextual conditions because the mean level at one frequency increases while the mean level at another changes to the same extent but in the opposite direction? This design was used by Marks (1988, 1992, 1993, 1994) in an extensive series of studies of what came to be labeled Differential Context Effects.

As shown in Fig. 3.5, the experimental design contains two conditions, A and B. The overall stimulus range extends by 50 dB in condition A and by 60 dB in condition B (the slightly lower values at 2,500 Hz compensate for the greater sensitivity at 2,500 Hz), but the means of the levels at 500 and 2,500 Hz both change across conditions. In condition A, the SPLs at 500 Hz are relatively low (mean = 52.5 dB) and the SPLs at 2,500 Hz are relatively high (mean = 67.5 dB), whereas in condition B, the SPLs at 500 Hz are relatively high (mean = 72.5 dB) and the SPLs at 2,500 Hz are relatively low (mean = 47.5 dB). A subset of four SPLs at each frequency is common to both contextual conditions (identified as bold in Fig. 3.5). The average judgments of loudness obtained in conditions A and B appear in Fig. 3.6, where the functions are grouped by frequency. Note that the four common stimuli are judged to have different values of loudness depending on the contextual condition. The common 500-Hz stimuli were judged louder in condition A and the common 2,500-Hz stimuli were judged louder in condition B. That is, at each frequency, loudness is smaller when the mean SPL is relatively high. The overall shift in loudness judgments summed across the two conditions is 17.7 dB, which constitutes 44% of the 40-dB physical shift between stimuli sets (20 dB at each frequency).

In this experimental paradigm, shifting contexts led to a relative increase in loudness at one frequency and a relative decrease at the other – constituting a Differential Context Effect (DCE). It is unlikely that DCEs arise from response biases operating in the task of numerical judgment. Similar results also have been reported in a direct-comparison task, where subjects were required to choose the louder of a pair of 500- and 2,500-Hz tones that could be drawn from stimuli in either condition A or condition B. The probability that a given 500-Hz tone was judged louder than a given 2,500-Hz tone was greater in condition A, whereas the probability that the 2,500-Hz tone was judged louder than the 500-Hz tone was greater in condition B (Marks).

It is also unlikely that DCEs arise from a more general decisional process that modifies the way subjects map loudness values across frequencies. According to

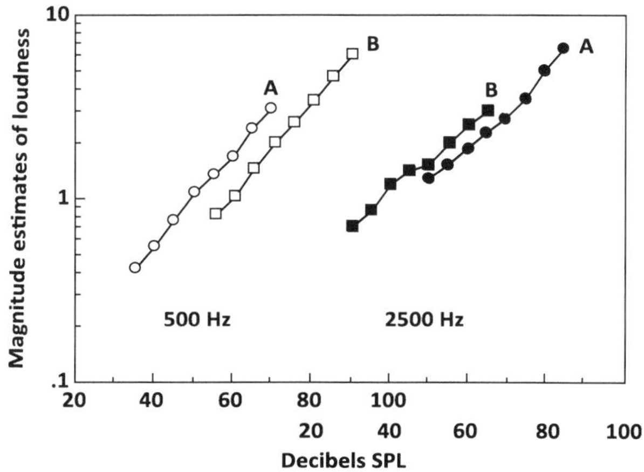

Fig. 3.6 Average magnitude estimates of loudness of 500 and 2,500-Hz tones, under two conditions of stimulus context. The common stimuli at each frequency were judged to be softer at the contextual condition where the mean SPL was higher (data from Marks 1988, reprinted by permission)

equal loudness functions, for example, those of Fletcher and Munson (1933), a 55-dB tone at 500 Hz is roughly as loud as a 50-dB tone at 2,500 Hz. But when the SPLs at 2,500 Hz are set relatively high contextually (as in condition A in Fig. 3.5), the loudness of a 55-dB tone at 500 Hz might map to the loudness of a 70-dB tone at 2,500 Hz. This could result, for instance, from a stimulus-centering bias. In many instances, people tend to center their response range at the center of the stimulus range (Hollingworth 1910; Poulton 1989). It is possible, therefore, that subjects can center their response ranges separately at different frequencies. If so, then the 55-dB SPL tone at 500 Hz and the 70-dB SPL tone at 2,500 Hz could fall at comparable positions in their respective ranges. Given this explanation, the implicit loudness matches would change with changes in context, but the underlying sensory representations of the sounds would be unaffected.

This explanation seems unlikely. Arieh and Marks (2003a) used a variant of the DCE paradigm depicted in Fig. 3.5 in order to measure response times for subjects to classify tones according to their frequency (high vs. low), while the intensity levels varied contextually across conditions. The results showed that the 2,500-Hz tones were classified more slowly in condition A and the 500-Hz tones were classified more slowly in condition B. The changes in response times mirrored the changes in loudness judgments that typify DCEs. More importantly, increases in response times were associated with increases in error rates; a positive correlation between response times and error rates is usually interpreted as a change in perceptual representation and not in decisional mapping (Wickelgren 1977).

From the data on DCEs reviewed so far, it is not possible to determine whether the changes in loudness at a given frequency are due to decrease in loudness when the mean SPL is relatively high, increase in loudness when the mean SPL is relatively low, or both. Subsequent research, however, determined that the shifts in loudness judgments across contextual conditions depend completely on reduction in the loudness of stimuli presented in high-SPL conditions. Marks (1993, Exp 15) exposed listeners to a series of repeated tones: 500 Hz at 53 dB, 500 Hz at 73 dB, 2,500 Hz at 48 dB, or 2,500 Hz at 68 dB. After exposure, listeners compared test tones at 500 and 2,500 Hz that in a baseline condition had been judged equally loud. Only exposure to the relatively high SPL at each frequency influenced the subsequent loudness comparisons. Exposure to the 500-Hz tone at 73 dB decreased the probability of judging a subsequent 500-Hz tone as louder than a 2,500-Hz tone, and exposure to the 2,500-Hz tone at 68 dB increased the probability. Exposure to the softer tones had essentially no effect.

This conclusion stands in contrast to the assimilative effects witnessed when the mean SPL varies at just one frequency; in that paradigm, greater mean SPL leads to greater judgments of loudness. When the stimulus ranges at two frequencies shift in a complementary manner, however, an adaptation-like effect reveals itself, in that loudness judgments are depressed at the frequency that contains relatively high sound levels. Because the overall range of SPLs, taken over both sound frequencies, is similar in the two contextual conditions, any assimilative effects that may arise are presumably equivalent in the different conditions and therefore not revealed. The presence of loudness reduction only in DCEs cannot be fully explained by Helson's adaptation-level theory either. If subjects set an AL independently at each frequency and judge each sound relative to its AL, then the theory predicts not only reduction in loudness when mean SPL (and AL) is high, but also enhancement of loudness when mean SPL (and AL) is low.

Other properties of the DCEs strengthen the inference that they do not originate in decisional processes but originate instead in an adaptation-like process that reduces loudness of suprathreshold tones by modifying their sensory representations. For example, DCEs show only partial transfer between the ears (Marks 1996). In Marks's study, DCEs equaled about 9.5 dB in the ear that was exposed to the relatively loud sounds but only 5.3 dB in the contralateral ear. According to a simple model that assumes independence and additive loudness reduction centrally and peripherally, the overall magnitude of the DCEs may be parceled into central and peripheral components of 5.3 and 4.2 dB, respectively. Although DCEs partly arise from or involve neural processes central to the locus of binaural integration, a substantial component appears to be peripheral and specific to the ear exposed to the relatively intense sounds.

The magnitude of the DCE depends on the difference between the two sound frequencies in a way that follows the predictions of filter (critical-band) models in hearing: the smaller the frequency difference, the smaller the changes in loudness (Marks and Warner 1991; Marks 1994). Roughly speaking, when the difference between the two frequencies is smaller than a critical bandwidth, the change in relative loudness is negligible. Presumably, when the two frequencies fall within a critical band, all of the signals are processed through a common channel, so that the intense sounds at one frequency reduce loudness equivalently at both frequencies, and thus changes in context produce no relative change in loudness, that is, no DCE.

Most studies of DCEs reported so far used a similar methodology. Using ME or its variants, subject judged the loudness of relatively short-duration tones presented one at a time and repeated many times within a block of trials. Each block of trials included two frequencies having different ranges of SPL. This method, while useful to detect the presence of DCEs, lacks the precise control over interstimulus intervals and size of stimulus steps to study in detail the processes that underlie the phenomenon, such as how changes in loudness depend on the temporal relation between sequences of tones and their intensity. Mapes-Riordan and Yost (1999) sought to distill the essence of the DCE paradigm into a series of discrete trials that consist of an inducer tone, a target tone, and a comparison tone, all of whose properties and relations can be precisely controlled. Results obtained under this new paradigm eventually received its own label: Induced Loudness Reduction (ILR, Scharf et al. 2002).

3.5 Induced Loudness Reduction

Figure 3.7 presents the basic stimulus sequences used in baseline and experimental trials to measure ILR. In baseline trials, subjects match the loudness of a target tone and a comparison tone of a different frequency. The comparison varies in intensity from trial to trial, usually by an adaptive method, to produce a level, I_B, that matches the loudness of the target. In experimental trials, an inducer, having the same frequency as the target and usually stronger than it, precedes the target and another

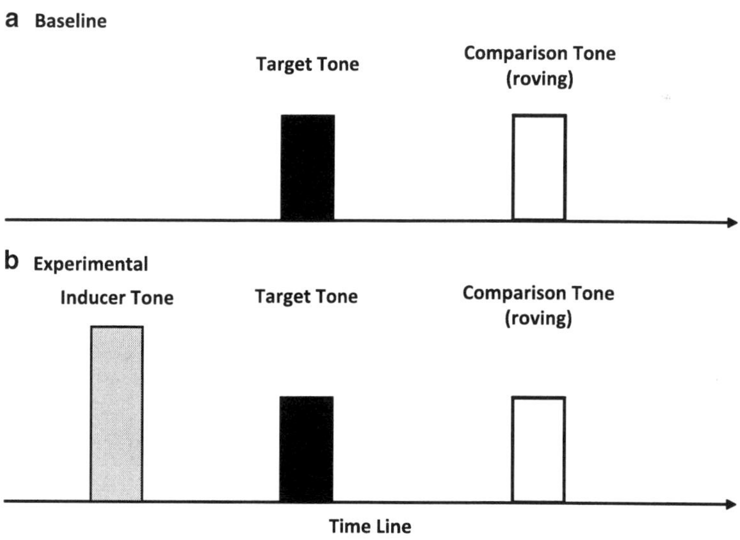

Fig. 3.7 Sequences of stimuli used to measure induced loudness reduction

match is determined between the target and the comparison, I_E. ILR is then computed as the dB difference between the two matches, $I_B - I_E$.

Optimal conditions to produce ILR are as follows: (1) the inducer is around 70–80 dB SPL and target is 10–20 dB SPL weaker; (2) the inducer and target have the same frequency, but the comparison differs from both in frequency by at least one critical band; and (3) the interstimulus interval (ISI) between inducer and target is about 1 s. Under these conditions, ILR is approximately 10 dB (Mapes-Riordan and Yost 1999; Arieh and Marks 2003a; Nieder et al. 2003). That is, the loudness of the target decreases by half (50% in sones) in the experimental condition compared to the baseline.

3.5.1 *Relation to Other Sequential Effects in Loudness*

Superficially, ILR bears an affinity to other kinds of loudness reduction (and threshold elevation) that depend on the sequence of stimulation. In forward masking, for instance, an inducing tone increases the threshold of a subsequent tone (Plomp 1964). But forward masking, unlike ILR, operates only over very short intervals, at most a few 100 ms, and may largely reflect short-term temporal integration (Fastl 1979). Longer time-courses are evident in auditory fatigue, produced by exposure to stimulation that is intense (sometimes exceeding 100 dB HL) and sustained (measured in minutes) (Hood 1950; Ward 1973b). Auditory fatigue is revealed primarily as increases in threshold (temporary threshold shift [TTS]: Mills et al. 1981), though also as decreases in loudness (temporary loudness shift, TLS: Botte and Mönikheim 1994). These consequences are consistent with the high levels of stimulus exposure; indeed, susceptibility to TTS has been tied to more permanent hearing loss wrought by years of exposure to environmental noise (Ward 1965). Whereas TTS and TLS both result from prolonged exposures to intense stimuli, ILR represents a reduction in loudness following inducing stimuli that may be as brief as 5 ms (Nieder et al. 2003) and only moderately intense (e.g., 80 dB). Moreover, fatiguing stimuli most strongly affect absolute threshold, whereas ILR appears to be absent at threshold, despite being clearly evident at suprathreshold levels (Mapes-Riordan and Yost 1999).

Another sequential effect characterized by reductions in loudness is loudness adaptation: the progressive decline in the loudness of an unchanging, continuous sound (e.g., Hellman et al. 1997; for a review of the early literature, see Scharf 1983). In loudness adaptation, as in ILR, loudness also decreases due to prior acoustic exposure, except that paradigms of loudness adaptation lack temporally discrete inducing and test signals. In the loudness-adaptation paradigm, stimulation is continuous in that no silent interval intervenes between the exposure stage and the test stage. In matching loudness, however, it is important to avoid interactions between test signal and comparison by separating the signals in time and presenting them to different ears (Hellman et al. 1997).

Surprisingly, and in contrast to other senses such as touch and taste, adaptation in loudness proved to be elusive. Initial reports of generous amounts of adaptation, over a wide range of levels and frequencies (e.g., Hood 1950), were later attributed not to adaptation per se but to interactions between the continuous tone whose loudness was putatively being measured and the probe tone presented at the end of the adapting stimulus (e.g., Elliott and Fraser 1970; Petty et al. 1970). Indeed, after the early probe procedures were abandoned in favor of methods such as magnitude estimation, measures of loudness adaptation suggested that it is a relatively small effect at all but low sensation levels. Near the absolute threshold, the decline in loudness is also known as "tone decay," a very weak tone becoming inaudible after prolonged exposure (Snyder 1973). Some loudness adaptation is observed, however, at levels as great as 60 dB above threshold (Hellman et al. 1997). Also, notably, loudness adaptation is most prominent at relatively high sound frequencies (Hellman et al. 1997).

In any case, loudness adaptation and ILR have strikingly different properties. As mentioned before, ILR is absent at low stimulus levels and as far as we know is not strongest at relatively high sound frequencies. Given the marked differences between the psychophysical behaviors of ILR, on the one hand, and forward masking, auditory fatigue, TLS and loudness adaptation, on the other, it is unlikely that ILR is mediated by any of the mechanisms that underlie the other temporal phenomena just reviewed.

3.5.2 Possible Mechanisms of ILR

One mechanism that has been offered to explain the reduction in loudness involves the descending efferent pathways that project from the medial olivocochlear neurons to the outer hair cells (Nieder et al. 2003). Feedback through the cochlear efferent system reduces vibration at the basilar membrane and activity in the auditory nerve. The efferent system responds more vigorously to ipsilateral than to contralateral stimulation, consistent with evidence that ILR is greater when induced ipsilaterally rather than contralaterally. The olivocochlear efferent system, however, produces maximum suppression near threshold (Stankovic and Guinan 1999), where ILR is virtually absent (Mapes-Riordan and Yost 1999). Thus, it seems unlikely that the olivocochlear feedback system is solely responsible for ILR.

Explanation in terms of another top-down process, one that is driven, in part, by the listeners' expectations, has been offered by Parker and Schneider (1994; see also Parker et al. 2002). They hypothesized that the presence or expectation of loud sounds at a particular sound frequency might activate, in an auditory filter centered at that frequency, a nonlinear attenuator – or gain control – that can reduce the amplitude of sounds to avoid "clipping" and enhance discrimination performance (see also Pynn et al. 1972). Parker et al. (2002) demonstrated the role of listeners' expectancies in a loudness identification task. In one condition, listeners were to identify four weak tones (e.g., 25, 30, 35, and 40 dB SPL), and in a second condition, a fifth tone, either 90 or 45 dB, was added to the stimulus set. Identification accuracy decreased only when the added tone was loud. Moreover, accuracy hardly changed when listeners were instructed to ignore a predictable loud tone, implying that the gain-control mechanism itself responds to the experimental contingencies. It is not clear, however, whether or how these results relate to ILR, where the reduction of loudness usually occurs even when the listeners are instructed to ignore the inducer (Arieh and Marks 2003b).

Monaural inducers not only reduce the loudness of subsequent tones, but also affect lateralization. Following monaural inducers, subjects tend to lateralize subsequent sounds to the contralateral ear (Arieh and Marks, 2007). The partial specificity of ILR to the ear of stimulation, coupled with the fact that relatively loud monaural inducers affect both loudness and lateralization, led Arieh and Marks to suggest that ILR reflects the operation of a general intensity-processing mechanism that reduces the magnitude of neural signals at both peripheral and central loci. This view implies that the effect of inducers on loudness is just one example of a much wider set of inducer-generated auditory aftereffects. Arieh and Marks (2003b) also suggested that inducers may set off two diametrically opposed processes: a short-term enhancement and long-term suppression.

3.5.3 Induced Loudness Reduction and Loudness Enhancement

It has been known for some time that when two tones are presented in close temporal proximity, with an ISI smaller than 100 ms, and the first tone is more intense than the second; the loudness of the second tone appears to be augmented, usually by about 10 dB. The effect has been labeled loudness enhancement and has been studied extensively, including its consequences for discrimination of loudness (Zwislocki and Sokolich 1974; Elmasian and Galambos 1975; Zeng 1994; Plack 1996). The time course of loudness enhancement is at odds, however, with the time course of ILR as mapped by Arieh and Marks (2003b). Arieh and Marks reported no change in loudness of the target tone at short ISIs and only suppression thereafter. Why the discrepancy?

Scharf et al. (2002) noted that most, if not all, studies of loudness enhancement used a three-tone paradigm similar to the one presented in Fig. 3.7, but with one crucial difference: in studies reporting loudness enhancement, the comparison tone had the same frequency as the inducer, thus making the comparison itself vulnerable to ILR. Scharf et al. hypothesized that because the comparison was usually presented a few 100 ms after the inducer (and the target), the comparison's loudness was reduced substantially by ILR. The subjects then would have to raise the level of the comparison so that its loudness will match that of the target, creating the appearance of enhancement. According to the strong version of this hypothesis, loudness enhancement does not exist at all, but is an artifact of an unfortunate choice of frequency for the comparison tone. If the comparison's frequency is changed, putative loudness enhancement disappears. A weaker version of the hypothesis of Scharf et al. would argue that most of the apparent enhancement may be attributed to ILR, but a small amount of loudness enhancement may be "real."

The general hypothesis was corroborated by Arieh and Marks (2003b), who were able to predict the amount of loudness enhancement from the time course of ILR. That is, they showed that the amount of loudness enhancement measured in the target tone was approximately equal to the predicted reduction in loudness of the comparison tone, as derived from its temporal relation to the inducer (but see Oberfeld 2007, for direct measurement of the changes in the comparison tone, showing that ILR alone may not fully explain enhancement). There still remains the puzzle, however, as to the fate of the target tone when the ISI is smaller than 200 ms.

Measures of the time course of ILR show that loudness begins to decline after about 200 ms, with little change evident before that point (Arieh and Marks 2003b). One possible explanation is that it takes about 200 ms for the inhibitory process underlying loudness reduction to develop. But this explanation is somewhat at odds with physiological evidence showing inhibitory neural processes at all levels of the auditory system occurring at ISIs shorter than 100 ms. A second possible explanation is that the inducer sets off two processes: short-term loudness enhancement and long-term loudness suppression. At short ISIs, the two processes offset each other, resulting in relatively constant loudness; at longer ISIs, however, only suppression remains, resulting in ILR.

The hypothesis that there are two contravening processes operating at short ISIs leads to the prediction, confirmed by Arieh et al. (2004), that the longer-term loudness suppression will show itself after repeated presentations of an inducer are discontinued, so there is no longer a trigger for the shorter-term enhancement. In the first series of trials, the loudness of a 60-dB target tone was unaffected when it followed an 80-dB inducer by 100 ms. When the inducer was subsequently omitted, however, the same target declined immediately in loudness by about 4–5 dB relative to baseline.

3.5.4 *Induced Loudness Reduction and Differential Context Effects: Mediated by the Same Mechanism?*

The preceding discussion of DCEs and ILR implicitly suggests that both phenomena reflect the operation of the same underlying auditory mechanism, a mechanism by which brief, relatively strong transient stimulation suppresses loudness (Epstein 2007). But is this really the case? Both effects indeed represent a reduction in loudness that is induced by transient stimulation, but do both effects reflect one and the same underlying mechanism? If so, then DCEs and ILR should reveal similar psychophysical properties.

A great deal of evidence suggests that ILR and DCEs arise largely from a common mechanism. Both ILR and DCEs require moderately intense inducers on the one hand, or context-inducing stimuli on the other. Both are frequency specific (Marks and Warner 1991; Marozeau and Epstein 2008). Both can last for relatively long periods of time, at least up to several minutes (Arieh et al. 2005; Epstein and Gifford 2006). And both show partial transfer to the contralateral ear (Marks 1996; Nieder et al. 2007).

But these communalities end when we consider the intensity-tuning properties of ILR and DCEs. In ILR, inducers of 80 dB reduce the loudness of subsequent test tones at 60–70 dB by as much as 10 dB, but reduce the loudness of test tones at 40 dB by only 4 dB, and reduce the loudness of test tones at 80 dB by only 1–3 dB (Mapes-Riordan and Yost 1999). Although results vary somewhat from study to study, the upshot is that ILR requires inducers of at least 65–80 dB and mainly affects test tones 10–20 dB lower, in the region of 50–70 dB. To better appreciate the intensity tuning of ILR, we recalculated the amount of loudness reduction from the data in Mapes-Riordan and Yost (1999) and show the result as the filled circles of Fig. 3.8. The curve clearly shows the nonuniform way that ILR depends on the level of the target tone.

The results of experiments that measure DCEs, however, imply reductions in loudness that are more or less constant (in dB) across all levels of the target (Marks 1988, 1993, 1994). In comparison to ILR, the open circles in Fig. 3.8 show the magnitude of loudness reduction calculated from data of Marks (1988); the calculations assume that DCEs are equal in both context-inducing conditions, so the reduction in loudness is half of the overall shift between the two conditions. As Fig. 3.8 makes clear, where ILR shows relatively sharp intensity tuning, DCEs do not.

Why the difference in intensity tuning? Let us assume that in fact ILR does occur at all levels of target intensity (except perhaps very close to absolute threshold). If so, then when the level of the inducer equals the level of the 80-dB target, the loudness of the target should be reduced substantially below its baseline level. Note, however, that, in the adaptive-matching procedure, the intensity of the comparison tone varies over trials. On those trials in which the level of the comparison lies near that of the 80-dB inducer, the comparison will itself produce ILR, affecting loudness on subsequent presentations of the comparison. To compensate for this reduction in its loudness, it becomes necessary to increase the SPL of the comparison to match the target. Thus, if a target of 80 dB undergoes substantial ILR, measuring it by the adaptive, three-stimulus method may underestimate it.

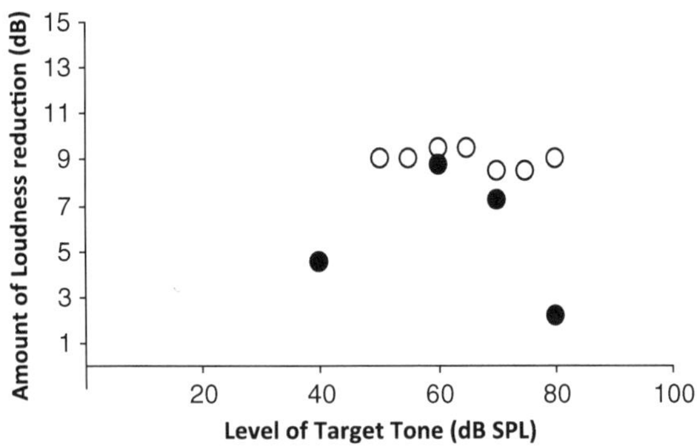

Fig. 3.8 The dependence of DCEs and ILR on the intensity of the target tone. DCEs (*open circles*) show a uniform reduction in loudness across all stimulus levels, whereas ILR (*filled circles*) show intensity tuning curve that begins at about 60–70 dB SPL (DCEs computed from data of Marks 1988 and ILR computed from data of Mapes-Riordan and Yost 1999)

Alternatively, it is also possible, of course, that the contextual effects reported by Marks and colleagues involve processes other than ILR. Other contextual effects are well known to operate in single-stimulus paradigms such as those used to measure DCEs. Chief among them are contrast and assimilation effects, discussed next.

3.6 Assimilation Effects and Contrast Effects

It may be convenient to assume that the psychophysical response made on a given trial n is determined solely by the physical intensity and other properties of the stimulus presented on that trial. Unfortunately, this is generally not true (Ward 1973a, 1990; Staddon et al. 1980; DeCarlo and Cross 1990). In fact, the subjects' responses typically depend, in systematic ways, on previous responses and on previous stimuli, even on trials as far as five back.

Many studies (e.g., Jesteadt et al. 1977; Ward 1990) have found the response on trial n to be positively correlated with the magnitude of the response on trial $n - 1$ and negatively correlated with the magnitude of the stimulus on trial $n - 1$. These coefficients have been generally interpreted as indicating two types of contextual effect. The positive response coefficients characterize assimilation, in that a high response on trial $n - 1$ increases the judgment of intensity on trial n, whereas a low response on trial $n - 1$ decreases the judgment on trial n. With assimilation, subjects tend to repeat their previous responses (see also Garner 1953). The negative stimulus coefficients, on the other hand, characterize the reverse effect, contrast, in that a more intense stimulus on trial $n - 1$ reduces the judgment given to a weaker stimulus on trial n, whereas a weaker stimulus on trial $n - 1$ increases the judgment on trial n.

Ward (1979, 1982) hypothesized that assimilation and contrast arise from different mechanisms. Assimilation to previous responses comes from a relatively high-level process of categorization, whereas contrast to previous stimuli comes from a lower-level sensory process. Ward's fuzzy judgment theory proposes that every stimulus generates an internal sensory representation, which is assumed to be a fuzzy subset of possible sensation levels, organized as an excitatory center plus an inhibitory surround. This organization causes mutual repulsion between the current representation and the traces of representations of earlier stimuli; repulsion in turn produces the observed contrast effect. Then, to choose a response, the fuzzy internal representation is compared to prototypes of sensation levels, also fuzzy, that are stored in long-term memory. The comparison of fuzzy representations often results in more than one adequate candidate for a response; the uncertainty is resolved by a heuristic process that chooses the response candidate closest to the response previously used – hence the assimilation of responses (but see DeCarlo and Cross 1990, for a different interpretation of assimilation and contrast effects).

Ward's theory predicts that assimilation will be largely unaffected by the nature of the stimulus (e.g., its acoustic frequency) because assimilation presumably arises from categorization processes that are independent of the nature of the sensory representations on which the processes operate. On the other hand, contrast will be sensitive to the make-up as well as the physical magnitude of the stimulus because it arises directly from the structure of the sensory representations. Indeed, this prediction is supported by several sets of findings using mixed-modality paradigms, in which stimuli on different trials are presented to different modalities (i.e., light and sound). In general, contrast of the current response to previous stimuli occurs only when the previous stimuli are presented to the same modality, whereas assimilation of the current response to previous responses occurs regardless of the modality of the previous stimuli (Ward 1982, 1985). Similar results obtain in mixed-frequency designs, in which sounds on different trials have different acoustic frequencies. Contrastive effects are strongest when successive sounds fall inside the same critical band. Because of the tonotopic representation of sound frequency in the auditory system, tones close in sound frequency overlap to a greater extent in their spatial representations than do tones farther apart in frequency, thereby leading to greater mutual repulsion among the former. On the other hand, assimilation is unaffected by the sound frequencies of the current and previous signals (Ward 1990).

As already mentioned, Ward (1982) found sequential effects when stimuli were presented on different trials to different sensory modalities, as well as to the same modality. DeCarlo and Cross (1990) later showed sequential dependencies in cross-modality matching, where subjects match the perceived intensity of stimuli in one modality to perceived intensity in another (see also Baird et al. 1977). Results of the studies by Ward and by DeCarlo and Cross show that sequential assimilation extends across modalities whereas contrast does not. These results locate the source of the cross-modal effects in loudness at the decisional phase of information processing. In the next section, we review pertinent research on cross-modal contextual effects on loudness, some of which lead to a different conclusion.

3.7 Cross-Modal Context Effects

A traditional way of assessing cross-modal interactions in hearing is to compare responses in two types of experimental conditions: a bimodal condition, where two stimuli (e.g., a target sound and an irrelevant light) are presented together, and a unimodal control condition, where the target sound is presented alone. Often, the subjects are asked to detect the sound or to judge its intensity, although in some cases they may be asked to report on both the sound and the light. Differences in performance between the bimodal and unimodal conditions provide evidence of cross-modal effects. By far, the most widely explored modalities in multisensory research are hearing and vision, with much of this research focusing on the effects of auditory stimuli on visual perception (Stein et al. 1996; Odgaard et al. 2003; Arieh and Marks 2008; see also Calvert et al. 2004, for a review), perhaps because complementary effects of visual stimuli on auditory perception have been elusive.

Early research on the effects of visual stimuli on auditory detection yielded a confusing array of results. Some studies reported that irrelevant visual stimuli enhanced auditory sensitivity, but only when the subjects were instructed to ignore the visual stimuli and not when they were asked to report both (Gregg and Brogden 1952; Seif and Howard 1975). Other studies revealed inconsistent effects, for example, some subjects showing enhanced auditory sensitivity and others not (Bothe and Marks 1970). Among the relatively early studies, however, only those by Bothe and Marks (1970) and Seif and Howard (1975) sought to measure or control the subjects' response criterion.

In a recent study, Lovelace et al. (2003) used signal-detection methodology to assess the contribution of sensory and decisional factors to the effect of a visual stimulus on the detection of weak sounds. The results of the first experiment revealed two effects of light flashes on auditory detection: an increase in the detection of sound (sensory effect), and a bias toward making "yes" responses (decisional effect). A second experiment reduced the temporal uncertainty of the stimulus presentation and consequently eliminated the difference in response bias, while leaving the sensory increase in sensory detection.

Lovelace et al. (2003) interpreted the improvement in sensory detection in terms of the operation of multisensory neurons in subcortical structures, such as the superior colliculus, that reach their peak firing rate only in the presence of joint auditory and visual inputs (Stein and Meredith 1993; Meredith 2002). Single-cell recording suggests that integration is optimal when the multisensory stimuli are spatially and temporally proximal. Accordingly, Lovelace et al. designed their study to maximize the conditions in which multimodal interaction could occur: they presented weak auditory stimuli simultaneously with the visual stimuli and at the same spatial location. The use of optimal stimulus conditions, together with a signal-detection method that enabled them to isolate the sensory and decisional components, may account for their positive findings.

The presence of even a small effect of a visual stimulus on auditory detection makes it at least plausible that visual stimuli might also affect the loudness of suprathreshold auditory stimuli. Odgaard et al. (2004) examined the effect of concurrently presented flashes of light on the perceived loudness of low-intensity bursts of white noise. The noise bursts were rated as louder in the presence of lights than in their absence. Importantly, the increases in loudness were unaffected by two manipulations known to modify response bias: changing the probability that the irrelevant light would accompany the sound, and changing the psychophysical task from rating the loudness of individual sounds to two-interval forced choice. Odgaard et al. interpreted the results as indicating that the light-induced enhancement of loudness may reflect an early-stage sensory interaction. It is worth noting that analogous manipulations did eliminate the enhancement of brightness judgments by irrelevant sounds, implying that the sound-induced increase in judgments of brightness reflected a late-stage decisional process

(Odgaard et al. 2003). Taken together, the results of the two studies suggest that the effect of light on loudness and the effect of sound on brightness are probably mediated by different mechanisms.

Auditory perception may also be affected by irrelevant tactile stimulation. Early reports focused on the effect of tactile stimulation on auditory thresholds and showed these effects to be largely inhibitory. Intense tactile stimulation increased auditory threshold – that is, effectively produced cross-modal masking (Gescheider and Niblette 1967). The effect was small, however, equaling about 1.6 dB under optimal conditions. Subsequent analysis using signal-detection methodology revealed that the cross-modal masking consists of a slight reduction in sensitivity coupled with an increase in the subjects' response criterion (Gescheider 1970).

Tactile stimulation may exert facilitatory effects, however, on the perception suprathreshold auditory stimuli. One study asked subjects to adjust the intensity of a weak probe tone to match the level of a standard tone. When an irrelevant vibrotactile stimulus accompanied the probe tone, subjects reduced its matching intensity, as if some of the energy of the tactile stimulus added to the energy of the probe tone (Schürmann et al. 2004). This result implies a facilitatory interaction between touch and hearing. The result is also consistent with evidence that tactile stimulation can reduce the response time to suprathreshold auditory stimuli (Murray et al. 2005). In a comprehensive study, Gillmeister and Eimer (2007) asked how a concurrent tactile stimulation affects both auditory detection and loudness. Using a two-interval forced-choice paradigm, Gillmeister and Eimer showed that the detection of near-threshold auditory sounds improved and the loudness of weak sounds increased, but only when the irrelevant vibratory stimulus was delivered simultaneously with the sounds, and not when the stimuli were asynchronous. The dependence of the facilitatory effect on temporal alignment of the sound and the vibration implied that the facilitation is rooted, at least in part, in a low-level sensory process.

The conclusion that cross-modal effects of visual and tactile stimuli on loudness can originate relatively early in information processing finds support in the rich body of physiological evidence documenting multisensory convergence in early cortical processing (Schroeder and Foxe 2004). Numerous studies have identified multisensory processes in higher-order associative parts of the brain, such as the parietal, temporal, and frontal lobes. Further, recent findings in both monkeys and humans suggest that multisensory activity could extend into the putatively unisensory stages of information processing (Giard and Peronnet 1999; Foxe et al. 2000). For example, recordings of event-related potentials (ERPs) have revealed visual and somatosensory modulation of neuronal activity in the auditory cortex, appearing just 40 ms after stimulus onset (Giard and Peronnet 1999; Foxe et al. 2000) and functional magnetic resonance imaging (fMRI) and magnetoencephalographic (MEG) recordings have revealed vibrotactile-induced activation of the auditory cortex (Levanen et al. 1998; Foxe et al. 2002). Taken together, the behavioral evidence and the physiological evidence point to machinery in the brain that could mediate early (predecisional) interactions between sound and vibration.

3.8 Summary

Contextual effects come in a variety of flavors. Even when we focus on just one perceptual dimension (loudness) and on just one type of context (stimulus context), we find evidence for two different classes of contextual effects. In many experimental paradigms, changing the distribution of the stimulus intensities at a single sound frequency effects leads to changes in loudness judgments that may be attributed largely to high-level decisional processes – more specifically, to the ways that subjects apply numbers to the magnitudes of sensation or to the ways that subjects map stimuli to predetermined response categories. Loudness judgments also change, however, from shifts in the ranges of sound intensities across two frequencies, from presenting loud sounds before relatively weak sounds, or from adding simultaneous visual or tactile stimulation. These later changes in judgment may reflect, at least in part, low-level changes in the underlying sensory representations of loudness.

It is likely that manipulations of stimulus context can evoke changes in both decisional and sensory processes. Increasing the mean intensity level of a set of stimuli can, in fact, have two opposing effects: an increase in loudness judgments due to assimilation of responses – there is a higher likelihood that a given sound will follow a strong sound and thus follow a large response – and a decrease in loudness judgments due to the adaptation-like effect of ILR, which depends on the presence of moderately intense sounds. Thus, the average judgments of loudness, such as those reported by Marks (1993) and presented in Fig. 3.2, presumably reflect the net effect of these two opposing tendencies. One might argue that the assimilation observed in Fig. 3.2 actually underestimates its magnitude because the judgments also reflect the opposing effects of ILR. One of the challenges of future research is to parcel out and quantify precisely the contributions of different sources of contextual effects – sensory and decisional – to judgments of loudness.

There has been, and continues to be, a predilection on the part of many researchers (though surely not all) to ascribe contextual effects, especially those resulting from changes in the ensembles of stimuli, to relatively late, response-based processes. Stevens (1975) treated contextual effects largely as biases in responding, as unavoidable nuisances – impediments to the discovery of general principles – that should be minimized and neutralized whenever possible. Others have sometimes sought to determine, one way or another, unbiased measures of sensory processing (e.g., Poulton 1989). By emphasizing how context can influence response processes, however, one may be blinded to the existence of changes in sensory representations.

That is not to say that non-sensory changes are less important than sensory ones. Gescheider (1988) noted that there are two fundamentally different approaches to research on psychophysical relationships. One "represents the approach of the sensory scientist whose goal is to obtain unbiased scales of sensory magnitude to study sensory processes" and the second "represents the approach of the cognitive scientist whose goal is to understand the process of judgment." (p. 183). We believe that the two approaches are complementary and are equally essential to understanding auditory processes in general and loudness in particular.

References

Algom D, Marks LE (1990) Range and regression, loudness scales, and loudness processing: toward a context-bound psychophysics. J Exp Psychol Hum Percept Perform 16:706–727.
Anderson NH (1975) On the role of context effects in psychophysical judgment. Psychol Rev 8:462–482.
Arieh Y, Marks LE. (2003a) Recalibrating the auditory system: a speed-accuracy analysis of intensity perception. J Exp Psychol Hum Percept Perform 29:523–536.
Arieh Y, Marks LE. (2003b) Time course of loudness recalibration: implications for loudness enhancement. J Acoust Soc Am 114:1550–1556.
Arieh Y, Marks LE. (2007) Posttransient shifts in auditory lateralization. Percept Psychophys 69:523–528.
Arieh Y, Marks LE (2008) Cross-modal interaction between vision and hearing: a speed–accuracy analysis. Percept Psychophys 70:412–421.
Arieh Y, Mailloux JR, Marks LE (2004) Loudness recalibration at short ISI: a closer look. J Acoust Soc Am 115:2600(A).
Arieh Y, Kelly K, Marks LE (2005) Tracking the time to recovery after induced loudness reduction. J Acoust Soc Am 117:3381–3384.
Baird JC, Green DM, Luce RD (1977) Variability and sequential effects in cross-modality matching of area and loudness. J Exp Psychol Hum Percept Perform 6:277–289.
Bothe GG, Marks LE (1970) Absolute sensitivity to white noise under auxiliary visual stimulation. Percept Psychophys 8:176–178.
Botte MC, Mönikheim S (1994) New data on the short-term effects of tone exposure. J Acoust Soc Am 95:2598–2605.
Braida LD, Durlach NI (1972) Intensity perception. II. Resolution in one-interval paradigms. J Acoust Soc Am 51:483.
Calvert G, Spence C, Stein BE (2004) The Handbook of Multisensory Processes. Cambridge, MA: MIT Press.
Chase S, Bugnacki P, Braida LD, Durlach NI (1983). Intensity perception. XII. Effect of presentation probability on absolute identification. J Acoust Soc Am 73:279–264.
DeCarlo LT, Cross DV (1990) Sequential effects in magnitude scaling: models and theory. J Exp Psychol Gen 119:375–396.
Durlach NI, Braida LD (1969) Intensity perception. I. Preliminary theory of intensity resolution. J Acoust Soc Am 46:372–383.
Elliott DN, Fraser WR (1970) Fatigue and adaptation. In: Tobias JV (ed.), Foundations of Modern Auditory Theory, Vol. 1. New York: Academic Press, pp. 115–156.
Elmasian R, Galambos R (1975) Loudness enhancement: monaural, binaural, and dichotic. J Acoust Soc Am 58:229–234.
Engen T, Levy N (1958) The influence of context on constant-sum loudness-judgments. Am J Psychol 71:731–736.
Epstein M (2007) An introduction to induced loudness reduction. J Acoust Soc Am 122:EL74–EL80.
Epstein M, Gifford E (2006) A potential carry-over effect in the measurement of induced loudness reduction. J Acoust Soc Am 120:305–309.
Fastl H (1979) Temporal masking effects: III. Pure tone masker. Acustica 43:282–294.
Fletcher H, Munson WA (1933) Loudness, its definition, measurement and calculation. J Acoust Soc Am 5:82–108.
Foxe JJ, Morocz IA, Murray MM, Higgins BA, Javitt DC, Schroeder CE (2000) Multisensory auditory-somatosensory interactions in early cortical processing revealed by high-density electrical mapping. Brain Res Cog Brain Res 10:77–83.
Foxe JJ, Wylie GR, Martinez A, Schroeder CE, Javitt DC, Guilfoyl D, Ritter W, Murray MM (2002) Auditory-somatosensory multisensory processing in auditory association cortex: an fMRI study. J Neurophysiol 88:540–543.

Garner WR (1953) An informational analysis of absolute judgments of loudness. J Exp Psychol 46:373–380.

Garner WR (1954) Context effects and the validity of loudness scales. J Exp Psychol 48:218–224.

Garner WR (1974) The Processing of Information and Structure. Potomac, MD: Erlbaum.

Gescheider GA (1970) Some comparisons between touch and hearing. IEEE Trans Man-Mach Syst. 11:28–35.

Gescheider GA (1988). Psychophysical scaling. Annu Rev Psychol 39:169–200.

Gescheider GA, Niblette RK (1967) Cross-modality masking for touch and hearing. J Exp Psychol 74:313–320.

Giard M, Peronnet E (1999) Audio-visual integration during multimodal object recognition humans: A behavioral and electrophysiological study. J Cog Neurosci 11:473–490.

Gillmeister H, Eimer M (2007) Tactile enhancement of auditory detection and perceived loudness. Brain Res 1160:58–68.

Grau JW, Kemler Nelson DG (1988) The distinction between integral and separable dimensions: evidence for integrality of pitch and loudness. J Exp Psychol Gen 117:347–370.

Gregg LW, Brogden WJ (1952) The effect of simultaneous visual stimulation on absolute auditory sensitivity. J Exp Psychol 43:179–186.

Hellman RP, Zwislocki JJ (1961) Some factors affecting the estimation of loudness. J Acoust Soc Am 33:687–694.

Hellman RP, Miskiewicz A, Scharf B (1997) Loudness adaptation and excitation patterns: effect of frequency and level. J Acoust Soc Am 101:2176–2185.

Helson H (1964) Adaptation-Level Theory: An Experimental and Systematic Approach to Behavior. New York: Harper & Row.

Hollingworth HL (1910) The central tendency of judgment. J Philos Psychol Sci Methods, 7:461–469.

Hood JD (1950) Studies in auditory fatigue and adaptation. Acta Otolaryngol Suppl 92:1–57.

Houtsma AJM, Durlach NI, Braida LD (1980) Intensity perception. XI. Experimental results on the relation of intensity resolution to loudness matching. J Acoust Soc Am 68:807–813.

Jesteadt W, Luce RD, Green DM (1977) Sequential effects in judgments of loudness. J Exp Psychol Hum Percept Perform 3:92–104.

Levanen S, Jousmaki V, Hari R (1998) Vibration-induced auditory cortex activation in a congenitally deaf adult. Curr Biol 8:869–872.

Lim LS, Rabinowitz WM, Braida LD, Durlach NI (1977) Intensity perception. VIII. Loudness comparisons between different types of stimuli. J Acoust Soc Am 62:1256–1267.

Lovelace CT, Stein BE, Wallace MT (2003). An irrelevant light enhances auditory detection in humans: a psychophysical analysis of multisensory integration in stimulus detection. Cog Brain Res 17:447–453.

Mapes-Riordan D, Yost WA (1999) Loudness recalibration as a function of level J Acoust Soc Am 106:3506–3511.

Marks LE (1988) Magnitude estimation and sensory matching. Percept Psychophys 43:511–525.

Marks LE (1992) The slippery context effects in psychophysics: intensive, extensive and qualitative dimensions. Percept Psychophys 51:187–198.

Marks LE (1993) Contextual processing of multidimensional and unidimensional auditory stimuli. J Exp Psychol Hum Percept Perform 19:227–249.

Marks LE (1994) Recalibrating the auditory system: the perception of loudness. J Exp Psychol Hum Percept Perform 20:382–396.

Marks LE (1996) Recalibrating the perception of loudness: interaural transfer. J Acoust Soc Am 100:473–480.

Marks LE, Aylor DE (1976) Perception of noise transmitted through barriers. J Acoust Soc Am 56:397–400.

Marks LE, Warner E (1991) Slippery context effect and critical bands. J Exp Psychol Hum Percept Perform 17:986–996.

Marozeau J, Epstein M (2008) Induced loudness reduction as function of frequency difference between test tone and inducer. Percept Psychophys 70:725–731.

Mellers BA, Birnbaum MH (1982) Loci of contextual effects in judgment. J Exp Psychol Hum Percept Perform 8:582–601.
Meredith MA (2002) On the neuronal basis for multisensory convergence: a brief overview. Cog Brain Res 14:31–40.
Mills JH, Gilbert RM, Atkins WY (1981) Temporary threshold shifts produced by wideband noise. J Acoust Soc Am 70:390–396.
Murray MM, Molholm S, Michel CM, Heslenfeld DJ, Ritter W, Javitt DC, Schroeder CE, Foxe JJ (2005) Grabbing your ear: rapid auditory–somatosensory multisensory interactions in low-level sensory cortices are not constrained by stimulus alignment. Cereb Cortex 15:963–974.
Nieder B, Buus S, Florentine M, Scharf, B (2003) Interactions between test-and inducer-tone durations in induced loudness reduction. J Acoust Soc Am 114:2846–2855.
Nieder B, Buus S, Cazals Y, Scharf, B (2007) Loudness reduction induced by contralateral tone (L). J Acoust Soc Am 122:35–37.
Oberfeld B (2007) Loudness changes induced by a proximal sound: loudness enhancement, loudness recalibration, or both? J Acoust Soc Am 121:2137–2148.
Odgaard EC, Arieh Y, Marks LE (2003) Cross-modal enhancement of perceived brightness: sensory interaction versus response bias. Percept Psychophys 65:123–132.
Odgaard EC, Arieh Y, Marks LE (2004) Brighter noise: sensory enhancement of perceived loudness by concurrent visual stimulation. J Cog Affect Behav Neurosci 4:127–132.
Parducci A (1965) Category judgment: a range-frequency model. Psychol Rev 72:407–418.
Parducci A, Perrett LF (1971) Category rating scales: effects of relative spacing and frequency of stimulus values. J Exp Psychol 89:427–452.
Parducci A, Thaler H, Anderson, NH (1968) Stimulus averaging and the context for judgments. Percept Psychophys 3:145–150.
Parducci A, Knobel, S, Thomas C (1976) Independent contexts for category ratings: a range-frequency analysis. Percept Psychophys 20:360–366.
Parker S, Schneider BA (1994) The stimulus range effect: evidence for top-down control of sensory intensity in audition. Percept Psychophys 56:1–11.
Parker S, Murphy DR, Schneider B (2002) Top-down gain control in the auditory system: Evidence from identification and discrimination experiments. Percept Psychophys 64:598–615.
Petty JW, Fraser WD, Elliott DN (1970) Adaptation and loudness decrement: a reconsideration. J Acoust Soc Am 47:1074–1081.
Plack CJ (1996) Loudness enhancement and intensity discrimination under forward and backward masking. J Acoust Soc Am 100:1024–1030.
Plomp R (1964) Rate of decay of auditory sensation. J Acoust Soc Am 36:277–282.
Poulton EC (1968) The new psychophysics: six models for magnitude estimation. Psychol Bull 69:1–19.
Poulton EC (1989) Bias in Quantifying Judgments. Hove, England: Lawrence Erlbaum.
Purks SR, Callahan DJ, Braida LD, Durlach NI (1980) Intensity perception. X. Effect of preceding stimulus on identification performance. J Acoust Soc Am 67:634–637.
Pynn CT, Braida LD, Durlah NI (1972). Intensity perception. III. Resolution in small-range identification. J Acoust Soc Am 51:559–566.
Robinson GH (1976) Biasing power law exponents by magnitude estimation instructions. Percept Psychophys 19:80–84.
Scharf B (1983) Loudness Adaptation. In: Tobias JV, Schubert ED (eds), Hearing Research and Theory, Vol. 2. New York: Academic Press, pp. 1–56.
Scharf B, Buus S, Nieder B (2002) Loudness enhancement: induced loudness reduction in disguise? J Acoust Soc Am 112:807–810.
Schneider B, Parker S (1990) Does stimulus context affect loudness or only loudness judgment? Percept Psychophys 48:409–418.
Schroeder CE, Foxe JJ (2004) Multisensory convergence in early cortical processing. In: Calvert G, Spence C, Stein BE (eds), The Handbook of Multisensory Processes. Cambridge, MA: MIT Press, pp. 295–310.

Schürmann M, Caetano G, Jousmäki V, Hari R (2004) Hands help hearing: facilitatory audiotactile interaction at low sound-intensity levels. J Acoust Soc Am 115:830–832.

Seif TF, Howard JH (1975) Stimulus compatibility effects of an accessory visual stimulus on auditory sensitivity. Percept Psychophys 17:504–510.

Snyder JM (1973) Threshold adaptation in normal listeners. J Acoust Soc Am 53:435–439.

Staddon JER, King M, Lockhead GR (1980) On sequential effects in absolute judgments experiments. J Exp Psychol Hum Percept Perform 6:290–301.

Stankovic KM, Guinan JJ (1999) Medial efferent effects on auditory-nerve responses to tail-frequency tones. I. Rate reduction. J Acoust Soc Am 106:857–869.

Stein BE, London N, Wilkinson LK, Price DD (1996) Enhancement of perceived visual intensity by auditory stimuli: a psychophysical analysis. J Cognitive Neurosci 8:497–506.

Stein BE, Meredith MA (1993) The Merging of the Senses. Cambridge, MA: MIT Press.

Stevens, JC (1958a) Stimulus spacing and the judgment of loudness. J Exp Psychol 56:246–250.

Stevens SS (1958b) Adaptation-level vs. the relativity of judgment. Am J Psychol 71:633–646.

Stevens SS (1975) Psychophysics: An Introduction to Its Perceptual, Neural, and Social Prospects. New York: John Wiley & Sons.

Stevens SS, Galanter EH (1957) Ratio scales and category scales for a dozen perceptual continua. J Exp Psychol 54:377–411.

Stevens SS, Greenbaum HB (1966) Regression effect in psychophysics. Percept Psychophys 1: 439–446.

Teghtsoonian R (1973) Range effects in psychophysical scaling and a revision of Stevens' law. Am J Psychol 86:3–27.

Teghtsoonian R, Teghtsoonian M (1978) Range and regression effects in magnitude scaling. Percept Psychophys 24:305–314.

Treisman M (1984) A theory of criterion setting: an alternative to the attention band and response ratio hypotheses in magnitude estimation and cross-modality matching. J Exp Psychol Gen 113:443–463.

Ward LM (1973a) Repeated magnitude estimation with a variable standard: sequential effects and other properties. Percept Psychophys 13:193–200.

Ward LM (1979) Stimulus information and sequential dependencies in magnitude estimation and cross-modality matching. J Exp Psychol Hum Percept Perform 5:444–459.

Ward LM (1982) Mixed-modality psychophysical scaling: sequential dependencies and other properties. Percept Psychophys 31:53–62.

Ward LM (1985) Mixed-modality psychophysical scaling: inter- and intramodality sequential dependencies as a function of lag. Percept Psychophys 38:512–522.

Ward LM (1990) Critical bands and mixed-frequency scaling: sequential dependencies, equal-loudness contours and power function exponents. Percept Psychophys 47:551–562.

Ward LM, Lockhead G (1971) Response system processes in absolute judgment. Percept Psychophys 9:73–78.

Ward WD (1965) The concept of susceptibility to hearing loss. J Occup Med 12:595–607.

Ward WD (1973b) Adaptation and fatigue. In: Jerger J (ed), Modern Developments in Audiology. New York: Academic Press, pp. 301–344.

Wedell DH, Parducci, A (2000) Social comparison: lessons from basic research on judgment. In: Suls J, Wheeler L (eds), Handbook of Social Comparison: Theory and Research. Dordrecht, the Netherlands: Kluwer, pp. 223–252.

Wickelgren WA (1977) Speed-accuracy tradeoff and information processing dynamics. Acta Psychol 41:67–85.

Zeng F-G (1994) Loudness growth in forward masking: relation to intensity discrimination. J Acoust Soc Am 96:2127–2132.

Zwislocki JJ, Sokolich WG (1974) On loudness enhancement of a tone burst by a preceding tone burst. Percept Psychophys 16:87–90.

Chapter 4
Correlates of Loudness

Michael J. Epstein

4.1 Introduction

This chapter reviews two issues related to responses correlated with loudness. First, the physiological effects of loud sounds are examined. Then, specific indirect measures, both perceptual and physiological that correlate with loudness growth, are summarized.

4.2 The Physiological Effects of Loud Sounds

High-intensity sounds can elicit reflexes, alter the physiological responses of the auditory system, and disrupt concentration, cognition, and sleep. This section explores some of the physiological effects of loud sounds along with the psychological and perceptual results of these effects. Most often, only the potential auditory harm of loud sounds is considered. However, the effects of sound exposure are not limited to the auditory system as will be explained in the following sections.

4.2.1 Nonauditory Effects

4.2.1.1 Acoustic Startle Reflex

The acoustic startle reflex is the response elicited by unexpected loud sounds. In human beings, the reaction includes physical movement away from the stimulus, a contraction of the muscles of the arms and legs, and often blinking. Saccadic eye movement toward

M.J. Epstein (✉)
Auditory Modeling and Processing Laboratory, Department of Speech-Language Pathology and Audiology, The Communications and Digital Signal Processing Center, Department of Electrical and Computer Engineering, Northeastern University, Boston, MA 02115, USA
e-mail: m.epstein@neu.edu

M. Florentine et al. (eds.), *Loudness*, Springer Handbook of Auditory Research 37,
DOI 10.1007/978-1-4419-6712-1_4, © Springer Science+Business Media, LLC 2011

the location of the stimulus accompanies these reactions. In addition, the reflex can sometimes include changes in blood pressure and respiration. The response typically begins with a blink and a facial grimace and ends with neck, shoulder, and back contraction to assume a shrunken position. These reactions are generally considered fast, reflexive responses to help avoid nearby danger during the brief period before fleeing is possible (Davis 1984). Human startle response is also affected by general emotional state. Abnormal reflexes are observed, for example, in posttraumatic stress disorders. In some cases, responses are heightened, in others lessened (Morgan et al. 1995, 1996, 1997a,b; Krystal et al. 1997; Klorman et al. 2003).

4.2.1.2 Short-Term Psychophysical Effects of Loud Sounds

A number of investigators have examined the nonauditory somatic response to loud sounds including early work by Davis et al. (1955) in which the complex of responses to sound was termed the N-response. The N-response includes: vasoconstriction of peripheral blood vessels coupled with increases in heart rate, deep breathing, and changes in skeletomuscular tension. These changes possibly also result in increased blood flow to the brain as the primary circulatory constriction occurs in the peripheral systems. Additional effects include increases in gastrointestinal motility (Davis and Berry 1964) and glandular activity leading to changes in blood chemical composition (Hale 1952).

4.2.1.3 Long-Term Psychophysical Effects of Loud Sounds

It is well established that exposure to undesired, loud sounds can cause significant annoyance (see Chap. 8), but the effects of exposure extend beyond simple nuisance. Numerous studies examining the effects of children's chronic exposure to loud sounds have shown clear and significant increases in resting blood pressure (see Evans and Lepore 1993 for a review of effects on children). Loud sounds have also been shown to influence the physiological and behavioral responses of fetuses (Uziel 1985; Chodynicki et al. 1986; Thurston and Roberts 1991).

In adults, there is some debate about whether there is habituation to noise, such that the physiological and psychological effects are less pronounced after extended exposure periods (Thompson 1993). There is, however, much evidence that continued annoyance, if present, may lead to stress responses that contribute to illness (van Dijk et al. 1987). Some studies have examined the effects of occupational exposure to loud sounds on the cardiovascular system (see Stansfeld and Matheson 2003 for a summary). Many of these studies found increases in blood pressure associated with noisy jobs. Additionally, it has been shown that industrial noise exposure results in increases in adrenaline release (Cavatorta et al. 1987). However, this evidence is somewhat tempered by the findings of Melamed et al. (1997, 1999), who provided evidence that the effects of noise exposure may be partly synergistic with the effects of job complexity and that without the presence of job-related stress, noise exposure may not have long-term nonauditory effects.

In addition to occupational noise, community noise exposure has been shown to systematically affect cardiovascular health. Several studies examining communities near aircraft noise have found increased prevalence of high blood pressure and provided evidence that aircraft noise exposure likely increases the risk of heart disease at least a small amount (Babisch et al. 1988, 1993a,b; Rosenlund et al. 2001). Surveys of communities in which excessive noise is present have also shown high percentages of symptomatic reports including headaches, difficulty sleeping, and general anxiety (Horne et al. 1994; Vlek 2005; Kjellstrom et al. 2007; Jarup et al. 2008).

The precise physiological effects of noise exposure, both short and long term, are still enough in debate that determining precise conclusions regarding general safety and predicting risk factors associated with specific exposures is quite challenging.

4.2.1.4 Cognitive Effects of Loud Sounds

The cognitive effects of loud sounds have been studied more thoroughly in children than in adults. In particular, background noise has been shown to inhibit learning in school-age children. This has been consistently shown in many diverse studies (see Evans and Lepore 1993 for review). Most of these studies, however, indicate that loudness has less of an effect than the information interference of the competing sounds, with competing "irrelevant" speech causing a greater disruption than other types of background sounds. This indicates that there is likely a cognitive component playing a role in addition to simple loss of auditory information (see Nelson and Soli 2000 for a summary).

Exposure to loud sounds is believed to also negatively influence memory physiologically and result specifically in changes in the morphology of Nissl bodies of neurons (Cui et al. 2009). This physiological, long-term change may be partly reflected in children who are exposed to chronic environmental noise. This population tends to have poorer auditory discrimination and speech perception as well as lower scores on standardized tests (Haines et al. 2001). However, much of the cognitive disruption caused by noise may not actually be auditory, but rather rooted in a central process. Glass and Singer (1972) presented listeners with background noise that, for some listeners, could be controlled, and for others, could not. Even when presented with the ability to control the noise, listeners infrequently did. When both groups were tested on another task afterward, the group who could not control the noise in the original condition performed significantly worse. Glass and Singer proposed that this resulted from stress caused by a feeling of helplessness. The listeners who previously could control the sound were then less affected by this stress.

4.2.1.5 Sleep Disruption

There is still some debate regarding the degree to which noise disrupts sleep. Although it is clear that indoor noise events can disrupt sleep, some investigators believe that listeners habituate to outside noise. Most studies find that aircraft noise, for example, does not increase sleep disturbance as measured by awakenings, but does increase blood pressure and perhaps alter the microstructure of sleep (Fidell et al. 2000; Carter et al. 2002; Franssen et al. 2004; Basner et al. 2006, 2008a,b; Schapkin et al. 2006; Aydin and Kaltenbach 2007; Haralabidis et al. 2008). This indicates that the physiological and psychological effects of noise may be more subtly presented than one might expect. [See Kryter (1970; 1972) for additional information on the non-auditory effects of noise exposure.]

4.2.2 Auditory Effects

Somewhat more is known about the auditory effects of loud sounds. These include temporary perceptual, and possibly protective phenomena, as well as morphological changes resulting in permanent damage to the auditory system.

4.2.2.1 Temporary Effects

There are several relatively short-term effects of exposures to loud sounds. Some of these are likely protective, others adaptive for optimal information receipt, and some likely result from cellular incapacitation.

Acoustic Reflex

The acoustic reflex occurs when the muscles of the middle ear contract in response to loud ipsilateral or contralateral stimuli. The contraction of the stapedius muscle results in a decrease in middle-ear sound transmission, primarily at low frequencies. The tensor tympani muscle may also be involved in the reflex, but it is not clear how the contraction of the tensor tympani mediates sound transmission. The elicitation of the acoustic reflex does not appear to occur by stimuli with fixed loudness as a function of frequency (Margolis and Popelka 1975). Therefore, it is not clear how loudness might be related to reflex threshold. Popelka et al. (1974) also found that the bandwidth of the elicitor affected reflex onset and strength, indicating that spectral summation or possibly spectral loudness summation has a role in the acoustic reflex.

Rabinowitz (1977) measured the attenuation as a function of frequency resulting from the acoustic reflex as a function of reflex-activator level (shown in Fig. 4.1). As activator level increased from 85 to 110 dB sound pressure level (SPL), the amount of attenuation at low frequencies varied from 0 dB to approximately 10 dB. For stimuli above 2 kHz, the reflex provided no attenuation regardless of reflex-activator level.

There are a number of theories regarding the function of the acoustic reflex. Some believe it to be a protective mechanism, but others question the value of a protective

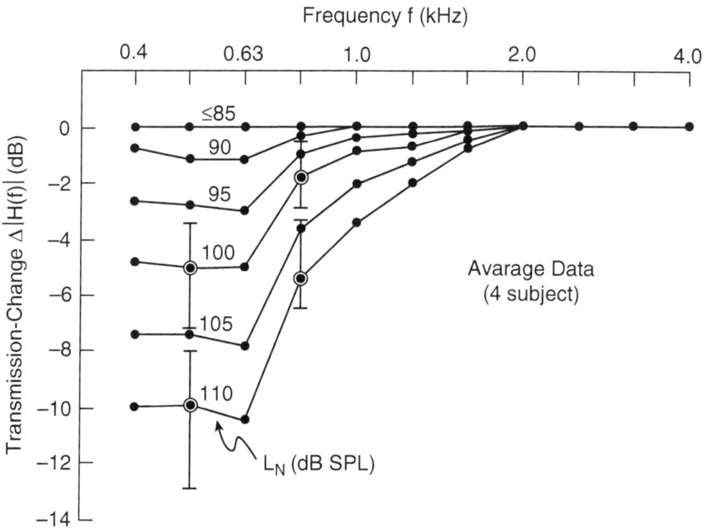

Fig. 4.1 Average results for the acoustic-reflex effects on middle-ear transmission. Average of curves fitted to the data for four subjects with elicitor level, L_N, as a parameter. Vertical bars indicate maximum and minimum (fitted) values over the four subjects for the circled data points (reprinted with permission from Rabinowitz 1977)

mechanism that begins some time after exposure to a loud stimulus. In some cases, an anticipatory acoustic reflex may occur as a result of the expectation of a loud sound. Bench (1971) found that young adults controlling the onset of a loud stimulus began to exhibit acoustic reflexes prior to the onset of the sound after some conditioning. When the sound onset was predictable, but not subject controlled, no anticipatory reflex was observed. Because the viability of a protective reflex is unclear, some investigators believe it may have multiple purposes (Sokolovski 1973) and have suggested that the reflex may serve to unmask speech information from the spread of excitation caused by high-level, low-frequency sounds.

Temporary Loudness Shift

Exposure to high-level sounds, particularly for extended periods, causes a reduction in the loudness of sounds. This is known as temporary loudness shift ((TLS), a.k.a. loudness fatigue) or specifically, in the case of changes to sound detectability at threshold, temporary threshold shift (TTS). Although exposure level has a significant effect on TTS, laboratory studies examining human TTS limit exposures to avoid the danger of possible permanent damage. Typically, the maximum temporary change in threshold reported from relatively modest noise exposure tends to be reported in the 20–40 dB range. Hirsh and Ward (1952) observed that after just 3 min of exposure to a 500-Hz tone set to 120 dB SPL, a maximum temporary threshold shift of about 20 dB occurred 2 min after the tone ended. The maximum shift also tends to occur for sounds around 4 kHz, even when exposure occurs at lower frequencies. This is also consistent with the greater likelihood of noise-induced hearing loss close to this frequency. Hirsh and Ward (1952) also found that recovery for the 3-min exposure took slightly over 10 min. The amount of TTS asymptotes after about 8 h of exposure and recovery begins rapidly, with near complete recovery after about 24 h. (See Fig. 4.2 for a data summary and Melnick (1991) for a thorough review of TTS in human subjects.)

In nonexperimental scenarios with dangerous noise exposures, it is likely that there may be greater maximum TTS and recovery times as long as 3–5 days. The few non-experimental TTS cases examined in the literature are often a result of noise exposures like gun fire (Bapat and Tolley 2006) or loud music (Sadhra et al. 2002). Rabinowitz (2000) examined the case of a young girl who attended a rock concert the night before an auditory evaluation. She showed a temporary 30-dB hearing loss at 4 kHz, which returned to normal after several days of recovery time.

It is not clear that loudness fatigue occurs in the same quantities as temporary threshold shifts. In other words, if threshold has changed 20 dB, it does not necessarily mean that the loudness of an 80 dB SPL sound will be equivalent to that of a 60 dB SPL sound in a healthy ear. In fact, Botte et al. (1993) hypothesized that loudness fatigue and TTS result from two independent, but correlated, mechanisms. Regardless, when TTS occurs, loudness fatigue is typically also present.

Physiologically, there is evidence that TLS may result from multiple mechanisms including changes in the hair-cell activity, possibly mediated by stereocilia rootlet length reduction (Liberman and Dodds 1987), reduction in neural activity, and temporary sensory-cell incapacitation due to the rapid production of metabolic waste products during heightened activity.

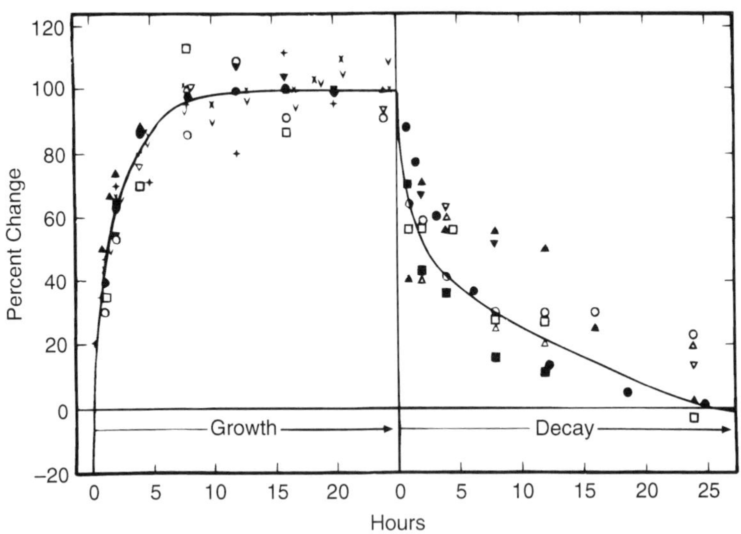

Fig. 4.2 Growth and decay of threshold shifts as a percent change where the asymptotic thresholds shift equals 100% (reprinted with permission from Melnick 1991)

Adaptation

Loudness adaptation is the decrease in loudness of a sound presented continuously over an extended period of time. Loudness adaptation differs from fatigue because it does not result from exposure to loud sounds and occurs only for sounds presented below about 30-dB sensation level (SL), with high-frequency tones adapting more than low-frequency tones. In fact, it is possible for continuous tones at high frequencies and low levels to become completely inaudible over time. If sounds are amplitude modulated, even relatively slightly, the effect tends to disappear (see Scharf 1983 for review). This indicates that loudness adaptation is useful for helping a listener ignore continuous background noise with constant level and spectral characteristics.

Induced Loudness Reduction

Induced loudness reduction (ILR), also known as recalibration, is a phenomenon by which a preceding higher-level tone (an inducer tone) reduces the loudness of a lower-level tone (a test tone). The amount of ILR depends on several factors including tone levels, frequency separation between inducer and test tones, the durations of inducer and test tones, the time separation between inducer and test tones, and the number of exposures to inducers. The effects of the ILR can last anywhere from several milliseconds to several minutes depending on the experimental scenario. Unlike similar phenomena, such as adaptation and fatigue, ILR primarily reduces the loudness of sounds at moderate levels. This could be a primary source of experimental loudness "biases" that cause discrepancies between estimates of loudness performed by listeners at the beginning of an experiment, when a listener has little prior sound exposure, and at the end of an experiment, after the listener has been exposed to higher-level sounds (see Epstein 2007 or Chap. 3 for a review). Because only some sound levels are affected, ILR alters the shape of a listener's loudness function.

4.2.2.2 Permanent Hearing Loss

Noise exposure can result in permanent threshold shift (PTS), a permanent reduction in auditory sensitivity. In addition to exposure time, sound kurtosis (i.e., peakedness), (Hamernik, and Qiu 2001) and sound level, a number of factors including age, genetics, and overall health contribute to the likelihood of sustaining a hearing loss due to noise exposure (Daniel 2007).

Cochlear Damage

Exposure to high-level sounds, both long and short term can result in permanent changes to cochlear integrity. Most commonly, damage initially begins with the outer hair cells. Outer-hair-cell stereocilia can be dislodged from the tectorial membrane or even torn from the cells. Tip links connecting the stereocilia may break or disconnect resulting in decreased synchronous movement of the stereocilia with hair cell movement (Husbands et al. 1999). Outer hair cells themselves may also break or detach from the organ of Corti. In extreme cases, the entire organ of Corti may either become dislodged or die from chemical irregularities resulting from the mixture of endolymph and perilymph when Reissner's membrane is damaged by vibration, removing the barrier between the scala media and scala vestibuli. This damage most often occurs around the first turn of the cochlea in the region that has a characteristic frequency of about 4 kHz. Figure 4.3 shows the flat preparations of the organ of Corti (OC) from a control (A) and a noise-exposed chinchilla (B).

Neuronal Loss or Reorganization

Until recently, most investigators remained concerned primarily about damage to the organ of Corti resulting from exposure to loud sounds. More recent research has indicated, however, that in many cases, neuronal losses are just as significant, if not more significant. Norena and Eggermont (2006) found that cats exposed to loud sounds for a prolonged period showed a reorganization of the tonotopic map in the auditory cortex and an increase in the synchrony of spontaneous activity. Interestingly, these changes were also found to be reversible with immediate enriched-sound stimulation. That is, after excessive exposure to noise, exposure to moderate-level sounds rich in high-frequency content helped prevent the development of some abnormalities in neural activity. Figure 4.4 shows an image from Kujawa and Liberman (2006), who found that mice exposed to loud sounds at a young age showed greater degeneration of neural auditory pathways over their lifespan, even when the young mice did not exhibit substantial damage to the organ of Corti.

Hearing Loss

Hearing loss resulting from exposure to loud sounds most often occurs around 4 kHz. In the early stages of noise-induced hearing losses, it is typical to see audiograms showing decreased hearing at 4 kHz with recovery above 6 kHz. As the loss worsens, the width of the notch increases and often high frequencies are all affected. For more information on hearing loss, see Chap. 9.

Tinnitus

Tinnitus is the perception of sound without any external stimulus present. Most tinnitus takes the form of tonal or narrowband sounds, but there are also many reports of other types of sounds including buzzing, hissing, clicking, roaring, and wind. The perception can be constant or intermittent. Often, the tinnitus is present constantly, but is only perceived when background noise subsides, frequently when laying down to sleep. Not all tinnitus is associated with exposure to loud sounds,

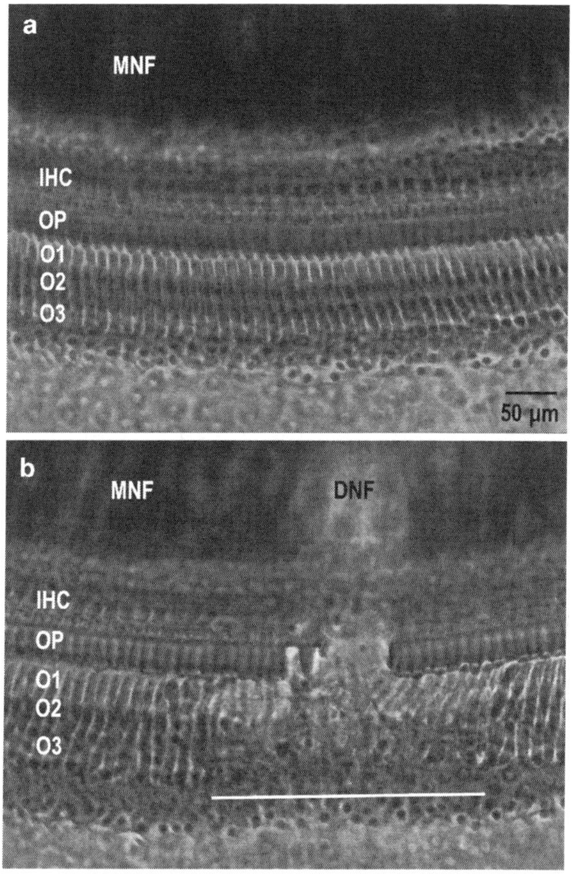

Fig. 4.3 Specially prepared image: Original photomicrographs of inner ears from experimental chinchillas provided by Bohne. Low-power photomicrographs of flat preparations of the organ of Corti (OC) from a control (**a**) and a noise-exposed chinchilla (**b**). (**a**) When undamaged, the cells of the OC (e.g., inner hair cells (IHC), outer hair cells (O1, O2, O3), outer pillars (OP)) have uniform appearances and orientations. The peripheral processes of the spiral ganglion cells (SGCs) are myelinated (MNF; appear *black* because of fixative) before they enter the OC. (**b**) Noise exposure (i.e., octave band of noise with a center frequency of 0.5 kHz, sound pressure level of 95 dB, 24 h) caused a focal loss of all OHCs, some IHCs and outer pillars (at horizontal white line) over a distance of 0.17 mm in the middle of the first cochlear turn. Note that the myelinated peripheral processes of the SGCs have degenerated (DNF) adjacent to the center of the OC lesion

but it is one very common cause. In these cases, it is possible that tinnitus results from abnormal neural synchrony (Norena and Eggermont 2006).

Studies on the loudness of tinnitus indicate that most tinnitus sufferers match their tinnitus loudness to the loudness of sounds at relatively low sensation levels (Vernon and Meikle 2003). However, the degree of annoyance often does not correspond with the loudness of the tinnitus (Hillera and Goebelb 2007). Judgments of loudness are also often confounded by annoyance and the meaning of the sound (see Chap. 8). Therefore, the interpretation of tinnitus loudness measures can be difficult.

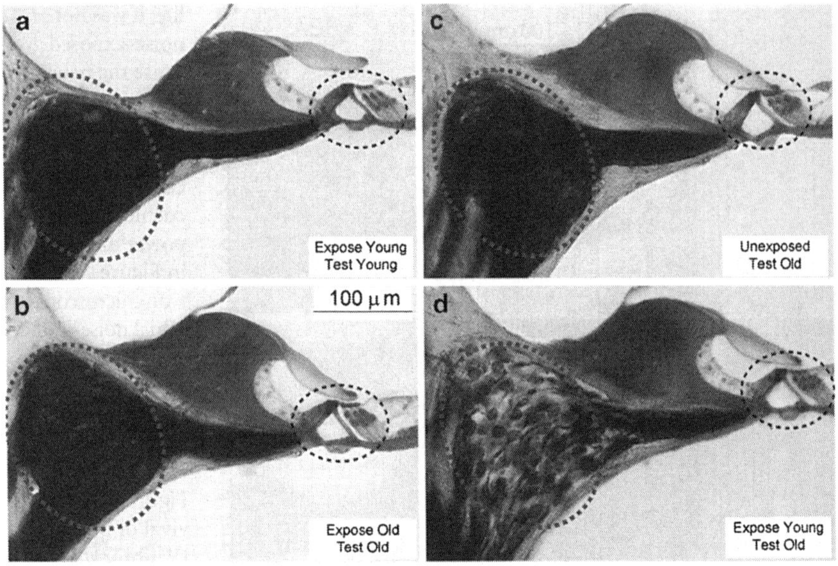

Fig. 4.4 Primary neuronal degeneration was seen in mice that were exposed and allowed to survive for many months. The degeneration, seen as decreased density of spiral ganglion cells (*heavy black circles*), although inner and outer hair cells (*light black circles*) are still present, is visible in cases exposed at 6 weeks and aged to 96 weeks (**d**) but not in cases exposed at 96 weeks and evaluated at 98 weeks (**b**) or in unexposed animals tested at 96 weeks (**c**) or in cases exposed at 6 weeks and tested at 8 weeks (**a**). All images are from the upper basal turn. Scale bar in (**b**) applies to (**a–d**) (reprinted with permission from Kujawa and Liberman 2006)

4.3 Laboratory Uses of Correlates of Loudness

4.3.1 Perceptual Correlates

4.3.1.1 Loudness Measurements

There are a number of different methodologies for measuring loudness. Listeners are often asked to express loudness as a number (magnitude estimation), a category (categorical scaling), or other psychophysical modality like the length of a line, a dynamometer squeeze pressure, or the magnitude of vibration. For further discussion of methods to measure loudness see Chap. 2.

4.3.1.2 Reaction Time

Measurements of simple reaction time – the duration between the presentation of a stimulus and the time at which a listener provides a response – correlate with loudness. Louder sounds yield faster reaction times, so reaction time is inversely correlated with loudness. In a typical reaction-time task, listeners are presented with sounds and asked to press a button when a sound is detected (see Wagner et al. 2004 for review).

In particular, this technique has been used to examine loudness near threshold, particularly in adults with hearing impairment (Florentine et al. 2004). In addition, reaction time has been used to examine loudness in infants (Leibold and Werner 2002).

4.3.2 *Physiological Correlates*

4.3.2.1 Otoacoustic Emissions

Otoacoustic emissions are low-level sounds recorded in the ear canal with a microphone. They result from the active mechanism of the auditory system and correlate with loudness. There are different types of otoacoustic emissions. Distortion-product otoacoustic emissions, which result from the interaction of two stimuli presented simultaneously and continuously, were measured as a function of level and found to exhibit many of the characteristics of loudness and basilar-membrane compression by a number of researchers in both normal-hearing and hearing-impaired listeners (Neely et al. 2003; Muller and Janssen 2004). Transient-evoked otoacoustic emissions, acoustic responses of the ear made after short stimuli are presented, have also been measured as a function of level and shown to correlate well with loudness functions estimated using a number of psychoacoustical tasks for normal-hearing listeners (Epstein and Florentine 2005; Epstein and Silva 2009). Figure 4.5 shows data from Epstein and Silva (2009) compared with the INEX loudness model, which is a modified power-function model that fits the loudness growth function better than the power function (Florentine and Epstein 2006).

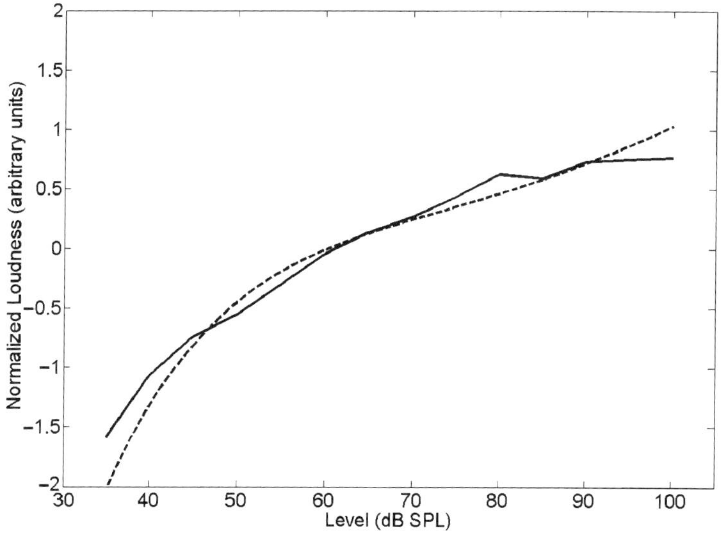

Fig. 4.5 A comparison of a loudness function estimated from six listeners' tone-burst otoacoustic emissions (*solid line*) and the Inflected Exponential (INEX) Loudness model of Florentine and Epstein (2006) (*dashed line*) (reprinted with permission from Epstein and Silva 2009)

4.3.2.2 Auditory Brainstem Response and Auditory Steady-State Response

The auditory brainstem response (ABR) is an objective neurological measure of the evoked potential that results from auditory stimulation. Several researchers have examined the relationship between the ABR and loudness (Howe and Decker 1984; Serpanos et al. 1997). These studies have typically used the most common ABR landmark, wave V, which is often used for clinical testing. Wave V latency decreases with increasing intensity and, much like reaction time, has a strong general inverse correlation with loudness. However, wave V latency asymptotes at a relatively low intensity, while loudness continues to grow with increasing intensity. Wave V latency is not the only ABR characteristic correlated with loudness. Pratt and Sohmer (1977) hypothesized that, rather than examining wave V latency, the first few components, approximately waves I + II of the ABR, are better suited for finding a correlate of loudness. Physiologically, it is likely that the early components of the ABR result from electrical activity from the cochlea and early neurons. Recent work also indicates it may be possible to use the sum of ABR coherent energy recorded over a relatively long time window to estimate loudness growth (Silva 2009, Silva and Epstein 2010).

Auditory steady-state response (ASSR) has also been used to estimate loudness growth in normal-hearing listeners (Menard et al. 2008; Zenker Castro et al. 2008). These studies found significant correlations between loudness growth and ASSR amplitude.

4.3.2.3 Basilar-Membrane Velocity

The relationship between sound intensity and basilar-membrane motion has been measured in animals using laser Doppler velocimetry (Ruggero et al. 1997). Buus and Florentine (2001) plotted human loudness functions modeled from temporal integration data and spectral summation data – known as the inflected exponential (INEX) loudness function (Florentine and Epstein 2006) – with basilar-membrane velocity measures made in a chinchilla (Ruggero et al. 1997) – shown in Fig. 4.6. They found that loudness was approximately proportional to basilar-membrane velocity squared. Other psychoacoustical measurements associated with basilar-membrane compression also correspond well with loudness growth functions (Oxenham and Plack 1997). This indicates that loudness for simple sounds is likely to be closely associated with the physical vibrations on the basilar membrane.

4.3.2.4 Loudness Coding in Neurons

Auditory-nerve fibers increase their firing rate as the intensity of a sound increases. Rate-level functions show the number of times a nerve fiber activates on average per second for a given stimulus. These functions show that the vast majority of auditory-nerve fibers tend to saturate (i.e., reach maximum firing rate) within only a 30–60 dB dynamic range. Because the auditory system as a whole is capable of processing a dynamic range of 120 dB, some mechanism other than general firing rate of the neurons must account for this ability. This is known as the dynamic-range problem.

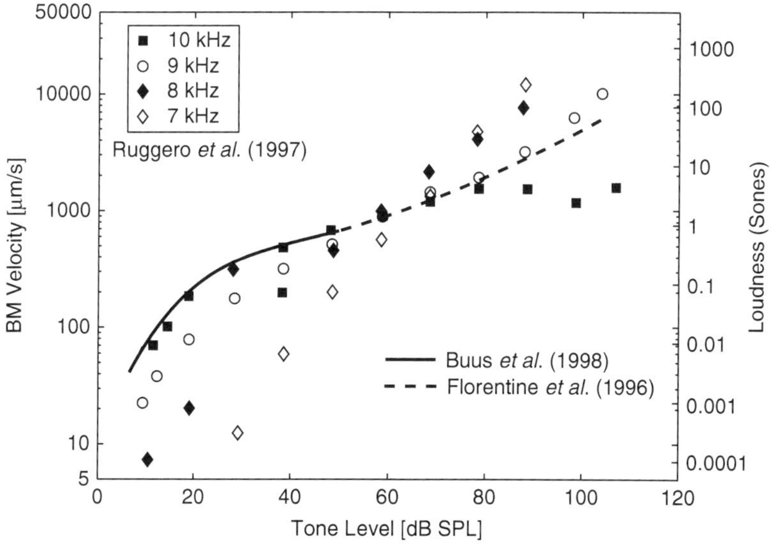

Fig. 4.6 Human loudness functions derived using two psychoacoustical methods plotted with basilar-membrane velocity measured in a chinchilla at a 10-kHz characteristic frequency location. The loudness data matches the basilar-membrane response to a tone close to characteristic frequency if it is assumed that loudness is linearly related to the square of basilar-membrane velocity. Responses at the same location to tones at lower frequencies are also shown for response comparison (replotted with permission from Buus and Florentine 2001)

Although our understanding of the physiology of loudness is far from complete, there are two compelling lines of thought to explain the dynamic range problem that are not mutually exclusive. The first is that loudness is related to the total amount of neural activity. The second is that loudness is related to the temporal properties of the neural activity. As a tone increases in level, it not only excites neurons with primary sensitivity (characteristic frequency or the frequency to which it responds best) near the tone frequency, it also excites an increasing number of neurons with adjacent characteristic frequencies. This is known as the "spread of excitation." Temporal theories are based on the fact that neurons that respond to a certain frequency tend to phase lock to it. In other words, they produce neural firings at precise times correlated with temporal properties of the sound wave. When the level of a tone increases, more neurons phase lock to it and the overall synchrony across the population of auditory nerve fibers increases. However, the timing theories have difficulty at high frequencies because the ability of the auditory nerve fibers to phase lock decreases at high frequencies, which is known as the roll-off in phase locking. It is possible that the amount and range of neural activity, as well as the timing and phase locking of neural activity, play roles in the perception of loudness. Despite how many data have been obtained, surprisingly little has been proven regarding the connection between physiological responses and our overall perception of loudness.

4.3.2.5 Brain Scans

Although the representation of loudness in the central auditory system is relatively poorly understood, several investigators have identified basic relationships between brain images and loudness. These relationships tend to be based on gross analysis of brain activations rather than fine timing or location information.

Langers et al. (2007) found that, for low-frequency stimuli, functional magnetic resonance imaging (fMRI) brain activation was more closely related to loudness measures than to stimulus presentation levels (i.e., level measures). They also observed a relatively constant growth in activation with changes in stimulus level at moderate sensation levels. At low sensation levels, growth was more rapid, perhaps indicative of the steeper loudness function observed at lower levels compared with moderate levels (For more information on loudness-growth functions, see Chap. 5.) High-frequency stimuli resulted in less consistent and clear results. Sigalovsky and Melcher (2006) found that fMRI activation increased with increasing broadband noise level at several stages of auditory processing, primarily the brainstem, thalamus, and cortex. They suggested that activation reflects the sum of activity over large neuronal populations, indicating both monotonic and nonmonotonic responses to tonal stimuli.

Bilecen et al. (2002) examined the amplitopicity (the relationship between sound amplitude and location of activation) of sound-level representation in the central auditory system. They found that activation volume and location changed as a function of level. Most notably, there was two-dimensional drift of cortical activation from the ventral to the dorsal edge and in the transverse temporal gyrus from the lateral toward the medial part. Figure 4.7 shows a scan of the amplitopic distribution of activity in the brain in response to different sound levels.

4.3.2.6 Direct Electrical Measurements and Stimulation

Cochlear nucleus recordings (Cai et al. 2009) have been used to search for correlates of an abnormally rapid growth of loudness (a.k.a. recruitment) with some types of hearing impairment (see Chap. 9). They observed steep slopes in non-primary-like neurons, but primary-like neurons showed recruitment-like behavior only when rates were summed across neurons of all best frequencies.

Researchers have also examined the loudness of the direct electrical stimulation used in cochlear implants. Loudness growth has been found to vary from patient to patient and even stimulus location to stimulus location. Some subjects exhibit linear loudness growth, others compressive growth (Blamey et al. 2000; Fu 2005; Sanpetrino and Smith 2006). This is in agreement with individual differences in loudness growth observed with psychoacoustical experiments with hearing-impaired listeners (Marozeau and Florentine 2007).

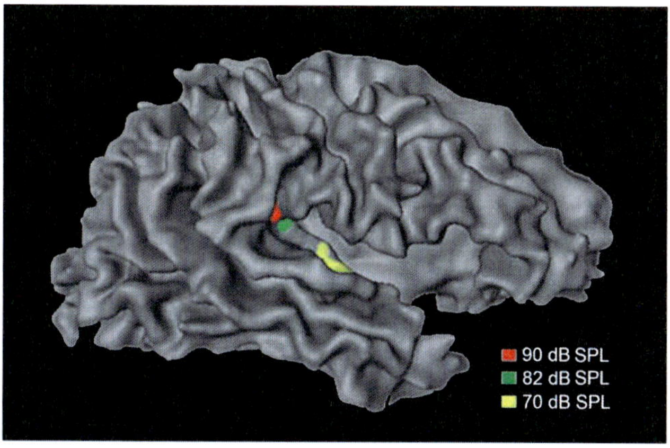

Fig. 4.7 A 3D-reconstructed surface of the right lateral hemisphere and its overlaid functional map for a single subject. For simplification, functional overlap regions are omitted and only activated areas occupying transverse temporal gyrus (TTG) or adjacent areas are displayed. The 90 dB SPL activated area (*red*) is centered at the most medial parts of the TTG, the 70 dB SPL activated area (*yellow*) is found most laterally of TTG, and the 82 dB SPL area is located in between (reprinted with permission from Bilecen et al. 2002)

4.4 Summary

Loud sounds result in a number of short- and long-term physiological responses. Some of these have undesirable effects including placing stress on health and wellness, causing damage to the auditory system, and possibly even disrupting educational development. Far more research needs to be done to gain a fuller understanding of these detrimental effects. In the laboratory, correlates of loudness are used to indirectly estimate auditory perception across a wide range of sound levels. These measures are potentially useful for supplementing direct subjective assessments for both learning more about how the auditory system functions and for clinical applications.

References

Aydin Y, Kaltenbach M (2007) Noise perception, heart rate and blood pressure in relation to aircraft noise in the vicinity of the Frankfurt airport. Clin Res Cardiol 96:347–358.

Babisch W, Gallacher JE, Elwood PC, Ising H (1988) Traffic noise and cardiovascular risk. The Caerphilly study, first phase. Outdoor noise levels and risk factors. Arch Environ Health 43:407–414.

Babisch W, Ising H, Elwood PC, Sharp DS, Bainton D (1993a) Traffic noise and cardiovascular risk: the Caerphilly and Speedwell studies, second phase. Risk estimation, prevalence, and incidence of ischemic heart disease. Arch Environ Health 48:406–413.

Babisch W, Ising H, Gallacher JE, Sharp DS, Baker IA (1993b) Traffic noise and cardiovascular risk: the Speedwell study, first phase. Outdoor noise levels and risk factors. Arch Environ Health 48:401–405.
Bapat U, Tolley N (2006) Temporary threshold shift due to recreational firearm use. J Laryngol Otol 121:927–931.
Basner M, Samel A, Isermann U (2006) Aircraft noise effects on sleep: application of the results of a large polysomnographic field study. J Acoust Soc Am 119:2772–2784.
Basner M, Glatz C, Griefahn B, Penzel T, Samel A (2008a) Aircraft noise: effects on macro- and microstructure of sleep. Sleep Med 9:382–387.
Basner M, Muller U, Elmenhorst EM, Kluge G, Griefahn B (2008b) Aircraft noise effects on sleep: a systematic comparison of EEG awakenings and automatically detected cardiac activations. Physiol Meas 29:1089–1103.
Bench J (1971) Anticipatory elicitation of the middle-ear muscle reflex. J Laryngol Otol 85:1161–1165.
Bilecen D, Seifritz E, Scheffler K, Henning J, Schulte AC (2002) Amplitopicity of the human auditory cortex: an fMRI study. Neuroimage 17:710–718.
Blamey PJ, Dooley GJ, James CJ, Parisi ES (2000) Monaural and binaural loudness measures in cochlear implant users with contralateral residual hearing. Ear Hear 21:6–17.
Botte MC, Charron S, Bouayad H (1993) Temporary threshold and loudness shifts: frequency patterns and correlations. J Acoust Soc Am 93:1524–1534.
Buus S, Florentine M (2001) Modifications to the power function for loudness. In Sommerfeld E, Kompass R, Lachmann T (eds), Fechner Day 2001. Berlin: Pabst, pp. 236–241.
Cai S, Ma WL, Young ED (2009) Encoding intensity in ventral cochlear nucleus following acoustic trauma: implications for loudness recruitment. J Assoc Res Otolaryngol 10:5–22.
Carter N, Henderson R, Lal S, Hart M, Booth S, Hunyor S (2002) Cardiovascular and autonomic response to environmental noise during sleep in night shift workers. Sleep 25:457–464.
Cavatorta A, Falzoi M, Romanelli A, Cigala F, Ricco M, Bruschi G, Franchini I, Borghetti A (1987) Adrenal response in the pathogenesis of arterial hypertension in workers exposed to high noise levels. J Hypertens Suppl 5:S463–S466.
Chodynicki S, Gindzienska E, Rogowski M (1986) Effect of noise and vibration on the ear during fetal development. Otolaryngol Pol 40:120–127.
Cui B, Wu M, She X (2009) Effects of chronic noise exposure on spatial learning and memory of rats in relation to neurotransmitters and NMDAR2B alteration in the hippocampus. J Occup Health 51:152–158.
Daniel E (2007) Noise and hearing loss: a review. J Sch Health 77:225–231.
Davis M (1984) The Mammalian Startle Response. Neural Mechanisms of Startle Behavior. New York: Springer.
Davis RC, Berry T (1964) Gastrointestinal reactions to response-contingent stimulation. Psychol Rep 15:95–113.
Davis RC, Buchwald AM, Frankman RW (1955) Autonomic and muscular responses and their relation to simple stimuli. Psychol Monographs 69:1–71.
Epstein M (2007) An introduction to induced loudness reduction. J Acoust Soc Am 122:EL74–80.
Epstein M, Florentine M (2005) Inferring basilar-membrane motion from tone-burst otoacoustic emissions and psychoacoustic measurements. J Acoust Soc Am 117:263–274.
Epstein M, Silva I (2009) Analysis of parameters for the estimation of loudness from tone-burst otoacoustic emissions. J Acoust Soc Am 125:3855–3864.
Evans GW, Lepore SJ (1993) Non-auditory effects of noise on children: a critical review. Child Environ 10:42–72.
Fidell S, Pearsons K, Tabachnick BG, Howe R (2000) Effects on sleep disturbance of changes in aircraft noise near three airports. J Acoust Soc Am 107:2535–2547.
Florentine M, Epstein M (2006) To honor Stevens and repeal his law (for the auditory system). In: Kornbrot DE, Msetfi RM, MacRae AW (eds), Fechner Day. St Albans: International Society for Psychophysics, pp. 37–42.

Florentine M, Buus S, Rosenberg M (2004). Reaction-time data support the existence of softness imperception in cochlear hearing loss. In Pressnitzer D, de Cheveigne A, McAdams S, Collet L (eds), Auditory Signal Processing: Physiology, Psychoacoustics, and Models. New York: Springer.

Franssen EA, van Wiechen CM, Nagelkerke NJ, Lebret E (2004) Aircraft noise around a large international airport and its impact on general health and medication use. Occup Environ Med 61:405–413.

Fu QJ (2005) Loudness growth in cochlear implants: effect of stimulation rate and electrode configuration. Hear Res 202:55–62.

Glass DC, Singer JE (1972) Behavioral aftereffects of unpredictable and uncontrollable aversive events. Am Sci 60:457–465.

Haines MM, Stansfeld SA, Job RF, Berglund B, Head J (2001) Chronic aircraft noise exposure, stress responses, mental health and cognitive performance in school children. Psychol Med 31:265–277.

Hale HB (1952) Adrenalcortical activity associated with exposure to low frequency sounds. Am J Psychol 171:732.

Hamernik RP, Qiu W (2001) Energy-independent factors influencing noise-induced hearing loss in the chinchilla model. J Acoust Soc Am 110:3163–3168.

Haralabidis AS, Dimakopoulou K, Vigna-Taglianti F, Giampaolo M, Borgini A, Dudley ML, Pershagen G, Bluhm G, Houthuijs D, Babisch W, Velonakis M, Katsouyanni K, Jarup L (2008) Acute effects of night-time noise exposure on blood pressure in populations living near airports. Eur Heart J 29:658–664.

Hillera W, Goebelb G (2007) When tinnitus loudness and annoyance are discrepant: Audiological characteristics and psychological profile. Audiol Neurotol 12:391–400.

Hirsh IJ, Ward WD (1952) Recovery of the auditory threshold after strong acoustic stimulation. J Acoust Soc Am 24:131–141.

Horne JA, Pankhurst FL, Reyner LA, Hume K, Diamond ID (1994) A field study of sleep disturbance: effects of aircraft noise and other factors on 5,742 nights of actimetrically monitored sleep in a large subject sample. Sleep 17:146–159.

Howe SW, Decker TN (1984) Monaural and binaural auditory brainstem responses in relation to the psychophysical loudness growth function. J Acoust Soc Am 76:787–793.

Husbands JM, Steinberg SA, Kurian R, Saunders JC (1999) Tip-link integrity on chick tall hair cell stereocilia following intense sound exposure. Hear Res 135:135–145.

Jarup L, Babisch W, Houthuijs D, Pershagen G, Katsouyanni K, Cadum E, Dudley ML, Savigny P, Seiffert I, Swart W, Breugelmans O, Bluhm G, Selander J, Haralabidis A, Dimakopoulou K, Sourtzi P, Velonakis M, Vigna-Taglianti F (2008) Hypertension and exposure to noise near airports: the HYENA study. Environ Health Perspect 116:329–333.

Kjellstrom T, Friel S, Dixon J, Corvalan C, Rehfuess E, Campbell-Lendrum D, Gore F, Bartram J (2007) Urban environmental health hazards and health equity. J Urban Health 84:i86–97.

Klorman R, Cicchetti D, Thatcher JE, Ison JR (2003) Acoustic startle in maltreated children. J Abnorm Child Psychol 31:359–370.

Krystal JH, Webb E, Grillon C, Cooney N, Casal L, Morgan CA, 3rd, Southwick SM, Davis M, Charney DS (1997) Evidence of acoustic startle hyperreflexia in recently detoxified early onset male alcoholics: modulation by yohimbine and m-chlorophenylpiperazine (mCPP). Psychopharmacology (Berl) 131:207–215.

Kryter KD (1970) The Effects of Noise on Man. Environmental Sciences. New York: Academic Press.

Kryter KD (1972) Non-auditory effects of environmental noise. Am J Public Health 62:389–398.

Kujawa SG, Liberman MC (2006) Acceleration of age-related hearing loss by early noise exposure: evidence of a misspent youth. J Neurosci 26:2115–2123.

Langers DR, van Dijk P, Schoenmaker ES, Backes WH (2007) fMRI activation in relation to sound intensity and loudness. Neuroimage 35:709–718.

Leibold LJ, Werner LA (2002) Relationship between intensity and reaction time in normal-hearing infants and adults. Ear Hear 23:92–97.

Liberman MC, Dodds LW (1987) Acute ultrastructural changes in acoustic trauma: serial-section reconstruction of stereocilia and cuticular plates. Hear Res 26:45–64.

Margolis RH, Popelka GR (1975) Loudness and the acoustic reflex. J Acoust Soc Am 58:1330–1332.

Marozeau J, Florentine M (2007) Loudness growth in individual listeners with hearing losses: a review. J Acoust Soc Am 122:EL81–87.

Melamed S, Froom P, Kristal-Boneh E, Gofer D, Ribak J (1997) Industrial noise exposure, noise annoyance, and serum lipid levels in blue-collar workers – the CORDIS Study. Arch Environ Health 52:292–298.

Melamed S, Kristal-Boneh E, Froom P (1999) Industrial noise exposure and risk factors for cardiovascular disease: findings from the CORDIS Study. Noise Health 1:49–56.

Melnick W (1991) Human temporary threshold shift (TTS) and damage risk. J Acoust Soc Am 90:147–154.

Menard M, Gallego S, Berger-Vachon C, Collet L, Thai-Van H (2008) Relationship between loudness growth function and auditory steady-state response in normal-hearing subjects. Hear Res 235:105–113.

Morgan CA, III, Grillon C, Southwick SM, Davis M, Charney DS (1995) Fear-potentiated startle in posttraumatic stress disorder. Biol Psychiatry 38:378–385.

Morgan CA, III, Grillon C, Southwick SM, Davis M, Charney DS (1996) Exaggerated acoustic startle reflex in Gulf War veterans with posttraumatic stress disorder. Am J Psychiatry 153:64–68.

Morgan CA, III, Grillon C, Lubin H, Southwick SM (1997a) Startle deficits in women with sexual assault-related PTSD. Ann N Y Acad Sci 821:486–490.

Morgan CA, III, Grillon C, Lubin H, Southwick SM (1997b) Startle reflex abnormalities in women with sexual assault-related posttraumatic stress disorder. Am J Psychiatry 154:1076–1080.

Muller J, Janssen T (2004) Similarity in loudness and distortion product otoacoustic emission input/output functions: implications for an objective hearing aid adjustment. J Acoust Soc Am 115:3081–3091.

Neely ST, Gorga MP, Dorn PA (2003) Cochlear compression estimates from measurements of distortion-product otoacoustic emissions. J Acoust Soc Am 114:1499–1507.

Nelson PB, Soli S (2000) Acoustical barriers to learning: children at risk in every classroom. Language Speech Hearing Serv Schools 31:356–361.

Norena AJ, Eggermont JJ (2006) Enriched acoustic environment after noise trauma abolishes neural signs of tinnitus. NeuroReport 17:559–563.

Oxenham AJ, Plack CJ (1997) A behavioral measure of basilar-membrane nonlinearity in listeners with normal and impaired hearing. J Acoust Soc Am 101:3666–3675.

Popelka GR, Karlovich RS, Wiley TL (1974) Letter: acoustic reflex and critical bandwidth. J Acoust Soc Am 55:883–885.

Pratt H, Sohmer H (1977) Correlations between psychophysical magnitude estimates and simultaneously obtained auditory nerve, brain stem and cortical responses to click stimuli in man. Electroencephalogr Clin Neurophysiol 43:802–812.

Rabinowitz PM (2000) Noise-induced hearing loss. Am Fam Physician 61:2749–2756, 2759–2760.

Rabinowitz WM (1977) Acoustic-Reflex Effects on the Input Admittance and Transfer Characteristics of the Human Middle-Ear (dissertation). Cambridge, MA: Massachusetts Institute of Technology.

Rosenlund M, Berglind N, Pershagen G, Jarup L, Bluhm G (2001) Increased prevalence of hypertension in a population exposed to aircraft noise. Occup Environ Med 58:769–773.

Ruggero MA, Rich NC, Recio A, Narayan SS, Robles L (1997) Basilar-membrane responses to tones at the base of the chinchilla cochlea. J Acoust Soc Am 101:2151–2163.

Sadhra S, Jackson CA, Ryder T, Brown MJ (2002) Noise exposure and hearing loss among student employees working in university entertainment venues. Ann Occup Hyg 46:455–463.

Sanpetrino NM, Smith RL (2006) The growth of loudness functions measured in cochlear implant listeners using absolute magnitude estimation and compared using Akaike's information criterion. Conf Proc IEEE Eng Med Biol Soc 1:1642–1644.

Schapkin SA, Falkenstein M, Marks A, Griefahn B (2006) Executive brain functions after exposure to nocturnal traffic noise: effects of task difficulty and sleep quality. Eur J Appl Physiol 96:693–702.

Scharf B (1983). Loudness adaptation. In: Tobias JV, Schubert ED (eds), Hearing Research and Theory, Vol. 2. New York: Academic Press, pp. 1–56.

Serpanos YC, O'Malley H, Gravel JS (1997) The relationship between loudness intensity functions and the click-ABR wave V latency. Ear Hear 18:409–419.

Sigalovsky IS, Melcher JR (2006) Effects of sound level on fMRI activation in human brainstem, thalamic and cortical centers. Hear Res 215:67–76.

Silva I (2009) Estimation of postaverage SNR from evoked responses under nonstationary noise. IEEE Trans Biomed Eng 56:2123–2130.

Silva I, Epstein M (2010) Estimation of loudness growth through tone-burst auditory brainstem responses. J Acoust Soc Am 127(6) 3629–3642.

Sokolovski A (1973) The protective action of the stapedius muscle in noise-induced hearing loss in cats. Arch Klin Exp Ohren Nasen Kehlkopfheilkd 203:289–309.

Stansfeld SA, Matheson MP (2003) Noise pollution: non-auditory effects on health. Br Med Bull 68:243–257.

Thompson SJ (1993) Review: extraaural health effects of chronic noise exposure in humans. Schriftenr Ver Wasser Boden Lufthyg 88:91–117.

Thurston FE, Roberts SL (1991) Environmental noise and fetal hearing. J Tenn Med Assoc 84:9–12.

Uziel A (1985) Non-genetic factors affecting hearing development. Acta Otolaryngol Suppl 421:57–61.

van Dijk FJ, Souman AM, de Vries FF (1987) Non-auditory effects of noise in industry. VI. A final field study in industry. Int Arch Occup Environ Health 59:133–145.

Vernon JA, Meikle MB (2003) Tinnitus: clinical measurement. Otolaryngol Clin North Am 36:293–305, vi.

Vlek C (2005) "Could we all be a little more quiet, please?" A behavioural-science commentary on research for a quieter Europe in 2020. Noise Health 7:59–70.

Wagner E, Florentine M, Buus S, McCormack J (2004) Spectral loudness summation and simple reaction time. J Acoust Soc Am 116:1681–1686.

Zenker Castro F, Barajas de Prat JJ, Larumbe Zabala E (2008) Loudness and auditory steady-state responses in normal-hearing subjects. Int J Audiol 47:269–275.

Chapter 5
Loudness in the Laboratory, Part I: Steady-State Sounds

Walt Jesteadt and Lori J. Leibold

5.1 Introduction

Many of the basic principles believed to govern the perception of loudness are based on data obtained from laboratory studies of the loudness of steady-state sounds. Although this chapter emphasizes the most recent publications in this area, much of the knowledge base comes from earlier work, some of it from the earliest studies with electrically generated sounds. Early studies continue to be of interest because there have been few changes in basic concepts or measurement techniques and because the precise stimulus control made possible by digital signal generation is not necessary for the production of most steady-state sounds of the type used in studies of loudness. Many of the concepts discussed in recent publications, such as the loudness of tones at threshold, the relation between growth of loudness and peripheral nonlinearity and the effect of masking on the loudness of broadband sounds can be found in papers published more than 70 years ago. The quality of the early work is remarkable and it is important to preserve that early history.

The material covered in this chapter has been organized in terms of the physical dimensions of steady-state sounds: level, frequency, duration, and bandwidth. The final two sections deal with the effect of partial masking on loudness and the literature on loudness as a function of age. Although methodology, models and some of the more complex issues such as context effects are reviewed in other chapters, some discussion of those issues is required here to provide a framework for discussion of the data.

Data on the loudness of steady-state sounds to be reviewed here have been obtained and reported with two different goals in mind. One is to use information concerning loudness to gain a better understanding of the function of the auditory system. The second is to provide methods of estimating loudness that can be used by engineers and others interested in loudness measurement and control. There has been a greater effort surrounding development of American National Standards

W. Jesteadt (✉)
Boys Town National Research Hospital, Omaha, NE 68131, USA
e-mail: walt.jesteadt@boystown.org

M. Florentine et al. (eds.), *Loudness*, Springer Handbook of Auditory Research 37,
DOI 10.1007/978-1-4419-6712-1_5,

Institute (ANSI) and International Standards Organization (ISO) standards and more reference to those standards in the literature on loudness than in most other areas of auditory research. Some of the data contributing to these standards have come from basic research on loudness obtained with the first goal in mind. Other data sets have been collected primarily to support development or improvement of standards. The standards for computation or estimation of loudness are based, in part, on assumptions concerning peripheral auditory function. These standards, however, always include additional assumptions primarily motivated by a need to approximate the known data. Data obtained with a third goal, accounting for loudness perception in listeners with hearing loss, are discussed by Chap. 9.

5.2 Loudness as a Function of Intensity

Given that loudness is generally defined as the subjective strength of a sound (Scharf 1978), it is not surprising that intensity is the physical dimension of steady-state sound that has the greatest impact on the perception of loudness. Loudness increases monotonically with sound intensity over the entire 120-dB dynamic range of the auditory system. An intensity range of 12 orders of magnitude is difficult to comprehend, but physical distances provide good examples. If the maximum of the 120-dB range is equated with the distance from New York to San Francisco (approximately 2,570 miles), the minimum intensity would be 4 μm, or 1/20 of the average diameter of human hair! Some would argue that it would be better to describe the dynamic range in terms of pressure, where it is only 6 orders of magnitude. If a sine wave at threshold can be displayed on an oscilloscope with a peak-to-peak amplitude of 1 cm (or approximately 0.4 in.), then a sine wave at 120 dB would have a peak-to-peak amplitude of 10 km (or approximately 6.2 miles). Either example demonstrates a remarkable range of audible intensities.

5.2.1 The Loudness Function

Four alternative functions describing the conversion from sound pressure level to loudness over a 120-dB range at 1 kHz are shown in Fig. 5.1. One is the ubiquitous power law function proposed by Stevens (1955). Two of the functions reflect current engineering standards (ISO 226 2003, ANSI S3.4 2007). The final function actually consists of raw data collected in 1-dB steps by Fletcher and Munson (1933). Although similar functions could be described at other frequencies, most data have been obtained at 1 kHz. Therefore, 1 kHz has been adopted as a reference for loudness at other frequencies. Results at other frequencies are summarized in Sect. 5.3.

The ordinate in Fig. 5.1 shows loudness in sones, plotted on a log scale. This unit of measurement, with 1 sone defined as the loudness of a 1 kHz tone presented in a free field at 40 dB sound pressure level (SPL), was proposed by Stevens (1936)

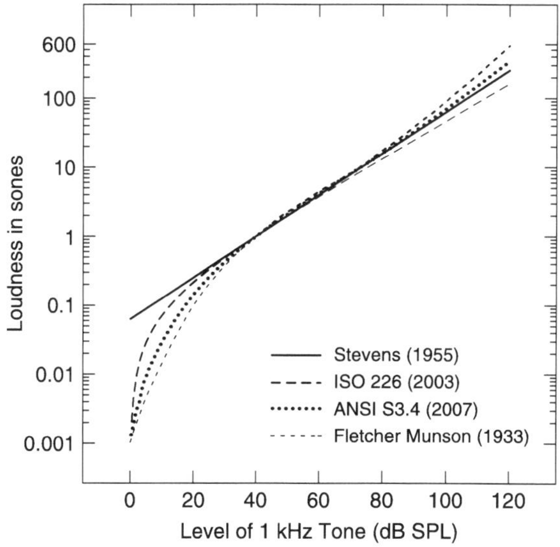

Fig. 5.1 Functions describing the relation between the level of a 1-kHz tone and its loudness in sones. The power law proposed by Stevens is compared to the functions described in the most recent loudness standards and to data reported by Fletcher and Munson (1933)

based on data summarized by Churcher (1935). The sone scale has the property that a sound with a loudness of 2 sones is twice as loud as a sound with a loudness of 1 sone. This relation may seem obvious, but many procedures used to obtain data concerning loudness do not lead to a true ratio scale that would allow one to make statements about one sound as being twice as loud or ten times as loud as another (e.g., Stevens 1946; Chaps. 1and 2). Data obtained when subjects adjust the level of one sound until it matches another sound in loudness, for example, do not generally allow conclusions about loudness ratios and some authors continue to question whether ratio scales of sensation are meaningful (Laming 1997). All four functions in Fig. 5.1 have been adjusted to pass through the 40-dB, 1-sone point. This corrects for the differences between binaural, free-field presentation and monaural or binaural presentation under headphones to the extent that such differences are independent of presentation level (see Chap. 7). Most studies of loudness in the laboratory have been conducted using headphones, generally with monaural presentation, but data are frequently corrected by a scale factor for plotting on a sone scale.

The oldest function in Fig. 5.1 (Fletcher and Munson 1933) was based on matching data. Fletcher and Munson converted the data to ratios by assuming that a tone presented binaurally would be twice as loud as the same tone presented monaurally and that a complex consisting of n equally loud tones widely spaced in frequency would be n times as loud as a single tone. They varied n from 1 to 10 and generated data in 1-dB steps. The approach to creating a loudness function with ratio properties by assuming that loudness is additive has been used in later studies (Buus et al. 1998; Neely et al. 2003). Although interaction among the components can reduce their contribution to the total loudness, violating the additivity assumption and leading to biased results, the original function has stood the test of time. The function in Fig. 5.1 was obtained by converting the data in Fletcher and Munson's Table 3 to sones by dividing all of the values in the table by the value for a 40-dB tone. Fletcher (1953) later fitted these data and data from other sources with an equation relating the physical level to loudness in sones. Above 40 dB SPL, the function was a straight line with a slope of 0.33.

The second function in Fig. 5.1 is the power law for loudness with an exponent of 0.3 proposed by Stevens (1955) on the basis of numerous studies in which subjects were asked to adjust the level of one tone until it was either half as loud or twice as loud as another or to provide other information, such as magnitude estimates (Chap. 2), that could be converted to loudness ratios. In one of his final publications, Stevens (1972) proposed a new standard for calculation of loudness based on a loudness function at 3.15 kHz with an exponent of 0.33, the value proposed by Fletcher (1953). His argument in favor of adopting a higher frequency as the standard was that judgments of loudness at high levels at 1 kHz were biased by spread of excitation to adjacent critical bands, a problem that could be avoided at a higher frequency where the critical bandwidth was wider. A similar argument has been made for use of frequencies higher than 1 kHz in masking experiments where spread of excitation to adjacent critical bands was a concern (e.g., Oxenham and Plack 1997). Most models of loudness assume spread of excitation, however, and later work found no significant difference between growth of loudness at 1 and 3 kHz (Hellman 1976). The effect of spread of excitation on loudness is more apparent when loudness functions from listeners with abrupt high-frequency losses are compared to those from listeners with more gradual losses (Florentine et al. 1997).

The third function shows the loudness growth assumed in the ISO 226 (2003) description of loudness contours, discussed in greater detail in Sect. 5.2.3. It assumes a somewhat more gradual growth of loudness at high levels, with an exponent of 0.27, and a steeper growth of loudness at low levels. The final function was generated using software distributed with the ANSI S3.4 (2007) standard for the loudness of steady-state sounds, based on a model described by Moore et al. (1997). The model is an update of a scheme for calculation of loudness proposed by Zwicker (Zwicker 1958; Zwicker and Scharf 1965; Chap. 10). It reflects our current understanding of peripheral auditory function and incorporates established procedures for estimation of excitation patterns (Glasberg and Moore 1990), but still includes some elements chosen to optimize the fit to the large body of existing loudness data. Glasberg and Moore (2006) have proposed modifications to further improve agreement with ISO 226 (2003) and ISO 389-7 (2005) and the function plotted in Fig. 5.1 includes those modifications. As shown in the figure, the relation between the latest model and the data obtained by Fletcher and Munson (1933) is striking.

5.2.2 *Variability in the Exponent*

Most studies of growth of loudness for the past 50 years have fitted data above 40 dB with a straight line, allowing growth to be characterized by a single power law exponent. The functions in Fig 5.1 reflect means across 11 subjects in the case of

Fletcher and Munson (1933) and efforts to summarize larger bodies of data for the others. The exponents range from 0.27 to 0.33. Individual studies show larger differences, even for laboratory subjects with normal hearing (Stevens 1955; Hellman 1991; Hellman and Meiselman 1993). Stevens (1955) presents a histogram showing the number of decibels required to halve or double loudness over a wide range of levels and a more limited range of frequencies as reported in 15 different studies. All of the 178 entries in the histogram represent means across subjects within a given condition. The numbers can be converted to estimates of power law exponents by dividing 10 log(2) by the number of decibel for doubling or halving. The estimates range from 0.125 to 1.2 with a mode of 0.33 and a median of 0.30. Values between 0.3 and 1.2 (there was a single estimate greater than 1.0) were typically obtained at low levels and reflect the steep portion of the loudness functions in Fig. 5.1, to be discussed in more detail later. Values less than 0.3 may reflect flatter regions in the function (Buus and Florentine 2002; Florentine and Epstein 2006), or these lower values might reflect individual differences. Logue (1976) reported exponents for 22 subjects with a mean of 0.26 and a standard deviation of 0.077. Hellman (1991) reviewed 78 studies and reported a mean exponent at 1 kHz of 0.3, with a standard deviation of 0.045. Hellman and Meiselman (1993) report exponents for 160 individual subjects with normal hearing with a mean of approximately 0.3 and a standard deviation of 0.065, not much larger than the standard deviation across studies found by Hellman (1991). Thus, there is reasonable agreement in the literature regarding the growth of loudness over the range from 40 to 90 dB SPL.

5.2.3 Loudness near Threshold

The data obtained by Fletcher and Munson (1933) and in many later data sets indicate that loudness at 1 kHz is not well approximated by a power function at levels below 40 dB SPL. Humes and Jesteadt (1991), Hellman (1997), and Buus et al. (1998) have reviewed the literature on this topic. The discussion here will be restricted to loudness near absolute threshold in subjects with normal hearing, but much of the interest in this topic is related to loudness in the presence of maskers (see Sect. 5.6) or loudness in the presence of hearing loss (Chap. 9).

Two key issues are the loudness of tones presented at threshold and the mathematical form of the loudness function near threshold. These issues are closely related because some equations intended to describe loudness below 40 dB predict that loudness at threshold is zero, while other equations predict loudness at threshold is greater than zero. Moore et al. (1997) revised an earlier model of loudness (Moore and Glasberg 1996) to incorporate the prediction that tones at threshold would have loudness greater than zero, reasoning that tones at threshold as defined in laboratory studies can be heard with a probability greater than chance and must therefore have loudness. Buus et al. (1998) used four- and ten-tone complexes made up of tones widely spaced in frequency that were below threshold when presented individually. The complexes were audible and yielded orderly loudness data. Under the assumption that loudness

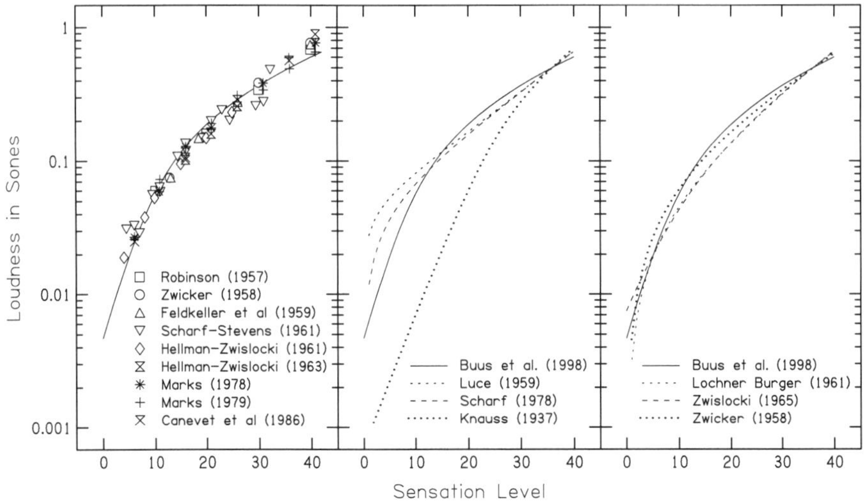

Fig. 5.2 Data and alternative functions describing the loudness of 1-kHz tones near threshold. A function proposed by Buus et al. (1998) appears in all three panels as a reference

is additive, loudness was greater than zero at threshold for five of the six subjects and arguably greater than zero even below threshold. Statements of this kind reflect the vagaries of defining threshold (e.g., Swets 1961), but there is a consensus that tones at threshold have measurable loudness.

Buus et al. (1998) reviewed six equations proposed by various authors as a description of loudness near threshold and proposed a seventh equation based on their own data. Data from nine previous studies reviewed by Buus et al. (1998) and the function they proposed to fit their data are shown in the left-hand panel of Fig. 5.2. The results of two sets of alternative equations are plotted in the middle and right-hand panels, with the Buus et al. (1998) equation as a reference. They proposed that the function describing the loudness of a tone presented at X dB above threshold for that tone was:

$$L = k\left(\left[1+\left(C\cdot 10^{x/10}\right)^{D}\right]^{B/D} - 1\right) \tag{5.1}$$

where B, C, and D are free parameters and k is a scale factor. The function that appears in all three panels of Fig. 5.2 was obtained with $k = 0.44$, $B = 0.11$, 10 log $C = -6$ dB, and $D = 1.4$. They note that 10 log C can be considered as the signal-to-noise ratio at threshold and that D is the asymptotic slope at low levels Although they describe B as the asymptotic slope at high levels, the function was not intended to fit loudness data above 40 dB SPL, where it generates a slope of 0.19, in good agreement with the shallow mid-level slopes of the loudness functions derived from temporal-integration data that are discussed later (Florentine et al. 1996; Buus et al. 1997). Given that many of the other equations make similar predictions, including all of those illustrated in the right-hand panel, not all seven will be reviewed here.

The concepts underlying the equations are of more general interest than the equations themselves. A full list can be found in Buus et al. (1998).

The earliest of these equations, shown in the middle panel of Fig. 5.2, was proposed by Knauss (1937) to fit the data published by Fletcher and Munson (1933), shown in Fig. 5.1. Knauss assumed that loudness was proportional to intensity at low levels and made a smooth transition from an exponent of 1 to an exponent of 0.33 in the region around 40 dB SPL. Stevens (1966b) later proposed a two-line approximation to growth functions for loudness and for scales in other sensory modalities that included a steeper power function near threshold and shallower function at higher levels. The approximation proposed by Stevens allowed exponents other than 1.0 at low levels and made no effort to describe a smooth transition between the regions with steep and shallow exponents. The remaining two functions shown in the middle panel both assume that loudness near threshold can be approximated by subtracting the threshold itself before applying the power law transformation. The first, proposed by Luce (1959) and by Stevens (1959, 1966b), corrects for threshold in units of intensity:

$$L = k\left(I_x - I_{\mathrm{th}}\right)^{0.3} \tag{5.2}$$

where I_x is the intensity of the tone whose loudness is to be estimated, I_{th} is the intensity at threshold, and 0.3 is the standard power law exponent. Other values of the exponent have been used, of course, but the standard value is used in the equations and in Fig. 5.2 for purposes of illustration.

The other function, proposed by Scharf (Scharf and Stevens 1961; Scharf 1978, 1997), makes the same type of correction, but in pressure rather than power:

$$L = k\left(P_x - P_{\mathrm{th}}\right)^{0.6} \tag{5.3}$$

where P_x is the pressure of the tone whose loudness is to be estimated, P_{th} is the pressure at threshold, and 0.6 is the standard power law exponent when expressing loudness as a function of pressure rather than power. Pressure and power are used interchangeably in power-law equations in the loudness literature, with a corresponding change in the exponent, but it is noteworthy that the two quantities are not interchangeable when making a threshold correction. Equation (5.3) results in a closer approximation to the data than (5.2).

Humes and Jesteadt (1991) compared the threshold correction given by (5.2) to an alternative suggested by Zwislocki and Hellman (1960) and Lochner and Burger (1961):

$$L = k\left(I_x^{\,0.3} - I_{\mathrm{th}}^{\,0.3}\right). \tag{5.4}$$

The application of a power law transformation to the threshold quantity in (5.4) suggests that the threshold itself is determined in part by more central factors, whereas (5.2) and (5.3) could be viewed as more peripheral corrections for threshold (see Marks, 1974, for a discussion of these issues).

The final equation of interest was developed by Zwislocki (1965) to account for the loudness of tones in noise:

$$L = k\left(I_x + I_{\mathrm{th}}\right)^{0.3} - I_{\mathrm{th}}^{\ 0.3} \tag{5.5}$$

The functions described by (5.4) and (5.5) appear in the right-hand panel of Fig. 5.2 along with the function described by (5.1) and an equation proposed by Zwicker (1958). All four make similar predictions below 40 dB, but only (5.1) and (5.5) predict loudness greater than zero at and even below threshold. The loudness function adopted in ISO 226 (2003), illustrated in Fig. 5.1, is based on the threshold correction described in (5.4), with an exponent of 0.27. Developers of the standard chose (5.4) despite its prediction of zero loudness at threshold, because it had fewer parameters and yielded more stable numerical solutions (see Suzuki and Takeshima 2004).

5.2.4 *Loudness at High Levels*

In contrast to the extensive literature available concerning loudness near threshold, there are few reports of efforts to establish the form of the loudness function at levels above 100 dB SPL. In some of the magnitude estimation data for 1 kHz tones reported by Stevens (1956), estimates for tones at the high end of the range, from 100 to 120 dB SPL, fell above the standard power-law function with an exponent of 0.3, suggesting more rapid growth of loudness at high levels. Stevens accounted for this effect as an artifact resulting from the choice of the standard. A small increase in the slope of the data obtained by Fletcher and Munson (1933) can be observed in Fig. 5.1. More recently, Viemeister and Bacon (1988), Florentine et al. (1996), and Buus et al. (1997) have obtained data suggesting steeper functions at high levels. Moore et al. (1997) cite the data from Viemeister and Bacon (1988) as the basis for assuming that loudness grows more rapidly above 100 dB, but the newly adopted standard (ANSI S3.4 2007) simply notes the lack of data at high levels.

5.2.5 *Loudness as a Reflection of Compressive Nonlinearity*

Many auditory masking phenomena are now recognized as reflections of the compressive nonlinearity of the peripheral auditory system (for reviews of data related to masking, see Moore and Oxenham 1998; Moore et al. 1999; Oxenham and Bacon 2003, 2004). The interpretation of loudness data as a reflection of peripheral compressive nonlinearity is relatively straightforward although the details have yet to be resolved. Early papers speculated that loudness was directly related to various physiological measures of the response of the peripheral auditory system (Fletcher and Munson 1933; Stevens 1936; Stevens and Davis 1936). This view was reinforced when it became clear that the peripheral response was highly nonlinear (Rhode 1971; Kim et al. 1973; Rhode and Robles 1974). More recent discussions of the relation of loudness to peripheral auditory physiology (e.g., Phillips 1987; Yates et al. 1990; Relkin and Doucet 1997; Schlauch et al. 1998; Neely et al. 2000; Buus and Florentine 2002) have been concerned with the specific nature of the transformation from basilar membrane or eighth nerve response to loudness. Goldstein (1974) and Hellman and Hellman (1975) summarized early data supporting the hypothesis that loudness was proportional to the total number of responses or spikes at the level of the auditory nerve. Yates et al. (1990) hypothesized that loudness was directly related to basilar membrane displacement, while both Schlauch et al. (1998) and Buus and Florentine (2002) argued that loudness is proportional to the square of displacement over the range from threshold through moderate levels. This difference of a factor of two reflects the choice of loudness data. Yates et al. (1990) compared their model of displacement to loudness growth functions obtained by Viemeister and Bacon (1988) that are shallower than other functions reported in the literature. Schlauch et al. (1998) compared basilar membrane displacement data published by Ruggero et al. (1997) to loudness functions obtained by Hellman (1976), while Buus and Florentine (2002) compared the same data from Ruggero et al. (1997) to loudness data obtained by Buus et al. (1998) and Florentine et al. (1998). Although the alternative sets of loudness data were in good agreement, Schlauch et al. (1998) speculated that the squaring operation occurred at the level of the inner hair cell, while Buus and Florentine (2002) argued that it occurs central to the auditory nerve. Relkin and Doucet (1997) questioned the view that loudness was a simple function of auditory nerve spike count, noting that the relation between loudness and their measure of spike count differed as a function of signal frequency and level.

The view that loudness is a reflection of peripheral compressive nonlinearity has had an impact on the interpretation of loudness functions of the type shown in Figs. 5.1 and 5.2. The data obtained by Fletcher and Munson (1933) are in the closest agreement with other measures of nonlinearity, showing a steep portion near threshold, a shallow, more compressive function in the mid intensities, and a less compressive function at high levels. Buus et al. (1998) proposed an equation, plotted in Fig. 5.2, that deviates from the other functions and from earlier loudness data at mid to high levels because the data in that study and data obtained in studies of temporal integration – to be discussed later – indicated a more gradual growth of loudness at mid levels, consistent with Fletcher and Munson (1933). One could argue, therefore, that the power law is an approximation that ignores meaningful features of the loudness function. Buus and Florentine (2002) proposed modifications to the power function for loudness to reflect shallower growth at mid levels than at low or high levels and suggested that the loudness function was a reflection of basilar membrane mechanics. Florentine and Epstein (2006) described the revised power law as an "inflected exponential" or INEX law. The hypothesis that growth of loudness is directly related to peripheral auditory function and that the shape of the loudness function should be consistent with other data reflecting differences in nonlinearity as a function of level runs counter to the assumption that similar functions govern the transform from growth of physical magnitude to growth of sensation in different sensory modalities (Stevens 1957, 1975; Marks 1974; Laming 1986).

5.3 Loudness as a Function of Frequency

The loudness of pure tones differs as a function of frequency as well as intensity, both because of the acoustics of the outer and middle ear and because the functions relating intensity to loudness, described in the previous section, differ as a function of frequency even when plotted relative to the threshold at each frequency. Differences in loudness as a function of frequency can be described in terms of loudness contours that map the physical levels required for equal loudness over a wide range of frequencies or in terms of growth of loudness with level at frequencies other than 1 kHz.

5.3.1 Loudness Contours

Information regarding loudness as a function of frequency is typically presented in the form of equal-loudness-level contours as illustrated in Fig. 5.3. The contours show the change in intensity required to maintain equal loudness as frequency shifts over the entire audible range. The lowest curve in Fig. 5.3 represents absolute threshold. Although all tones at threshold might be considered to be equally loud, the curve itself is not based on loudness judgments. By convention, loudness contours are referenced to the levels of 1-kHz tones. The contours marked 60, 40,

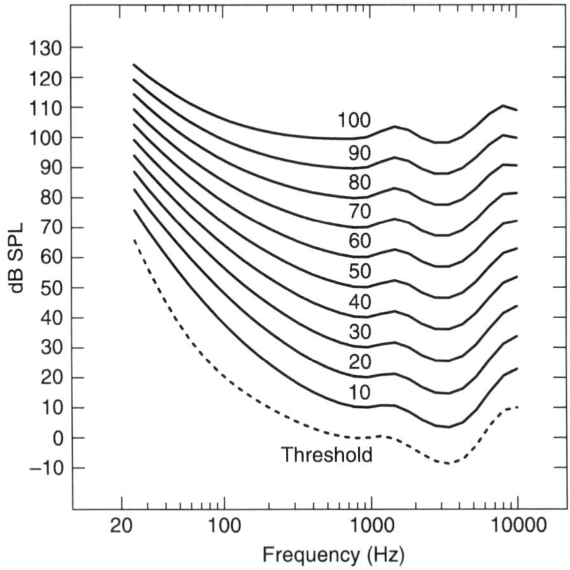

Fig. 5.3 Loudness contours as described by the ISO 226 (2003) standard. All points on a given contour are judged to be equally loud. The *dashed line* represents absolute threshold. The contours were generated using MATLAB code distributed by Jeff Tackett and modified by Harisadhan Patra

and 20, for example, pass through 60, 40, and 20 dB SPL at 1 kHz and points on those contours are said to have loudness levels of 60, 40, and 20 phons, respectively. A tone with a loudness level of *x* phons is as loud as a 1-kHz tone presented at x dB SPL. Note that equal-loudness-level contours do not specify the actual loudness of the tones on a given contour. We can convert to loudness, however, using a function relating physical level at 1 kHz to loudness in sones. The loudness contours in Fig. 5.3 are based on the recently revised ISO standard (ISO 226 2003).

The earliest loudness contours were based on data obtained at Bell Labs by Kingsbury (1927). Contours reported by Fletcher and Munson (1933) were used as the standard for many years, but later norms were based on data obtained by Robinson and Dadson (1956). Fletcher and Munson (1933) obtained data under headphones with binaural presentation, and then corrected the data to be equivalent to presentation in a free field. They used a procedure in which the subject listened to the variable frequency and level, followed by a 1-kHz standard with this sequence presented twice before the subject indicated whether the reference tone was louder or softer than the variable tone. Robinson and Dadson (1956) used a similar procedure, but the stimuli were presented by a loudspeaker. Because the psychophysical task in loudness-contour studies is to match the loudness of two tones by the method of adjustment or by paired comparisons, there has been more agreement concerning procedures and interpretation of the data than there has been in the case of functions relating level to loudness in sones. Nonetheless, there were significant differences between the data reported by Fletcher and Munson (1933) and Robinson and Dadson (1956), and the most recent data differ substantially from both of the classic studies.

One obvious difference between the two classic studies is that Robinson and Dadson (1956) obtained their data in a free field. Fletcher and Munson (1933) applied a headphone to free-field correction to their data, but comparisons to data actually collected in a free field are limited by the accuracy of that correction. The ISO standard assumes free-field conditions because that information is of greatest interest for engineering purposes, given that most people do not use headphones in everyday listening situations. Data obtained using headphones may be of more interest, however, when the goal is to relate loudness data to other psychoacoustics data obtained under headphones. A third category is insert phones, which are used in hearing aid receivers, and in many audiology clinics and hearing research laboratories. This category has increased in importance with the widespread use of insert phones by people when listening to music. Keidser et al. (2000) asked 11 subjects to adjust the level of octave bands of speech-babble noise centered on either 0.5 or 3 kHz to equate them in loudness to a reference at 1.5 kHz, using stimuli presented in a free field or through a hearing aid receiver. Even though the levels were corrected based on probe tube measurements of the level at the eardrum in each subject, the 0.5-kHz noise band was adjusted to a level 10 dB higher in an occluded ear than in a free field. The source of this difference is unclear.

The work leading up to the latest revision of the ISO standard for loudness contours is reviewed by Suzuki and Takeshima (2004). Beginning with a study by Fastl and Zwicker (1987), a series of studies reported significant deviations from the Robinson and Dadson contours at frequencies below 1 kHz, in some cases as large as 15 dB. This prompted development of the new ISO standard illustrated in Fig. 5.3, based on data from 12 studies from 1983 forward, with most data obtained in free-field conditions. The new contours are higher at low frequencies than those reported by Robinson and Dadson, so sounds at low frequencies must be more intense to be equal in loudness to those at 1 kHz and above. This means that the contribution of low frequencies to the overall loudness of complex sounds was overestimated by standards based on earlier data.

The general form of the loudness contours reported in recent studies is similar to those reported by Fletcher and Munson (1933) and Robinson and Dadson (1956). The lowest contour represents quiet threshold, with an elevation in the low frequencies attributed to attenuation by the middle ear. The exact form of the threshold function in Fig. 5.3 is more similar to the function reported by Robinson and Dadson (1956) than that of Fletcher and Munson (1933). At frequencies below 1 kHz, loudness contours are more closely spaced, so that high-level contours are more uniform in dB SPL than the threshold contour. At frequencies above 1 kHz, loudness contours are nearly parallel to one another (Hellman et al. 2000; Takeshima et al. 2002). The new ISO contours are based on the most recent data, but differ from contours based on the ANSI standard, as shown in Fig. 5.4. The functions in Fig. 5.4 were generated using software distributed with the ANSI S3.4 standard. The software implements a version of the loudness model described by Glasberg and Moore (2006), but that model is very similar to the model described by Moore et al. (1997) on which ANSI S3.4 is based. Differences between the contours in Fig. 5.3 and predictions of the model proposed by Moore et al. (1997) are discussed by Suzuki and Takeshima (2004).

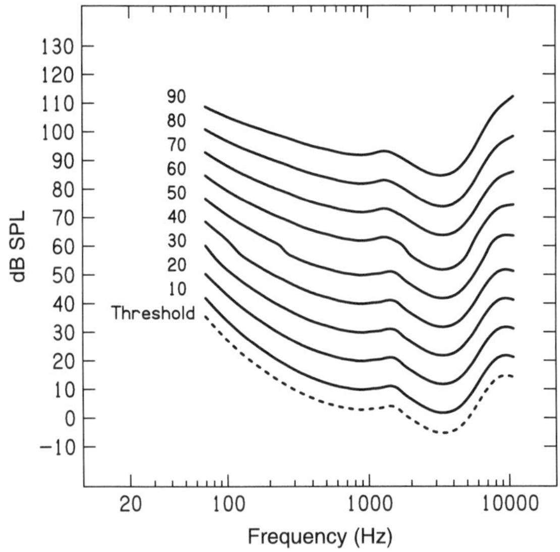

Fig. 5.4 Loudness contours predicted by the ANSI S3.4 (2007) standard, as modified by Glasberg and Moore (2006). The contours were generated using software developed by Brian Glasberg distributed with the ANSI standard

The ISO functions in Fig. 5.3 provide a more accurate description of loudness as a function of frequency than the ANSI functions in Fig. 5.4, because the ISO functions were fitted to many sets of data where tones at different frequencies were matched in loudness. The ANSI standard, however, may provide a more accurate representation of the loudness growth function that provides a conversion from loudness level in phons to loudness in sones.

5.3.2 Loudness Functions at Low and High Frequencies

Most comparisons of loudness as a function of frequency have been made in the context of efforts to develop equal-loudness-level contours. Relatively few studies have obtained growth of loudness functions directly at frequencies other than 1 kHz. Growth functions can be constructed at any frequency, however, from knowledge of loudness contours. In the left panel of Fig. 5.5, for example, loudness functions for tones at different frequencies have been estimated from the functions in Fig. 5.3, using the ISO 226 loudness function at 1 kHz shown in Fig. 5.1 to convert from phons to sones. Similar figures have been constructed by Scharf (1978, 1997) based

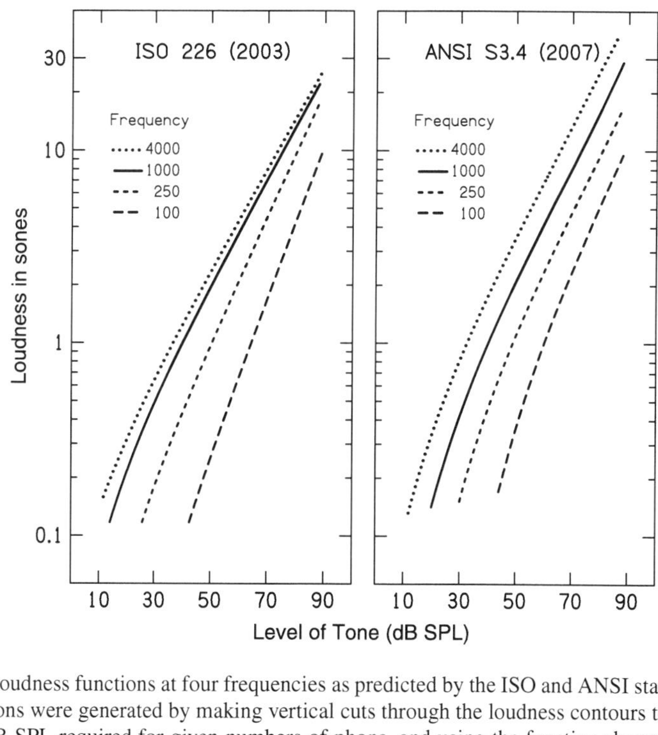

Fig. 5.5 Loudness functions at four frequencies as predicted by the ISO and ANSI standards. The ISO functions were generated by making vertical cuts through the loudness contours to obtain the levels in dB SPL required for given numbers of phons, and using the function shown in Fig. 5.1 to convert from phons to sones. The ANSI functions were generated directly by the model

on earlier data. In the right panel of Fig. 5.5, loudness functions have been estimated for the same four frequencies using the model incorporated in the ANSI S3.4 standard (Moore et al. 1997; Glasberg and Moore 2006). Takeshima et al. (2003) note that when fitting families of equal-loudness-level contours to matching data, it is important to consider the form of the loudness function that underlies the spacing of contours obtained at different levels. If loudness contours were fitted to matching data for one level at a time, measurement errors might result in discontinuities when the data were replotted as loudness growth functions.

Several studies have obtained loudness functions at low or high frequencies as a means of exploring theoretical issues concerning the measurement of loudness. In an early study, Hellman and Zwislocki (1968) used a combination of magnitude estimation and production to obtain loudness growth functions at 0.1, 0.25, and 1 kHz. Functions at the lower frequencies were steeper than at 1 kHz. The use of magnitude production as well as estimation made it possible to relate the loudness of points on one function to points on another. Hellman and Zwislocki (1968) verified this by converting their functions from loudness to loudness level in phons and demonstrating good agreement with data reported in studies of loudness contours. Additional support for translation between loudness and loudness level was provided by Schneider et al. (1971), who obtained loudness functions using a magnitude estimation task in which subjects were presented with tones selected at random from 7 levels at any of 11 frequencies. In agreement with Hellman and Zwislocki (1968), they found steeper power-law functions at frequencies below 0.4 kHz. Schneider et al. (1971) then used the 11 loudness growth functions to reconstruct loudness contours, which approximated those in the literature.

Hellman (1974) reported additional data on the growth of loudness at 0.25 kHz as part of a study on the effect of high-pass noise on the growth of loudness. In a later study, Hellman (1976) noted that data obtained at low frequencies show a more pronounced deviation from a strict power function than is typically observed at 1 kHz, with a mid-level bulge and plateau. Loudness functions at 1 and 3 kHz were found to have identical form, with the exception of an offset attributable to the difference in threshold for the two frequencies (Hellman 1976). This result is consistent with results of many studies indicating that loudness contours are parallel above 1 kHz, but it does not provide support for Stevens' contention that loudness functions at higher frequencies might be steeper and better approximated by a power law as a result of spread of excitation being confined to a single critical band (Stevens 1972).

5.4 Loudness as a Function of Duration

The increase in loudness as a function of duration, sometimes referred to as temporal integration of loudness, has elements in common with the improvement in absolute threshold as a function of duration. The latter has been modeled as energy summation with a leaky integrator (Plomp and Bouman 1959; Zwislocki 1960) or as a result of improved accuracy associated with multiple looks (Viemeister and

Wakefield 1991). The temporal integration of loudness has also been modeled as a leaky integrator (Munson 1947; Zwislocki 1969), with the assumption that integration occurred at a neural level. Munson's (1947) model was the first use of the concept of a leaky integrator in the hearing research literature. The more recent multiple-looks model of improvement in absolute threshold at longer durations cannot explain the temporal integration of loudness.

The increase in loudness with increasing duration has been measured by asking subjects to compare two stimuli differing in duration and to adjust the level of one stimulus to equate both stimuli in loudness (Munson 1947; Miller 1948) or to make similar comparisons in a task where the level of one stimulus was controlled by an adaptive procedure (Florentine et al. 1996). More recently, Epstein and Florentine (2005, 2006) have made direct measurements of the loudness of stimuli differing in duration. The effect of duration is greater for mid-level tones than for tones lower or higher in level. This effect is shown for both loudness matches and magnitude estimation in figure 5.6 from Epstein and Florentine (2006). This is one of a series of papers that have provided detailed data on the effect of duration obtained with a number of measurement paradigms (Florentine et al. 1996, 1998; Buus et al. 1997; Epstein and Florentine 2005, 2006). These articles provide support for the hypothesis, first formulated by Florentine et al. (1996), that a given increase in duration increases loudness by the same ratio at all levels. The implication of the equal-loudness-ratio hypothesis is that the large amount of temporal integration at mid levels shown in Fig. 5.6 reflects a shallow portion of the loudness function rather than

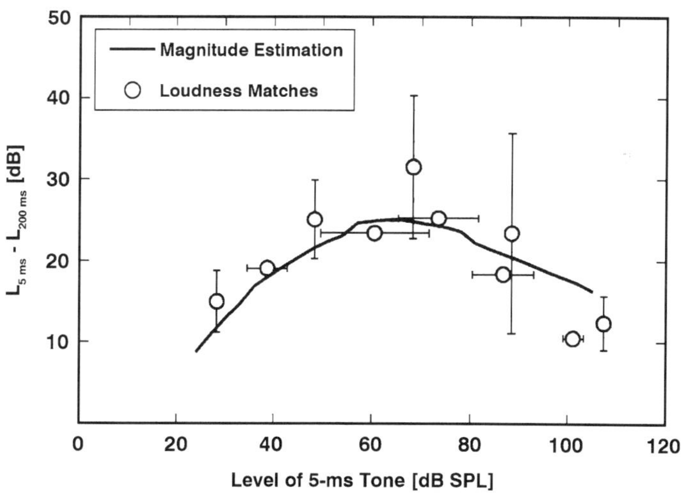

Fig. 5.6 When the level of tones is adjusted to equate a 5 and a 200-ms tone in loudness, the level of the 5-ms tone must be higher and this difference is greatest at mid intensities. Epstein and Florentine (2006) demonstrated that this effect can be predicted from the difference between magnitude estimation functions for 5 and 200-ms tones. The data for loudness matches were obtained in an earlier study (Epstein and Florentine 2005). (Reprinted with permission from Epstein and Florentine 2006, Copyright 2006, Acoustical Society of America)

an increase in the ratio between the loudness of long sounds and the loudness of short sounds. The equal-loudness-ratio hypothesis makes it possible to derive the shape of the loudness function from loudness matching data. Loudness functions derived from matching data like those shown in Fig. 5.6, but from the original Florentine et al. (1996) study, are shown in Fig. 5.7.

Loudness matching data cannot be used to test the hypothesis that an increase in duration changes loudness by the same ratio at all levels because the matching data do not provide a measure of loudness, per se. More recent studies have addressed this issue by using cross-modality matching (Epstein and Florentine 2005) and magnitude estimation (Epstein and Florentine 2006) to measure loudness for short- and long-duration tones. The cross-modality data, shown in Fig. 5.8, provide clear support for the equal-loudness-ratio hypothesis. Functions for short- and long-duration tones, obtained with a direct scaling procedure and plotted on an ordinate proportional to loudness, are parallel and show the curvature of the loudness functions in Fig. 5.7 that were reconstructed from loudness matching data. These data provide support in turn for loudness functions that are consistent with other measures of peripheral nonlinear as described in Sect. 5.2.5. In summary, recent studies of loudness as a function of duration have assumed that the effect of duration was constant at all levels and have used observed differences as a function of level to derive a loudness function that deviates in important ways from a simple power law.

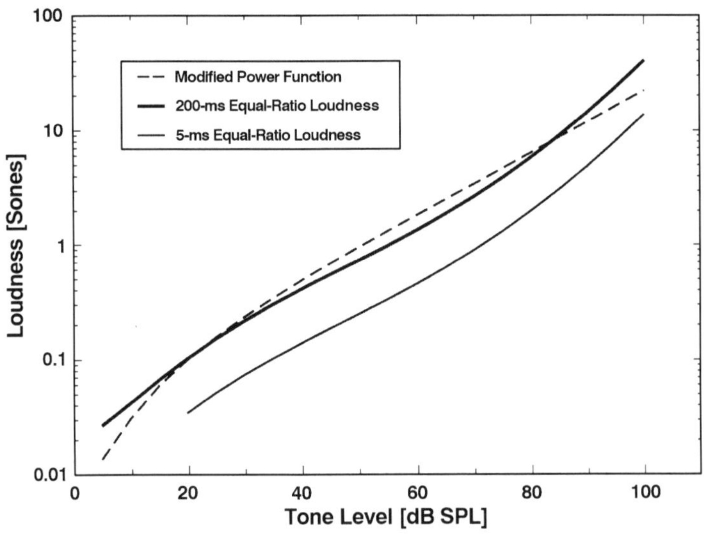

Fig. 5.7 Functions showing the relation between the level of a tone and loudness for short- and long-duration tones reconstructed from matching data by fitting polynomials to data of the type shown by circles in Fig. 5.6, then assuming that the mid-level bulge in the matching data reflects a shallow portion of the loudness function. The *dashed line* shows the loudness function proposed by Zwislocki (1965), presented here as (5.5). (Reprinted with permission from Florentine et al. 1996, Copyright 2006, Acoustical Society of America)

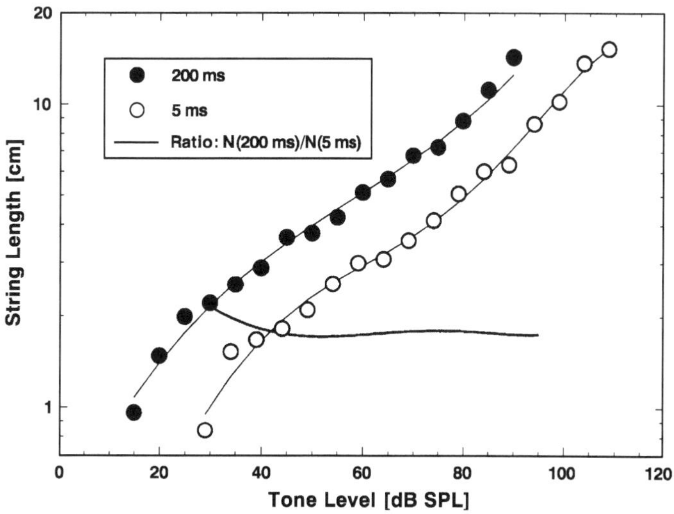

Fig. 5.8 Functions showing the relation between the level of a tone and loudness for short- and long-duration tones by cross-modality matching to string length. In this task, the subject cuts a piece of string to a length proportional to the loudness of the tone. This allows construction of loudness functions without reference to the loudness-ratio hypothesis by fitting polynomials to the string-length values. Loudness ratios can then be estimated from the vertical distance between the functions for short and long tones over the range of levels where both functions are represented. The ratios, shown by the *heavier line*, are nearly constant. (Reprinted with permission from Epstein and Florentine 2005, Copyright 2005, Acoustical Society of America)

5.5 Loudness as a Function of Bandwidth

The loudness of a pure tone depends on the physical dimensions of level, frequency, and duration, described in previous sections. Although many studies have examined the loudness of pure tones, most natural sounds (including speech) are comprised of multiple frequency components and/or bands of noise that can span across a wide range of frequencies. For these complex sounds, it has been well documented that the frequency range, or spectral bandwidth, plays an important role in determining loudness. For example, many studies have demonstrated that the overall level of a narrow-band sound must be higher than the level of a broadband sound in order to be equally loud (e.g., Fletcher and Munson 1933; Zwicker and Feldtkeller 1955; Zwicker et al. 1957; Scharf 1970). This effect is typically referred to as spectral loudness summation.

5.5.1 Measures of the Critical Band

Loudness increases with increasing frequency separation for bands of noise (e.g., Zwicker et al. 1957) and for multitonal complexes (e.g., Scharf 1961). However, loudness summation is observed only when the spectral bandwidth exceeds a minimum value. This minimum value, which serves as one of many measures of the critical band, can be estimated from the breakpoint in the function that relates bandwidth to perceived loudness while holding total power constant (e.g., Zwicker and Feldtkeller 1955; Zwicker et al. 1957; Scharf 1961, 1970; Leibold et al. 2007). In their classic study, Zwicker et al. (1957) had listeners match the loudness of a four-tone complex to the loudness of a single pure tone (or vice versa). Across conditions, the center frequency of the complex was 0.5, 1, or 2 kHz and the bandwidth was varied. For conditions in which the frequency separation of the complex was less than a single critical band, loudness was roughly independent of bandwidth. In contrast, loudness increased as frequency separation began to exceed the critical band. These results were confirmed in later studies (e.g., Scharf 1959, 1962; Florentine et al. 1978; Cacace and Margolis 1985; Schneider 1988; Verhey and Kollmeier 2002; Leibold et al. 2007). Data from the recent study by Leibold et al. (2007), who used a five-tone complex centered on 1 kHz, are shown in Fig. 5.9. The results of these studies have often been summarized by stating that energy sums for components that are in the same critical band, whereas loudness sums for components that are in different critical bands. Note, however, that complete loudness summation has not been observed in most studies. If complete loudness summation occurred, further separation of the tonal components would not lead to an increase in loudness. Recent studies specifically address this point (e.g., Hübner and Ellermeier 1993; Leibold et al. 2007).

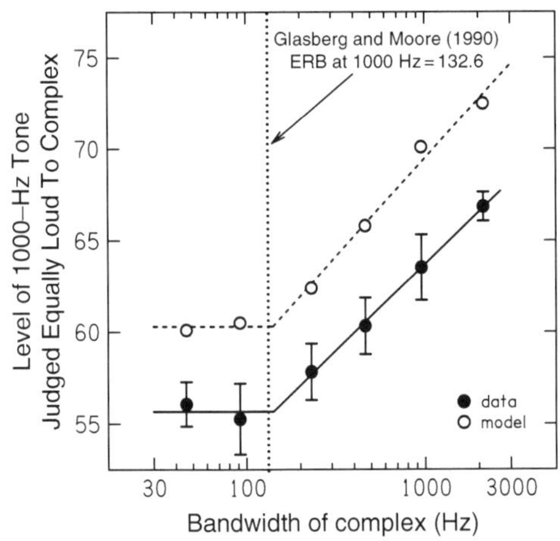

Fig. 5.9 Results of an experiment measuring the level of a 1-kHz tone judged equal in loudness to the level of a 60-dB, five-tone complex centered on 1 kHz as a function of the bandwidth of the complex. The mean level of the 1-kHz tone, averaged across subjects, is shown by the filled circles for each bandwidth (±1 SE). The *open circles* represent estimates provided by the Moore et al. (1997) loudness model. The subjects in this study showed a bias in matching the loudness of a tone to the loudness of narrowband complexes that were higher in level. The *dotted vertical line* indicates the ERB at 1,000 Hz (Glasberg and Moore, 1990). Note that the best two-line fit to both the data and the model are in agreement with the estimate of the ERB (from Leibold et al. 2007)

Beginning with Zwicker (1960), models of loudness summation have assumed that loudness is determined separately in each critical band via spectral filtering. Peripheral compressive nonlinearity and masking effects are reflected in the transform from the intensity in a given critical band to specific loudness. Finally, loudness is summed across critical bands (e.g., Fletcher and Steinberg 1924; Zwicker and Scharf 1965; Moore et al. 1997; Zwicker and Fastl 1999). Currently, the most widely used model of loudness summation is the excitation-pattern model proposed by Moore et al. (1997). The Moore et al. (1997) model accounts for spectral loudness summation by representing sounds as excitation patterns that are summed and then converted to specific loudness, as proposed in earlier models (e.g., Zwicker and Scharf 1965). Thus, the Moore et al. (1997) model predicts masking between components, and a resulting reduction in loudness, even when the components are widely separated in frequency.

Estimates of the critical band obtained from studies of loudness summation have been shown to vary with the center frequency of the noise band or multitonal complex. For center frequencies at or below about 0.2 kHz, the width of the critical band is approximately 90 Hz. As the center frequency increases, the critical band becomes wider. For example, the width of the critical band at 8 kHz is approximately 800 Hz. These critical band estimates are proportional to those obtained from the breakpoint in other procedures, including masking functions (e.g., Spiegel 1981) and estimates based on critical ratios (Fletcher, 1940).

5.5.2 Contributions of Individual Components to Total Loudness

An unresolved question in the study of loudness is how the individual frequency components of a complex sound contribute to its overall loudness. Several studies have examined the loudness of complex sounds as a function of bandwidth while varying component spacing, the number of components or the relative level of components. Component spacing appears to influence loudness, at least for conditions in which the individual components of a multitonal complex are located in multiple critical bands. Zwicker et al. (1957) examined the effect of uniform and irregular component spacing for four-tone complexes, maintaining the same overall bandwidth. Listeners were asked to match the loudness of a standard tone at 0.5, 1, or 2 kHz to the loudness of a four-tone complex. The relative spacing of the two intermediate tones in the complex was manipulated. The results indicated greater loudness when components were evenly spaced with respect to critical bands compared to when component spacing was irregular. Moreover, loudness appeared to be greater for complexes with irregular spacing when the frequencies of the two intermediate components were closer to the lowest frequency component in the four-tone complex than when the intermediate components were closer in frequency to the highest frequency component.

At least two studies have varied the number of components within a multitonal complex of fixed bandwidth. In one study, Scharf (1959) asked listeners to match the loudness of multitonal complexes with 2, 3, 4, or 8 equally-intense components to that of a 1.5-kHz tone. In one series of conditions, the bandwidth was 0.175 kHz. In another series of conditions, the bandwidth was 1.6 kHz. Within each series, components were uniformly spaced and the level of the standard tone was varied. No relation between number of components and loudness was found for either the narrow or wide series of fixed bandwidths. Scharf (1959) suggested that the greater number of critical bands contributing to total loudness for conditions in which components were widely spaced may offset the effect of increased partial masking for conditions in which components are narrowly spaced. In a subsequent study, however, Florentine et al. (1978) examined loudness summation as a function of the number of components of a multitonal complex with a wide frequency range. Results for multitonal complexes with two, three, and four tones indicated that loudness increased as the number of components increased. Florentine et al. (1978) suggested that the discrepancy between their results and those reported by Scharf (1959) reflect differences in the stimuli used. Whereas Florentine et al. (1978) used moderate level complexes (65 dB SPL) expected to produce the greatest loudness summation, Scharf presented stimuli at levels of 25, 50, 75, and 90 dB SPL. The relation between loudness summation and level is discussed in the following section.

The earliest studies of loudness summation used equally intense or equally loud components. However, Scharf (1962) argued that natural sounds typically contain components that differ in level and manipulated the relative levels of a three-tone complex. Complexes with both narrow and wide bandwidths were examined. The results indicated that loudness was independent of spectral shape when all components were located within the same critical band. In contrast, relative spacing influenced loudness when components were separated by more than a single critical band. For these more widely spaced conditions, loudness was greatest when all three components were equally intense. Although reducing the relative intensity of either side tone resulted in a decrease in overall loudness, this effect was largest when energy was shifted towards the low-frequency side tone. This finding suggests that partial masking among components contributes to the reduced loudness observed for complexes with components of unequal level, at least when the overall level of the complex is high. Recent data by Leibold et al. (2007) provide additional evidence that the relative level of individual components can influence the overall loudness of a multitonal complex. Perceptual weights were measured as a function of component position for five-tone complexes of varying bandwidth centered on 1 kHz. As shown in Fig. 5.10, weights were similar across components for conditions in which all components fell within the same critical band. In contrast, the range of weights increased with increasing bandwidth. Specifically, greater weight was assigned to the lowest and highest frequency components as the spectral bandwidth increased. Weights were related to masked thresholds in all but the widest bandwidth conditions. Similar weights were observed in all but the widest spacing condition when the Moore et al. (1997) model of loudness summation was used to estimate the loudness of each multitone complex on each trial and decisions based on those estimates were substituted for the decisions made by an actual subject. This is a strong test of the assumption that the contribution of individual components of a multitone complex to total loudness is determined by the masking between components.

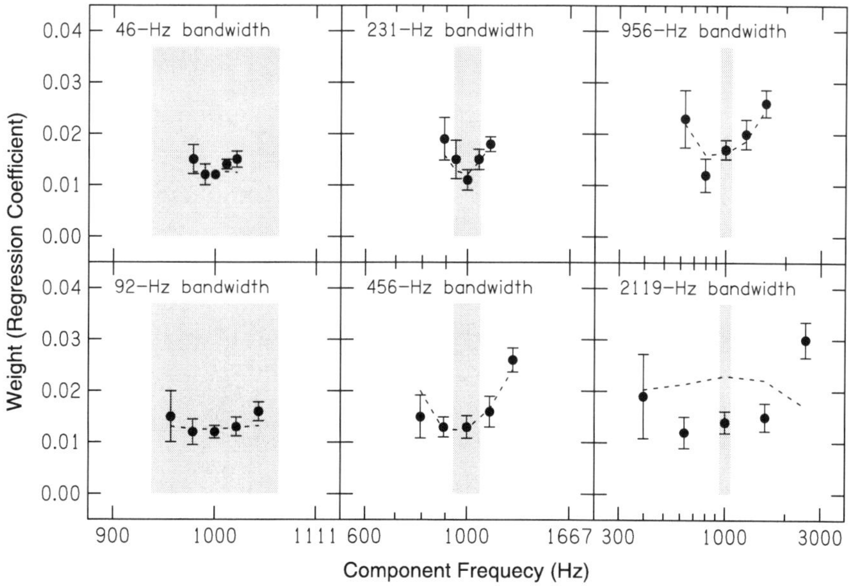

Fig. 5.10 Data from Leibold et al. (2007) showing the relative contributions of the individual tones in a five-tone complex to the overall loudness judgment with a different bandwidth in each panel. The *symbols* and *error bars* show means and standard errors for seven subjects. The *dashed lines* show predictions of the Moore et al. (1997) loudness model. The *shaded area* represents the critical band centered on 1 kHz. The model predicts the relative contributions of individual tones in all but the widest bandwidth condition (from Leibold et al. 2007)

5.5.3 Effect of Overall Level

Several studies have demonstrated that loudness summation differs as a function of overall level, with greater changes in loudness as a function of bandwidth for moderate levels than for either low or high levels (Zwicker and Feldtkeller 1955; Zwicker et al. 1957; Scharf 1959). An example of this effect is shown in Fig. 5.11, where the 1-kHz data from Zwicker et al. (1957) have been plotted and reanalyzed. Scharf (1961) observed a similar effect in measurements of the loudness of a four-tone complex as a function of bandwidth in the presence of filtered white noise. When all components fell within the same critical band, loudness summation did not change when the level of either the complex or the noise was increased. Loudness summation increased, however, when components fell within more than one critical band. For loudness measured in quiet or in noise, loudness summation was greater at moderate sensation levels (SLs) than either high or low levels. Scharf (1961) suggested that the reduced loudness summation observed at high levels reflects partial masking between components.

Overall level also appears to play a role in loudness additivity across critical bands. Hübner and Ellermeirer (1993) observed that loudness additivity does not

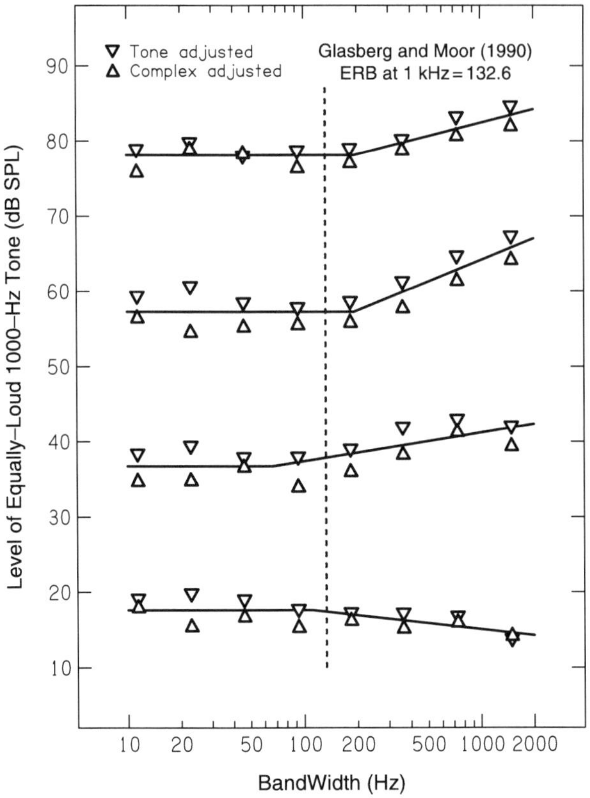

Fig. 5.11 Data from Zwicker et al. (1957) showing loudness matches between a 1-kHz tone and a four-tone complex as a function of the bandwidth of the complex for stimuli presented at four different overall levels. The *lines* shown by Zwicker et al. (1957), based on an assumed critical bandwidth, have been replaced by lines obtained by least squares fits to the data, with the assumption that loudness was independent of bandwidth for bandwidths of 100 Hz or less. The breakpoints determined by the data have a geometric mean of 128.5 Hz, close to the predicted ERB of 132.6 Hz. The rate of change in loudness with bandwidth is clearly dependent on overall level, with a decrease observed at low levels

hold under all conditions for a two-tone complex. It is likely that factors such as suppression and partial masking play a larger role at higher levels, even for complexes that are widely spaced in frequency.

The increase in loudness summation at moderate levels is comparable to the increased effect of duration observed at moderate levels (e.g., Florentine et al. 1996) and it is reasonable to assume that the form of the loudness function contributes to this effect as well. Because the interaction among components varies with level, however, it would be difficult to use the change in loudness summation with level to solve for the form of the loudness function. Given an assumed loudness function of the type shown in Fig. 5.7, it should be possible to correct for the contribution of the slope of the loudness function to loudness summation data.

5.6 Partial Masking of Loudness

The effect of bandwidth on the loudness of multitone stimuli discussed in the preceding section is determined in part by the masking of each tone by the other tones. The interaction between components and its effect on the loudness of the multi-tone complex is well predicted by models based on excitation patterns (Scharf 1964; Zwicker and Scharf 1965; Moore et al. 1997). Rather than considering the effects of such interactions on the loudness of the entire stimulus, it is sometimes of greater interest to consider the effect of one component on the loudness of another component. Tones presented at a certain level in broadband noise, for example, may be clearly audible, but may not be as loud as a tone of the same level presented in quiet. Such effects are referred to as partial masking of loudness (for review, see Scharf 1964).

The first study of partial masking in normal-hearing subjects compared loudness matching functions for tones presented in quiet and in noise to loudness matching functions for tones presented to the two ears of subjects with unilateral cochlear hearing loss (Steinberg and Gardner 1937). Data were obtained from three or four subjects with normal hearing at octave frequencies from 0.25 to 8 kHz using noise levels chosen to produce 10, 20, and 40 dB of masking. Tones presented in quiet alternated with tones presented in noise and the subject's task was to adjust the level of one set of tones to equate them in loudness to the other set of tones. The data for 1 kHz reported by Steinberg and Gardner are shown in Fig. 5.12. A similar

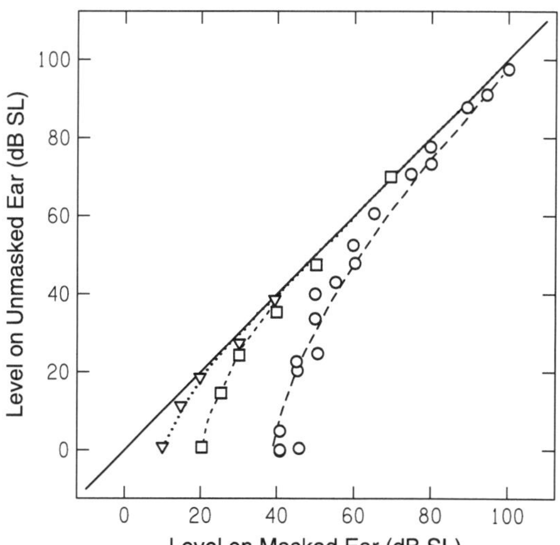

Fig. 5.12 Data showing loudness matches between 1 kHz tones in masked and unmasked conditions. The reference in both cases is the absolute threshold for a 1 kHz tone (adapted from Steinberg and Gardner 1937)

pattern is observed in subjects with cochlear hearing loss (see Chap. 9) showing maximum effect of the noise on the loudness of the tone at threshold with progressively less effect at higher levels. Steinberg and Gardner (1937) compared the data to similar matching functions obtained from subjects with unilateral hearing loss and presented a physiologically based excitation pattern model to account for the results. Their excitation patterns differ in form from those used by Moore et al. (1997), but the basic account of partial masking and its relation to the form of loudness functions in cochlear hearing loss has remained unchanged.

Many later studies have used a similar matching procedure with tone and noise stimuli (Jerger and Harford 1960; Lochner and Burger 1961; Zwicker 1963; Hellman and Zwislocki 1964; Stevens 1966b; Stevens and Guirao 1967). Hellman and Zwislocki (1964) obtained data showing the effect of an octave band of noise centered on 1 kHz on the loudness of a 1-kHz tone, using the matching of loudness levels and also through direct measurement of loudness using a combination of magnitude estimation and magnitude production. One of their goals was to demonstrate the value of direct measurement procedures by using loudness functions of the type shown in Fig. 5.13 to predict matching data of the type shown in Fig. 5.12. They therefore drew smooth functions through the data as shown in Fig. 5.13 and from those functions determined the levels of tones in noise that would be judged to be

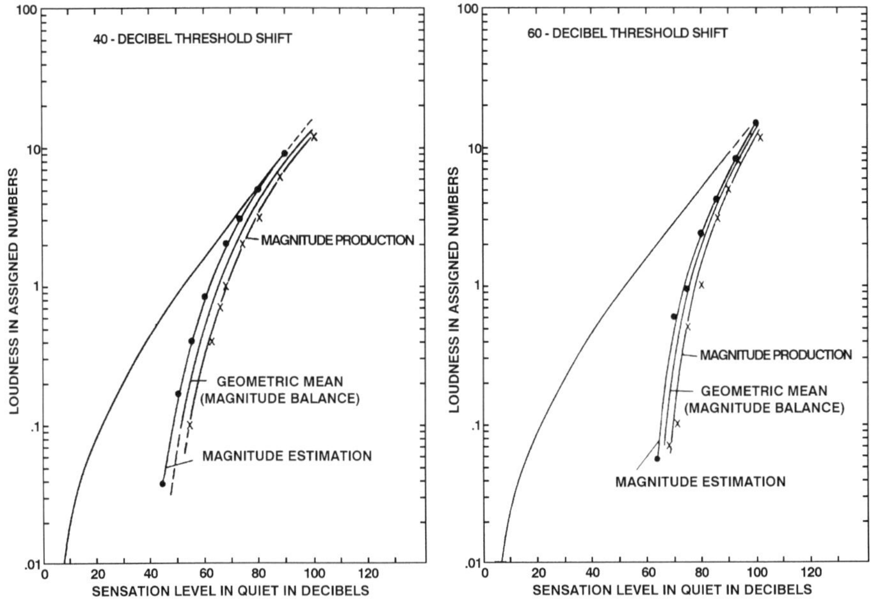

Fig. 5.13 Two figures from Hellman and Zwislocki (1964) have been placed side-by-side to show the effect of two levels of noise on a direct measure of the loudness of a 1-kHz tone obtained from a combination of magnitude estimation and magnitude production. The leftmost function in each panel showing the growth of loudness in the absence of noise was obtained in an earlier study. (Reprinted with permission from Hellman and Zwislocki 1964, Copyright 1964, Acoustical Society of America)

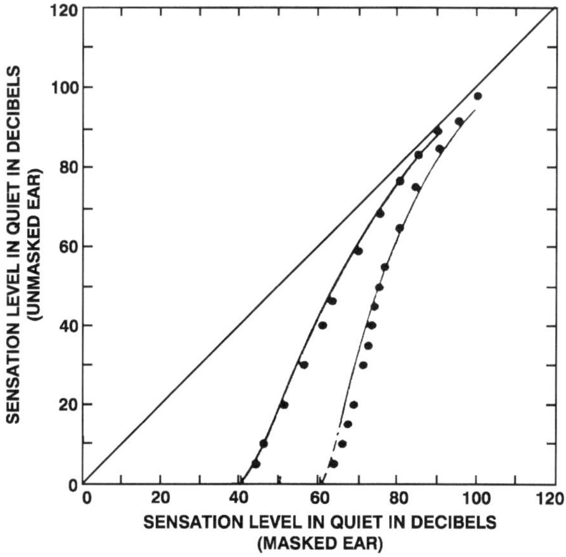

Fig. 5.14 Hellman and Zwislocki (1964) obtained loudness matches for tones in quiet and in noise as well as direct measures of loudness, then used the difference between the loudness functions shown in the two panels of Fig. 5.13 to predict the results obtained with loudness matches. The predictions, shown by *curved lines*, provide a good fit to the data. (Reprinted with permission from Hellman and Zwislocki 1964, Copyright 1964, Acoustical Society of America)

equal in loudness to tones in quiet. They then used those pairs of levels to construct the matching functions shown in Fig. 5.14, demonstrating a good fit to loudness matching data obtained independently. These authors address a wide range of other methodological issues including potential biases in matches based on whether the masked or unmasked tone is varied and the variability of matches. They noted that a tone appears less loud when it is varied in a method-of-adjustment procedure than when it is constant in level. They also noted that matches are generally more variable in the mid levels when the unmasked tone is varied than when the masked tone is varied. They attributed the difference in variability to the fact that the loudness function was shallower for the unmasked tone. This study is unique in that the relation between matching data and numerical magnitude balance data was demonstrated without the use of an equation to represent the form of the loudness function, although Zwislocki (1965) later used the data as support for (5.5).

The equations describing loudness near threshold reviewed in Sect. 5.2.2 have typically been applied to masked thresholds as well as quiet thresholds and most were developed to fit data obtained in studies of partial masking. Lochner and Burger (1961), for example, obtained loudness matching data for 1-kHz tones presented in quiet and in three levels of an octave band of noise centered on 1 kHz. They fitted the data by assuming the loudness function described by (5.4) in Sect. 5.2.2.

Partial masking data obtained by matching the loudness of tones presented in quiet and in noise have typically been fitted by using a loudness function with an assumed threshold correction to convert the data to loudness in sones, then predicting the levels in quiet and noise required for equal loudness. Pavel and Iverson (1981) took a different approach to the problem by developing a model of the relations among the matching functions themselves that in turn places some constraints on the form of threshold corrections in loudness functions. They assumed that any given shift in the level of the masker would cause all tone levels to shift by a fixed amount to preserve a loudness match. They noted that matching data from Lochner and Burger (1961), Hellman and Zwislocki (1964), and Stevens and Guirao (1967) all satisfy this shift invariance requirement.

The partial masking data reviewed up to this point have been obtained using bands of noise an octave wide or wider with a tone presented near the center of the noise. Under these conditions, the excitation pattern of the tone increases with tone level much as it would in quiet, except that the higher absolute levels result in more rapid growth. Data obtained using narrower bands of noise and, in some cases, tones presented at frequencies outside the noise bands provide additional information about excitation patterns and loudness and demonstrate that the threshold corrections discussed in Sect. 5.2.2 are not valid in all cases. Gleiss and Zwicker (1964) reviewed data reported by Zwicker (1963) for the partial masking produced by broadband noise and by noise a critical band wide, centered on 1 kHz. They noted that the loudness of the tone in critical-band wide noise grows more rapidly than the Lochner and Berger formula (5.4) would predict, while the loudness of a tone in broadband noise grows somewhat more slowly than Lochner and Berger observed for a tone in an octave band of noise. Gleiss and Zwicker (1964) concluded that these effects are consistent with an excitation pattern model, but rule out use of any single formula to describe masked loudness.

Another study completed at about the same time presents problems even for excitation pattern models. Scharf (1964) demonstrated asymmetric effects in the partial masking produced by a narrow band of noise, one critical band wide, using tones located at frequencies one and two critical bands below and above the noise as well as at the center of the noise. He found that the noise reduced the loudness of tones lower in frequency more than the loudness of tones higher in frequency, even though masked thresholds were elevated more on the high frequency side. Threshold correction formulas reviewed in Sect. 5.2.2 would provide a poor description of these data. Although Scharf provided a visual account of the data in terms of the asymmetric growth of excitation patterns with level for both the noise and the tones, it is noteworthy that models of loudness summation and partial masking based on excitation patterns (Zwicker and Scharf 1965; Moore et al. 1997) have provided poorer fits to these data than to any other data sets they have considered. Moore et al. (1997) noted that fits to data of this type are more sensitive to assumptions concerning auditory filter shape and that the fluctuations of the narrowband masker may have influenced both the loudness judgments and the threshold measurements in conditions where the tones were located outside the band of noise.

Two other studies have considered cases where masking is asymmetric. Hellman (1972) compared masked thresholds and partial masking for 1-kHz tones and narrow bands of noise centered on 1 kHz. She found that the threshold for the tone in the presence of a one-critical-band wide noise masker (0.925–1.080 kHz) was about 20 dB higher than the threshold for the noise in the presence of a 1-kHz tone masker presented at the same overall level. The masked loudness matching functions for tones in the presence of noise were steeper than for noise in the presence of tones, but functions for the two combinations were similar when the levels of the maskers were chosen to produce equal amounts of masking. Gockel et al. (2003) compared masked thresholds and partial masking for maskers consisting of periodic complex tones and noise bands chosen to have the same excitation patterns as the complex tones. They found that the noise was 12–16 dB more effective than the complex tone as a masker and as a partial masker. As noted in the preceding text, Gleiss and Zwicker (1964) summarized data showing that it was necessary to take excitation patterns into account to predict the form of masked loudness functions rather than relying solely on a threshold correction. Gockel et al. (2003) showed that excitation patterns are insufficient and that temporal properties of the waveforms play a significant role.

5.7 Loudness as a Function of Age

Most studies that have examined loudness perception have focused on understanding the performance of young adult listeners. These listeners enter the laboratory with a wide range of auditory experiences and a mature auditory system. The interaction between loudness and age has received less attention; despite mounting evidence of substantial changes in intensity processing during infancy and childhood (reviewed by Werner 2007) and an emerging literature describing effects of advanced aging on hearing (e.g., Fitzgibbons et al. 2007). This section describes the small number of studies that have examined loudness for steady-state sounds as a function of age, studies primarily focused on understanding the relation between sound level and loudness.

5.7.1 Loudness During Infancy and Childhood

Studies describing the early development of loudness are limited, reflecting the difficulties associated with obtaining reliable subjective responses to sound from infants and young children. In previous chapters, methods have been described that rely on verbal instructions and require the listener to match, order or scale loudness (Chaps. 2 and 3). These sophisticated responses are not feasible for use with infants and young children. Thus, few data describing loudness during the infant and toddler years are available.

Although subjective loudness measurements have not been obtained from infants, at least two studies have examined loudness during infancy using nonverbal correlates (see Chap. 4). Bartoshuk (1964) examined the relation between the intensity of a 1-kHz tone and the magnitude of cardiac acceleration in newborns. The data were reported to be well fitted by a power function with an exponent of 0.53, similar to the exponent of 0.6 reported for the relation between numerical magnitude estimates and pressure in adults (e.g., Stevens 1959). Note, however, these data have not been replicated and the relation between cardiac acceleration and loudness is not well established in adults.

Leibold and Werner (2002) examined reaction time (RT) to sound for 6–9-month-old infants and adults in response to 1- and 4-kHz pure tones. The mean RT decreased systematically with increasing intensity for both age groups. In addition, RT-intensity functions for most infants were steeper than for most adults. It is difficult to attribute these findings to age differences in sensory processing (Werner 2007). Instead, infants' steeper functions are consistent with the idea that infants and adults differ in how they listen to sounds. In support of this argument, central processes appear to influence adults' perception of loudness under some conditions (e.g., Schlauch 2004; Leibold et al. 2007).

The data describing the development of loudness are more complete for older children. By the time children enter school they can reliably perform magnitude estimation and cross-modality matching (Bond and Stevens 1969; Teghtsoonian 1980; Collins and Gescheider 1989; Serpanos and Gravel 2000). Using these methods, studies of loudness in children ages 4 years and older have consistently reported that loudness grows with increasing intensity in a similar way for children and adults (Dorfman and Megling 1966; Bond and Stevens 1969; Collins and Gescheider 1989; Serpanos and Gravel 2000). For example, Bond and Stevens (1969) found no evidence that 5-year-olds differed from adults when asked to match the perceived magnitude of light to sound intensity. Collins and Gescheider (1989) later reported similar estimates of loudness growth for 4–7-year-olds compared to adults for both absolute magnitude estimation (Fig. 5.15) and cross-modality matching to line length. Despite large individual differences, loudness data for individual children and adults tested by Collins and Gescheider were consistent across time. More recently, Serpanos and Gravel (2000) examined cross-modality matching in 4–10-year-old children and adults. As in the earlier studies, estimates of loudness growth were similar for children and adults.

An additional line of evidence consistent with the idea that loudness is adult-like by at least 4–5 years of age comes from studies that have examined the upper limit of the dynamic range. This upper limit is often referred to as the loudness discomfort level (LDL). Kawell et al. (1988) obtained reliable estimates of LDLs from 7 to 14-year-old children with sensorineural hearing loss. No systematic differences in LDLs were found between the children and a group of hearing-impaired adults. Macpherson et al. (1991) extended the lower age limit for reliably measuring LDLs to 4 years and reported LDL estimates comparable to previous adult data (e.g., Hawkins 1980) for both normal-hearing and hearing-impaired children. In summary, data describing loudness during infancy are limited. Preliminary data comparing RTs to sound across infants and adults may indicate increased loudness growth during infancy, but additional studies are required to establish the course of development during infancy and into the toddler years. Few developmental effects have

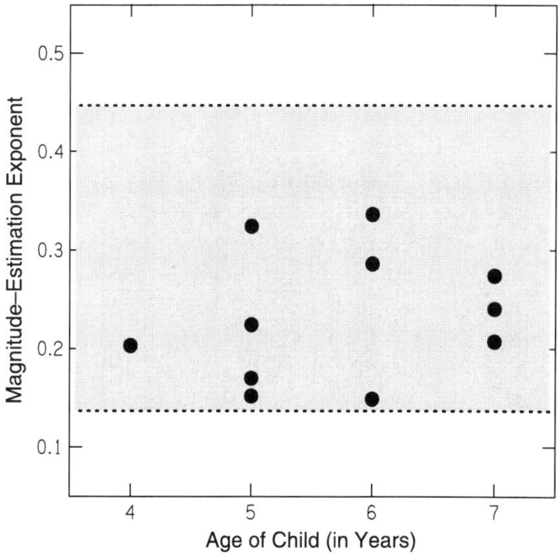

Fig. 5.15 Loudness exponents for absolute magnitude estimation (AME) of a 1-kHz tone are shown for individual children. The *shaded area* shows ±2 SD around the average AME exponent for adults. Consistent with exponents obtained for cross-modality matching, the data indicate similar loudness perception across 4–7-year-old children and adults (adapted from Collins and Gescheider 1989)

been reported for older children. Converging evidence suggests that the perception of loudness is mature by the time children enter school.

5.7.2 Loudness and the Aging Auditory System

Mounting evidence indicates that auditory processing declines with increasing age. For example, advanced age has been associated with decreased auditory temporal processing, changes thought to be largely independent of changes in absolute sensitivity (e.g., Fitzgibbons et al. 2007). Although age-related changes have been observed for auditory processing in other domains, few studies have examined loudness perception in elderly listeners.

The limited number of studies investigating age-related changes in loudness perception reflects difficulties associated with accounting for the many other changes associated with aging. As Marshall (1981) noted, a number of factors are likely to affect auditory performance for elderly listeners, including changes in general cognitive functioning, declines in central auditory processing and age-related changes in hearing sensitivity. Of perhaps the most importance for the study of senescent changes in loudness perception is age-related hearing loss (presbycusis), because the presence of hearing loss has been shown to influence loudness

(Hellman and Meiselman 1993; Chap. 9). Elderly listeners tend to have elevated audiometric thresholds compared to young adults, even if their thresholds are within the range considered normal on an audiogram. Interpreting these differences in hearing across older and younger adults is more complex when hearing loss is present, because it can be difficult to separate age-related hearing loss from hearing loss caused by other factors such as noise.

The few data that have been reported in the literature have primarily focused on the relation between age-related hearing loss and rapid loudness growth. Rapid loudness growth, or recruitment, is typically observed for cochlear hearing loss, but is often absent for retrocochlear hearing loss. The data are limited, but early studies by Pestalozza and Shore (1955) and Harbert et al. (1966) suggested that many elderly listeners do not show rapid loudness growth. In a later study, Knight and Margolis (1984) tested this hypothesis by examining rapid loudness growth in elderly listeners. The results were consistent with the earlier suggestions, indicating that the effects of advanced age on loudness is related to neural, rather than cochlear, pathology.

In summary, the effects of advanced aging on loudness are not well understood. In part, this lack of information reflects difficulties associated with differentiating peripheral from central factors, and also equating the degree of hearing loss across older and younger listeners. An important unresolved question is whether age-related changes in loudness occur independently of age-related changes in hearing loss.

5.8 Summary

It was noted at the beginning of the chapter that the literature describing studies of the loudness of steady-state sounds covers a period of more than 70 years and that many early studies were of high quality and still relevant. Some of the earliest work cited here, including Fletcher and Steinberg (1924), Fletcher and Munson (1933), Stevens (1936), Stevens and Davis (1936) and Steinberg and Gardner (1937), includes procedures and theories that are still current. Nonetheless, this chapter has identified several areas where significant changes have occurred in recent years and other areas with unresolved issues that warrant additional work.

The most significant advance in the loudness literature of steady-state sounds in recent years is the clear demonstration that the loudness function is a reflection of peripheral nonlinearity, consistent with other measures of peripheral nonlinearity (Florentine et al. 1996; Buus et al. 1997; Buus and Florentine 2002; Epstein and Florentine 2005, 2006; Florentine and Epstein 2006). This interpretation of the data emerged from the use of polynomial fitting procedures to reveal orderly deviations from the simple power functions that had been used to summarize large amounts of data in the literature. The difference was not in the stimuli or the procedures used for data collection, but in the data analysis. The new approach, not generally available until the widespread use of computerized data analysis, emphasized the complexity of the data rather than providing the simplest possible summary. In doing so, it increased the links between measures of loudness and other measures of peripheral nonlinearity based on masking. At the same time, it may have decreased the value of comparisons across sensory continua. A loudness function with the complex form shown in Figs. 5.7 and 5.8 is a poor device for measuring other sensory continua (Stevens 1966a) and it is likely that the growth functions for many of those continua are equally complex.

Another significant change is reflected in the most recent ISO (2003) loudness contours, provided in Fig. 5.3, showing a significant shift at low frequencies compared to previous standards. The origin of this shift to higher levels at low frequencies is unclear, but the new contours include data from a large number of laboratories around the world and were developed using data analysis procedures that were not available to Robinson and Dadson (1956). The data concerning loudness at frequencies other than 1 kHz have been greatly expanded by this effort.

The new ISO standard was followed by an ANSI (2007) standard based on a model of loudness that has been widely distributed as a stand-alone computer program (Moore et al. 1997). Although models of loudness are the purview of another chapter in this volume (see Chap. 10), the model proposed by Moore et al. (1997) was cited frequently here because it is a powerful tool for the evaluation and interpretation of laboratory data. The widespread availability of this model will move the field forward by making it convenient to compare results obtained with arbitrary combinations of tones and noises to predictions generated by a consensus model. Use of the model to predict masked thresholds based on partial loudness (e.g., Buss 2008) will lead to further connections between the literature on loudness and the literature on masking.

A final area of recent progress is the exploration of the effects of both infant development and advancing age on the perception of loudness. Further progress in this area will require additional efforts to separate peripheral and central factors contributing to loudness.

Two unresolved issues discussed in this summary of the literature suggest areas for future work. The first is the result reported by Keidser et al. (2000) that low frequency stimuli do not sound as loud when presented in occluded ear canals even though levels have been equated using probe-tube measurements. Given increased use of hearing aids and personal music systems, it is surprising that this result has not received more attention. The second, and considerably larger issue, is that there are many reports of results that cannot be predicted by the Moore et al. (1997) model and thus run counter to our understanding of the rules governing excitation patterns and loudness. These reports include Scharf's (1964) classic study of partial masking as well as later studies (e.g., Gockel et al. 2003), demonstrating that temporal properties of steady-state sounds can influence loudness judgments.

Acknowledgments Harisadhan Patra, Melissa Krivohlavek, Jessica Messersmith, Abbey Correll, Skip Kennedy, and Barbara Olmedo contributed to the preparation of this chapter. Their help is gratefully acknowledged. This work was supported by NIH grants R01 DC006648, R01 DC006616, and R03 DC008389.

References

American National Standards Institute (2007) ANSI S3.4. Procedure for the Computation of Loudness of Steady Sounds. New York: American National Standards Institute.

Bartoshuk AK (1964) Human neonatal cardiac responses to sound: a power function. Psychon Sci 1:151–152.

Bond B, Stevens SS (1969) Cross-modality matching of brightness to loudness by 5-year-olds. Percept Psychophys 6:337–339.

Buss E (2008) The effect of masker level uncertainty on intensity discrimination. J Acoust Soc Am 123:254–264.

Buus S, Florentine M (2002) Modifications to the power function for loudness. In: Summerfield E, Kompass R, Lochmann T (eds), Fechner Day 2001. Berlin: Pabst, pp. 236–241.

Buus S, Florentine M, Poulsen T (1997) Temporal integration of loudness, loudness discrimination, and the form of the loudness function. J Acoust Soc Am 101:669–680.

Buus S, Musch H, Florentine M (1998) On loudness at threshold. J Acoust Soc Am 104:399–410.

Cacace AT, Margolis RH (1985) On the loudness of complex stimuli and its relationship to cochlear excitation. J Acoust Soc Am 78:1568–1573.

Churcher BG (1935) A loudness scale for industrial noise measurements. J Acoust Soc Am 6:216–225.

Collins AA, Gescheider GA (1989) The measurement of loudness in individual children and adults by absolute magnitude estimation and cross-modality matching. J Acoust Soc Am 85:2012–2021.

Dorfman DD, Megling R (1966) Comparison of magnitude estimation of loudness in children and adults. Percept Psychophys 1:239–241.

Epstein M, Florentine M (2005) A test of the equal-loudness-ratio hypothesis using cross-modality matching functions. J Acoust Soc Am 118:907–913.

Epstein M, Florentine M (2006) Loudness of brief tones measured by magnitude estimation and loudness matching. J Acoust Soc Am 119:1943–1945.

Fastl H, Zwicker E (1987) Loudness level at 400 Hz: Psychoacoustical measurement and calculation by ISO 532B. Fortschritte der Akustik, DAGA (87, pp. 189–193.

Fitzgibbons PJ, Gordon-Salant S, Barrett J (2007) Age-related differences in discrimination of an interval separating onsets of successive tone bursts as a function of interval duration. J Acoust Soc Am 122:458–466.

Fletcher H (1940) Auditory patterns. Res Mod Phys 12:47–65.

Fletcher H (1953) Speech and Hearing in Communication. New York: D. Van Nostrand.

Fletcher H, Munson WA (1933) Loudness, its definition, measurement, and calculation. J Acoust Soc Am 5:82–108.

Fletcher H, Steinberg JC (1924) The dependence of the loudness of a complex sound upon the energy in the various frequency regions of the sound. Phys Rev 24:306–317.

Florentine M, Epstein M, (2006) To honor Stevens and repeal his law (for the auditory system). In Kornbrot, D, Msetfi, R, MacRae A. (eds), Fechner Day 2006. England: St. Albans, pp. 37–42.

Florentine M, Buus S, Bonding P (1978) Loudness of complex sounds as a function of the standard stimulus and the number of components. J Acoust Soc Am 64:1036–1040.

Florentine M, Buus S, Poulsen T (1996) Temporal integration of loudness as a function of level. J Acoust Soc Am 99:1633–1644.

Florentine M, Buus S, Hellman RP (1997) A model of loudness summation applied to high-frequency hearing loss. In: Jesteadt W (ed), Modeling Sensorineural Hearing Loss. Mahwah, NJ: Lawrence Erlbaum, pp. 187–198.

Florentine M, Buus S, Robinson M (1998) Temporal integration of loudness under partial masking. J Acoust Soc Am 104:999–1007.

Glasberg BR, Moore BC (1990) Derivation of auditory filter shapes from notched-noise data. Hear Res 47:103–138.

Glasberg BR, Moore BC (2006) Prediction of absolute thresholds and equal-loudness contours using a modified loudness model. J Acoust Soc Am 120:585–588.

Gleiss N, Zwicker E (1964) Loudness function in the presence of masking noise. J Acoust Soc Am 36:393–394.

Gockel H, Moore BC, Patterson RD (2003) Asymmetry of masking between complex tones and noise: partial loudness. J Acoust Soc Am 114:349–360.

Goldstein JL (1974) Is the power law simply related to the driven spike response rate from the whole auditory nerve? In: Moskowitz HR, Scharf B, Stevens JC (eds), Sensation and Measurement: Papers in Honor of S. S. Stevens. Dordrecht-Holland: D. Reidel, pp. 223–229.

Harbert F, Young IM, Menduke H (1966) Audiologic findings in presbycusis. J Aud Res 6:297–312.

Hawkins DB (1980) The effect of signal type on the loudness discomfort level. Ear Hear 1:38–41.

Hellman RP (1972) Asymmetry of masking between noise and tone. Percept Psychophys 11:241–246.

Hellman RP (1974) Effect of spread of excitation on the loudness function at 250 Hz. In: Moskowitz HR, Scharf B, Stevens JC (eds), Sensation and Measurement: Papers in Honor of S. S. Stevens. Dordrecht-Holland: D. Reidel, pp. 241–249.

Hellman RP (1976) Growth of loudness at 1000 and 3000 Hz. J Acoust Soc Am 60:672–679.

Hellman RP (1991) Loudness measurement by magnitude scaling: implications for intensity coding. In: Bolanski Jr SJ, Gescheider GA (eds), Ratio Scaling of Psychological Magnitude. Hillsdale, NJ: Lawrence Erlbaum, pp. 215–228.

Hellman RP (1997) Growth of loudness in sensorineural impairment: experimental results and modeling implications. In: Jesteadt W (ed), Modeling Sensorineural Hearing Loss. Mahwah, NJ: Lawrence Erlbaum, pp. 199–212.

Hellman WS, Hellman RP (1975) Relation of the loudness function to the intensity characteristic of the ear. J Acoust Soc Am 57:188–192.

Hellman RP, Meiselman CH (1993) Rate of loudness growth for pure tones in normal and impaired hearing. J Acoust Soc Am 93:966–975.

Hellman RP, Zwislocki JJ (1964) Loudness function of a 1000–cps tone in the presence of a masking noise. J Acoust Soc Am 36:1618–1627.

Hellman RP, Zwislocki JJ (1968) Loudness determination at low sound frequencies. J Acoust Soc Am 43:60–64.

Hellman RP, Takeshima H, Suzuki Y, Ozawa K, Sone T (2000) Equal-loudness contours at high frequencies. Proc InterNoise 2000 Nice:3653–3658.

Hübner R, Ellermeier W (1993) Additivity of loudness across critical bands: a critical test. Percept Psychophys 54:185–189.

Humes LE, Jesteadt W (1991) Models of the effects of threshold on loudness growth and summation. J Acoust Soc Am 90:1933–1943.

International Organization for Standardization (2003) ISO 226:2003 Acoustics. Normal equal-loudness contours. Geneva: International Organization for Standardization.

International Organization for Standardization (2005) ISO 389–7:2005 Acoustics. Reference zero for the calibration of audiometric equipment. Part 7: Reference threshold of hearing under free-field and diffuse-field listening conditions. Geneva: International Organization for Standardization.

Jerger JF, Harford ER (1960) Alternate and simultaneous binaural balancing of pure tones. J Speech Hear Res 3:15–30.

Kawell ME, Kopun JG, Stelmachowicz PG (1988) Loudness discomfort levels in children. Ear Hear 9:133–136.

Keidser G, Katsch R, Dillon H, Grant F (2000) Relative loudness perception of low and high frequency sounds in the open and occluded ear. J Acoust Soc Am 107:3351–3357.

Kim D, Molnar CE, Pfeiffer RR (1973) A system of nonlinear differential equations modeling basilar-membrane motion. J Acoust Soc Am 54:1518–1529.
Kingsbury BA (1927) A direct comparison of the loudness of pure tones. Phys Rev 29:588–600.
Knauss HP (1937) An empirical formula for the loudness of a 1000-cycle tone. J Acoust Soc Am 9:45–46.
Knight KK, Margolis RH (1984) Magnitude estimation of loudness II: loudness perception by presbycusic listeners. J Speech Hear Res 27:28–32.
Laming D (1986) Sensory Analysis. San Diego, CA: Academic.
Laming D (1997) The Measurement of Sensation. New York: Oxford University Press.
Leibold LJ, Werner LA (2002) Relationship between intensity and reaction time in normal-hearing infants and adults. Ear Hear 23:92–97.
Leibold LJ, Tan H, Khaddam S, Jesteadt W (2007) Contributions of individual components to the overall loudness of a multitone complex. J Acoust Soc Am 121:2822–2831.
Lochner JPA, Burger JF (1961) Form of the loudness function in the presence of masking noise. J Acoust Soc Am 33:1705–1707.
Logue AW (1976) Individual differences in magnitude estimation of loudness. Percept Psychophys 19:279–280.
Luce RD (1959) On the possible psychophysical laws. Psychol Rev 66:81–95.
Macpherson BJ, Elfenbein JL, Schum RL, Bentler RA (1991) Thresholds of discomfort in young children. Beh Aud 12:184–190.
Marks LE (1974) Sensory Processes: The New Psychophysics. New York: Academic.
Marshall L (1981) Auditory processing in aging listeners. J Speech Hear Dis 46:226–240.
Miller GA (1948) The perception of short bursts of noise. J Acoust Soc Am 20:160–170.
Moore BC, Glasberg BR (1996) A revision of Zwicker(s loudness model. Acustica 82:335–345.
Moore BC, Oxenham AJ (1998) Psychoacoustic consequences of compression in the peripheral auditory system. Psychol Rev 105:108–124.
Moore BC, Glasberg BR, Baer T (1997) A model for the prediction of thresholds, loudness, and partial loudness. J Audio Eng Soc 45:224–240.
Moore BC, Vickers DA, Plack CJ, Oxenham AJ (1999) Inter-relationship between different psychoacoustic measures assumed to be related to the cochlear active mechanism. J Acoust Soc Am 106:2761–2778.
Munson WA (1947) The growth of auditory sensation. J Acoust Soc Am 19:584–591.
Neely ST, Gorga MP, Dorn, PA (2000) Distortion product and loudness growth in an active, nonlinear model of cochlear mechanics. In: Wada, H, Takasaka T, Ikeda K, Ohyama K, Koike T (eds), Proceedings of the International Symposium on Recent Developments in Auditory Mechanics. Singapore: World Scientific, pp. 237–243.
Neely ST, Schairer KS, Jesteadt W (2003) Estimates of cochlear compression from measurements of loudness growth. In: Pressnitzer D, de Cheveigné A, McAdams S, Collet L (eds), Auditory Signal Processing, Physiology, Psychoacoustics, and Models. New York: Springer, pp. 42–56.
Oxenham AJ, Bacon SP (2003) Cochlear compression: perceptual measures and implications for normal and impaired hearing. Ear Hear 24:352–366.
Oxenham AJ, Bacon SP (2004) Psychophysical manifestations of compression: normal-hearing listeners. In: Bacon SP, Fay RR, Popper AN (eds), Compression: From Cochlea to Cochlear Implants. New York: Springer, pp. 62–106.
Oxenham AJ, Plack CJ (1997) A behavioral measure of basilar-membrane nonlinearity in listeners with normal and impaired hearing. J Acoust Soc Am 101:3666–3675.
Pavel M, Iverson GJ (1981) Invariant characteristics of partial masking: implications for mathematical models. J Acoust Soc Am 69:1126–1131.
Pestalozza G, Shore I (1955) Clinical evaluation of presbycusis on the basis of different tests of auditory function. Laryngoscope 65:1136–1163.
Phillips DP (1987) Stimulus intensity and loudness recruitment: neural correlates. J Acoust Soc Am 82:1–12.

Plomp R, Bouman MA (1959) Relation between hearing threshold and duration for tone pulses. J Acoust Soc Am 31:749–758.
Relkin EM, Doucet JR (1997) Is loudness simply proportional to the auditory nerve spike count? J Acoust Soc Am 101:2735–2740.
Rhode WS (1971) Observations of the vibration of the basilar membrane in squirrel monkeys using the Mössbauer technique. J Acoust Soc Am 49:1218–1231.
Rhode WS, Robles L (1974) Evidence from Mössbauer experiments for nonlinear vibration in the cochlea. J Acoust Soc Am 55:588–596.
Robinson DW, Dadson RS (1956) A re-determination of the equal-loudness relations for pure tones. Br J Appl Phys 7:166–181.
Ruggero MA, Rich NC, Recio A, Narayan SS, Robles L (1997) Basilar-membrane responses to tones at the base of the chinchilla cochlea. J Acoust Soc Am 101:2151–2163.
Scharf B (1959) Loudness of complex sounds as a function of the number of components. J Acoust Soc Am 31:783–785.
Scharf B (1961) Loudness summation under masking. J Acoust Soc Am 33:503–511.
Scharf B (1962) Loudness summation and spectral shape. J Acoust Soc Am 34:228–233.
Scharf B (1964) Partial masking. Acustica 14:17–23.
Scharf B (1970) Critical bands. In: Tobias JV (ed), Foundations of Modern Auditory Theory. New York: Academic, pp. 159–202.
Scharf B (1978) Loudness. In: Carterette EC, Friedman MP (eds), Handbook of Perception: Vol. 4. Hearing. New York: Academic, pp. 187–242.
Scharf B (1997) Loudness. In: Crocker MJ (ed), Encyclopedia of Acoustics, Vol. 3. New York: Wiley, pp. 1481–1495.
Scharf B, Stevens JC (1961) The form of the loudness function near threshold. Proceedings of the Third International Congress on Acoustics. Amsterdam: Elsevier, pp. 80–82.
Schlauch RS (2004) Loudness. In: Neuhoff JG (ed), Ecological Psychoacoustics. New York: Elsevier, pp. 318–345.
Schlauch RS, DiGiovanni JJ, Ries DT (1998) Basilar membrane nonlinearity and loudness. J Acoust Soc Am 103:2010–2020.
Schneider B (1988) The additivity of loudness across critical bands: a conjoint measurement procedure. Percept Psychophys 43:211–222.
Schneider B, Wright AA, Edelheit W, Hock P, Humphrey C (1971) Equal loudness contours derived from sensory magnitude judgments. J Acoust Soc Am 51:1952–1959.
Serpanos YC, Gravel JS (2000) Assessing growth of loudness in children by cross-modality matching. J Am Acad Audiol 11:190–202.
Spiegel MF (1981) Thresholds for tones in maskers of various bandwidths and for signals of various bandwidths as a function of signal frequency. J Acoust Soc Am 69:791–795.
Steinberg JC, Gardner MB (1937) The dependence of hearing impairment on sound intensity. J Acoust Soc Am 9:11–23.
Stevens SS (1936) A scale for the measurement of a psychological magnitude: loudness. Psychol Rev 43:405–416.
Stevens SS (1946) On the theory of scales of measurement. Science 103:677–680.
Stevens SS (1955) The measurement of loudness. J Acoust Soc Am 27:815–829.
Stevens SS (1956) The direct estimation of sensory magnitudes-loudness. Am J Psychol 69:1–25.
Stevens SS (1957) On the psychophysical law. Psychol Rev 64:153–181.
Stevens SS (1959) Tactile vibration: dynamics of sensory intensity. J Exp Psychol 59:210–218.
Stevens SS (1966a) Matching functions between loudness and ten other continua. Percept Psychophys 1:5–8.
Stevens SS (1966b) Power-group transformations under glare, masking, and recruitment. J Acoust Soc Am 39:725–735.
Stevens SS (1972) Perceived level of noise by mark VII and decibels (E). J Acoust Soc Am 51:575–601.

Stevens SS (1975) Psychophysics: Introduction to Its Perceptual, Neural and Social Prospects. New York: Wiley.
Stevens SS, Davis H (1936) Psychophysiological acoustics: pitch and loudness. J Acoust Soc Am 8:1–13.
Stevens SS, Guirao M (1967) Loudness functions under inhibition. Percept Psychophys 2:459–465.
Suzuki Y, Takeshima H (2004) Equal-loudness-level contours for pure tones. J Acoust Soc Am 116:918–933.
Swets JA (1961) Is there a sensory threshold?: when the effects of the observer(s response criterion are isolated, a sensory limitation is not evident. Science 134:168–177.
Takeshima H, Suzuki Y, Ashihara K, Fujimori T (2002) Equal-loudness contours between 1 kHz and 12.5 kHz for 60 and 80 phons. Acoust Sci Tech 23:106–109.
Takeshima H, Suzuki Y, Ozawa K, Kumagai M, Sone T (2003) Comparison of loudness functions suitable for drawing equal-loudness-level contours. Acoust Sci Tech 24:61–68.
Teghtsoonian M (1980) Children(s scales of length and loudness: a developmental application of cross-modality matching. J Exp Child Psychol 30:290–307.
Verhey JL, Kollmeier B (2002) Spectral loudness summation as a function of duration. J Acoust Soc Am 111:1349–1358.
Viemeister NF, Bacon SP (1988) Intensity discrimination, increment detection, and magnitude estimation for 1–kHz tones. J Acoust Soc Am 84:172–178.
Viemeister NF, Wakefield GH (1991) Temporal integration and multiple looks. J Acoust Soc Am 90:858–865.
Werner LA (2007) Issues in human auditory development. J Com Dis 40:275–283.
Yates GK, Winter IM, Robertson D (1990) Basilar membrane nonlinearity determines auditory nerve rate- intensity functions and cochlear dynamic range. Hear Res 45:203–219.
Zwicker E (1958) Über psychologische und methodische grundlagen der lautheit (On psychological and methodological bases of loudness). Acustica 8:237–258.
Zwicker E (1960) Ein verfahren zur berechnung der lautstärke. Acustica 10:304–308.
Zwicker E (1963) Über die lautheit von gedrosselten und ungedrosselten schallen. Acustica 13:194–211.
Zwicker E, Fastl H (1999). Psychoacoustics: Facts and Models. Berlin: Springer.
Zwicker E, Feldtkeller R (1955) Über die lautstärke von gleichförmiger geräuschen. Acustica 5:303–316.
Zwicker E, Scharf B (1965) A model of loudness summation. Psychol Rev 72:3–26.
Zwicker E, Flottorp G, Stevens SS (1957) Critical band width in loudness summation. J Acoust Soc Am 29:548–557.
Zwislocki JJ (1960) Theory of temporal auditory summation. J Acoust Soc Am 32:1046–1060.
Zwislocki JJ (1965) Analysis of some auditory characteristics. In: Luce RD, Bush RR, Galanter E (eds), Handbook of Mathematical Psychology. New York: Wiley, pp. 1–97.
Zwislocki JJ (1969) Temporal summation of loudness: an analysis. J Acoust Soc Am 46:431–441.
Zwislocki JJ, Hellman RP (1960) On the "Psychophysical law." J Acoust Soc Am 32:924.

Chapter 6
Loudness in the Laboratory, Part II: Non-Steady-State Sounds

Sonoko Kuwano and Seiichiro Namba

6.1 Introduction

The human auditory system conveys information about temporal variations in music, speech, and environmental sounds. Most of the sounds in daily life are temporally varying and non-steady state. Sounds in a temporal stream within a certain time interval are integrated by the auditory system, and information is extracted from these data (Fraisse 1978). An object can be recognized and differentiated in because it has its own specific pattern of temporal variation, different from that of other objects. Jones (1978) described that an auditory perception of a pattern is a meaningful succession within an event and conveys information about the outer world. Music, speech, and noise have unique patterns of temporal variation of sound energy. In daily life, the physical variation of a sound itself is seldom paid much attention: people generally listen to sound to grasp the information that it conveys, such as the meaning of the speech, the sources of machinery sound, the location of the sound sources, the melody of music, and so forth (Garner 1974).

Auditory perception researchers have conducted many experiments under physically well controlled conditions using synthesized steady-state sounds such as pure tones and white noise. They thereby discovered many basic and important phenomena, such as thresholds of hearing, masking, and critical band (e.g., Fastl and Zwicker 2007; Chap. 5).

In general, as Neisser (1976) noted, stimuli used in laboratory perception experiments are extremely restricted and different from those experienced in daily life. In the study of hearing, temporally varying phenomena in real life cannot be examined via experiments using steady-state sounds. Steady-state sounds are easily controlled in experimental situations, but experiments using such sounds may not measure crucial information that affects hearing in real-life situations. Problems with using steady-state sounds in psychoacoustical experiments were recently

S. Kuwano (✉)
Osaka University, Toyonaka, Osaka 560–0083, Japan
e-mail: kuwano@see.eng.osaka-u.ac.jp

M. Florentine et al. (eds.), *Loudness*, Springer Handbook of Auditory Research 37,
DOI 10.1007/978-1-4419-6712-1_6, © Springer Science+Business Media, LLC 2011

discussed from the aspect of ecological validity in books by Plomp (2002) and Neuhoff (2004).

It is important that both the precision of the experiment and ecological validity are established. Hearing researchers have made progress toward this aim by establishing a relation to problems in daily life, such as telephone communication, radio broadcasting, and noise control engineering. In these fields, complex non-steady-state sounds, such as speech, music, and noise, are often used as stimuli. Applied researchers have attempted to simulate situations in daily environments in various experiments, and to maintain ecological validity in hearing research (Neisser 1976; Chap. 8).

It is necessary to develop methods that enable examination of the perception of non-steady-state sounds while maintaining high experimental precision. To achieve such precision, the physical properties of the stimulus must be carefully controlled. The stimuli are, for example, synthesized sounds whose intensity, frequency, and temporal conditions are varied in a systematically programmed order (Namba et al. 1971, 1972). When sounds with clear physical properties are used, it is possible to obtain a difference limen and the point of subjective equality using conventional psychophysical methods (see Chap. 2). However, it is important to develop new methods to find the complex effect of temporal variation on the subjective impression. Two such methods are the "acoustic menu" (Molino et al. 1979) and "the method of continuous judgment by category" (Namba and Kuwano 1980; Kuwano and Namba 1985; Kuwano 1996; see Chap. 2).

In the acoustic menu method of Molino et al. (1979), participants listen to long-duration real-life sounds while they are seated in a laboratory and engaged in daily activities, such as reading books. The participants are presented with a pair of alternating sounds and given a control switch. They are asked to switch to the alternate sound if they perceive the sound presented at that moment as unpleasant. The total duration of the sound is used as an index of the unpleasantness of the sound. From an ecological viewpoint, this is an interesting method, as it can be applied to long-duration non-steady-state sounds because verbal responses are not used (Molino 1974), and adverse effects of sounds can be measured by avoidance behavior in response to the sounds. The method of continuous judgment by category is introduced later in this chapter.

The study of the perception of non-steady-state sounds ranges widely from (1) basic studies, such as discovering the mechanism of how the auditory system identifies the sound pattern, follows the temporal stream, and processes the information included in the sound, to (2) applied studies, such as finding an appropriate metric for noise evaluation and finding clues to synthesize sounds of musical instruments and speech as naturally as possible. In Sect. 6.2, the definition and classification of non-steady-state sounds are introduced. The subsequent sections discuss the loudness of regular and irregular time-varying sounds, non-steady-state sounds and duration, and continuous judgments of loudness along a temporal stream.

6.2 Classification of Time-Varying Sounds

6.2.1 Classification in ISO 2204

The classification of time-varying sounds can be found in ISO 2204 (1973). Sounds are classified as follows:

A. Steady noise: A noise with negligibly small fluctuations of level within the period of observation
B. Non-steady noise: A noise whose level shifts significantly during the period of observation
   - B-1. Fluctuating noise: A noise whose level varies continuously and to an appreciable extent during the period of observation.
   - B-2. Intermittent noise: A noise whose level suddenly drops to the ambient level several times during the period of observation, the time during which the level remains at a constant value different from that of the ambient, being of the order to magnitude of 1 s or more.
   - B-3. Impulsive noise: A noise consisting of one or more bursts of sound energy, each of a duration less than about 1 s.
     - B-3-1. Quasi-steady impulsive noise: A series of noise bursts of comparable amplitude with intervals shorter than 0.2 s between the individual bursts.
     - B-3-2. An isolated burst of sound energy: The envelope waveform of the burst may be of constant or nearly constant amplitude, or it may be that of a decaying transient.

Examples of each category of the classifications are shown in Fig. 6.1a–e (Namba 1984). As indicated in these figures, various non-steady-state sounds in daily life are easily classified into these categories, which seem practical because they can be related to noise measurement. However, these classifications are not without problems.

6.2.2 Problems in the Classification of Non-Steady-State Sounds

In the classification scheme of ISO 2204, sounds are defined only on the basis of sound level patterns regardless of variations in frequency and phase. However, the temporal variations in frequency and phase play an important role in daily life situations. For example, minute temporal variations of frequencies greatly contribute to the timbre of musical instruments.

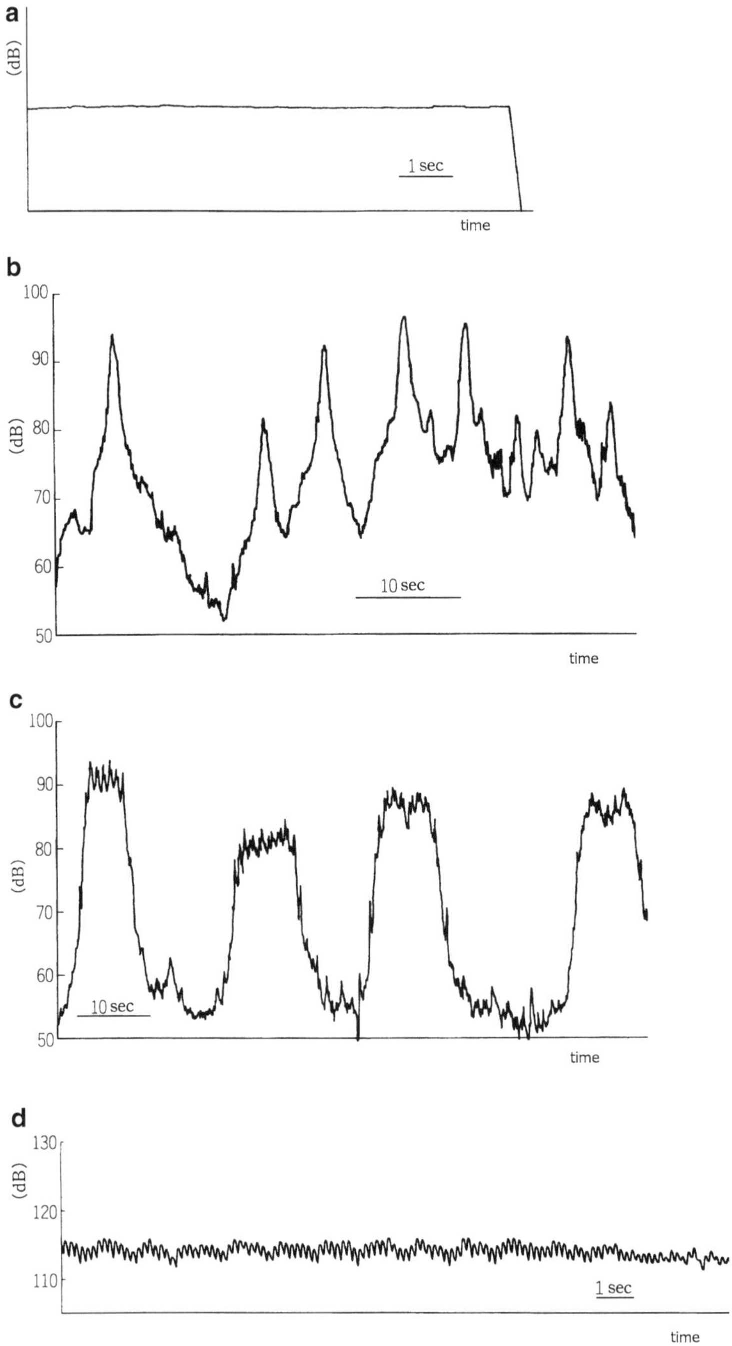
a
(dB)
1 sec
time
b
100
90
80
70
60
50
(dB)
10 sec
time
c
100
90
80
70
60
50
(dB)
10 sec
time
d
130
120
110
(dB)
1 sec
time

e

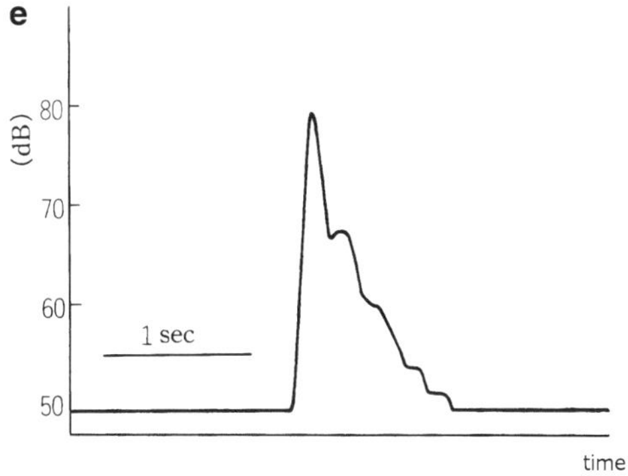

Fig. 6.1 Examples of (**a**) steady noise (cooling fan noise); (**b**) fluctuating noise (road traffic noise); (**c**) intermittent noise (train noise); (**d**) quasi-steady-state impulsive noise (neumatic hammer); (**e**) an isolated burst of sound energy (clap of hands produced in a room) (from Namba 1984, with permission from Nakanishiya Shuppan)

Even if only sound level patterns are considered, the differentiation between steady-state sounds and non-steady-state sounds is not clear. For example, in the classifications of ISO 2204, steady-state sound is defined as a sound whose level fluctuation is negligibly small, and non-steady-state sound is defined as a sound whose level varies significantly. There is no criterion to differentiate quantitatively between them. This may suggest that sounds cannot be physically differentiated and that whether a sound is steady state or non-steady state should be subjectively differentiated on the basis of hearing. Non-steady-state sounds are not always subjectively perceived as non-steady-state sounds. White noise, for example, fluctuates at random in sound level, frequency, and phase. However, it is perceived as being steady state. When the speed of temporal variation is fast and the mean of the temporal variation in sound level is constant along the temporal stream, the sound is subjectively perceived as being steady state, even though it is physically non-steady state.

There is no description about off-time and the number of events in the case of intermittent sounds in the classifications of ISO 2204. According to the results of the experiment by Symmes et al. (1955), the limit where intermittent white noise is assimilated to be continuous depends on the number of periodicity, ratio between on- and off-time in a cycle, and sound level. Scharf (1978) reviewed several studies on the loudness of pulse trains and concluded that it becomes difficult to differentiate between intermittent and continuous sounds when a pulse with the duration r (s) is repeated with a frequency of $1/r$. For example, when a pulse of 0.01 s is repeated at the frequency of 100/s, it is perceived as being continuous. It is necessary to find the limit where an intermittent sound is perceived as continuous when considering the definition and measurement of intermittent sound and quasi-steady impulsive noise.

According to the classifications of ISO 2204, the envelope patterns of single bursts are much simpler than those of impulsive sounds in daily life. The envelope patterns of impulsive sounds are different from each other depending on the sound sources. Moreover, the sound level patterns as well as the frequency components of the sounds, such as gunshots (Rice and Zepler 1967) and sonic booms (Zepler and Harel 1965), are complex. It is necessary to consider whether these sounds can be discussed on the same basis as other sounds.

6.2.3 Classification of Non-Steady-State Sounds on the Basis of Hearing

Rapidly changing non-steady-state sounds may be perceived as being steady state. However, sounds whose time-averaged sound level gradually increases or decreases would be perceived as non-steady-state sounds even if the sound level rapidly changes in a short time range, and each individual change cannot be discriminated by hearing. This is exemplified by the noise from an approaching and leaving aircraft.

The conclusion of Scharf (1978) mentioned in Sect. 6.2.2 is helpful for the differentiation between intermittent sounds and level-fluctuating sounds because it depends on whether the sound is perceived as intermittent or not. This is also applicable to the limit of temporal condition where quasi-steady intermittent sounds are perceived as being steady state.

It is difficult to define isolated bursts because their temporal structures (or patterns) are different depending on the sound source. The role of pattern in the perception of a single event is discussed in Sect. 6.3. In daily life, a single burst is seldom generated only once; it is usually repeatedly generated. The intermittent sounds often consist of such single bursts.

As mentioned in the preceding text, it is important to take perceptual aspects into consideration in the definition and classification of non-steady-state sounds. However, the classification of ISO 2204 is useful in its simplicity and it may be valuable because it suggests that there are various points to be considered in finding the psychological laws in the perception of non-steady-state sounds.

6.3 Loudness of Regular Time-Varying Sounds

6.3.1 Modulation Frequency and Loudness

Terhardt (1974) reported that amplitude modulation is detected in terms of the fluctuation of loudness when the amplitude is modulated with a modulation frequency of 5 Hz. When the modulation frequency is higher than 20 Hz, loudness becomes constant and an unpleasant impression of “roughness” is perceived. Further, Zwicker (1952) reported that the difference limen for amplitude-modulated sounds is at a minimum when the amplitude changes three to four times in a second. This confirms the results of pioneering experiments in which Riesz (1928) measured the difference limen using beats of sounds. These results suggest that the speed of fluctuation plays an important role in the differentiation between steady-state and non-steady-state sounds.

Precise information about the temporal structure of a sound is lost in the classification of ISO 2204. Is this information really irrelevant in sound evaluation? The effect of envelope patterns, such as rise time and temporal position of the intensity increment, on loudness is discussed in Sect. 6.3.2.

6.3.2 The Effect of Rise Time of a Sound on Loudness

Several researchers have conducted psychological and physiological studies to examine the dynamic characteristics of hearing using sounds with rise time (abbreviated as rising sound; Vigran et al. 1964; Davis and Zerlin 1966; Gjaevenes and Rimstad 1972; Carter 1972). Steady-state sounds with gradual rise times were used in most of these studies. For example, in the study of Gjaevenes and Rimstad (1972), the duration of the sound was 0.7 s and the rise time varied from 0.04 to 0.48 s. Although the conclusions in their paper are not clear, it is obvious that rise time affects the total energy of rising sounds depending on the envelope pattern.

Namba et al. (1974) examined loudness and timbre of rising sounds with controlled sound energy. In a series of experiments, the loudness of rising sounds was matched to that of steady-state sounds, as shown in Fig. 6.2. When the loudness of a rising sound is compared with the loudness of a steady-state sound with the same maximum level and duration, the rising sound is perceived to be softer than the steady-state sound

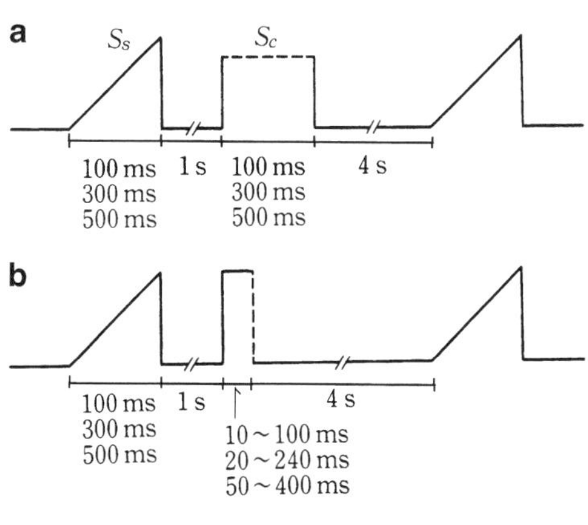

Fig 6.2 Stimulus patterns of Exp. 1 in *upper figure* and Exp. 2 and 3 in *lower figure*. In Exp. 1 the loudness of standard stimulus (Ss) was matched to that of comparison stimulus (Sc) by changing the amplitude of Sc. In Exp. 2 the point of subjective equality (PSE) was obtained by the method of limit by changing the duration of Sc. In Exp. 3 the loudness of Ss was matched to that of Sc by changing the duration of Sc (from Namba et al. 1974, with permission from the Acoustical Society of Japan)

because the rising sound has one third of the total energy as the steady state sound. The results of the experiments are shown in Fig. 6.3, in which the loudness of rising sounds was matched to that of steady-state sounds using the method of adjustment (Exp. 1, 3) and the method of limits (Exp. 2) by varying the amplitude (Exp. 1) and duration (Exp. 2, 3) of steady-state sounds. The abscissa in Fig. 6.3 is the sound exposure level (L_{AE}), which is the total energy level using a 1-s reference duration.

Figure 6.3 shows that the perception of equal loudness (the point of subjective equality, PSE) correlates well with L_{AE}. When the total energy is the same, there is no systematic difference in loudness between rising sounds and steady-state sounds. Kuwano et al. (2007) investigated the loudness of rising, decaying, and steady-state sounds whose duration varied from 80 to 1,280 ms. The loudness of these sounds was judged using magnitude estimation. The geometric means of the 28 judgments were calculated for each stimulus and compared to L_{AE}. Results are shown in Fig. 6.4. A high correlation can be seen between L_{AE} and loudness. L_{AE} seems to be a good descriptor of the loudness of impulsive sounds. It seems that there is no systematic effect of envelope patterns on the loudness of these sounds.

Kuwano and Namba (1983) examined the loudness of various kinds of impulsive sounds with different durations and envelope patterns in a series of eight experiments. The duration of the sounds used in these experiments was shorter than 1 s. Figure 6.5 summarizes the results and shows a good correspondence between loudness and L_{AE}, with a coefficient of correlation of 0.970. The correlation

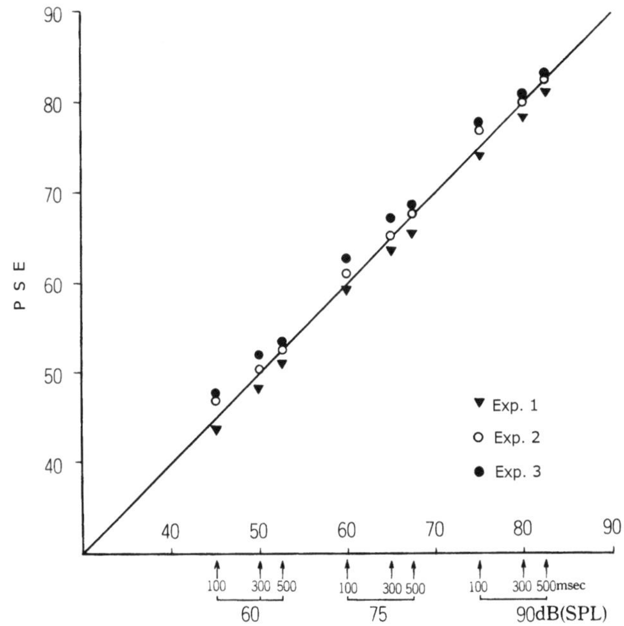

Fig. 6.3 Relation between the point of subjective equality (PSE) for loudness and sound exposure level (L_{AE}) of rising sounds (from Namba et al. 1974, with permission from the Acoustical Society of Japan)

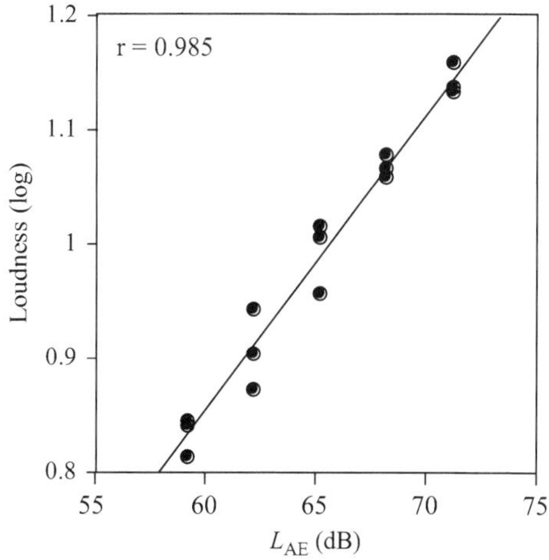

Fig. 6.4 Relation between the loudness of steady-state sounds, rising sounds, and decaying sounds ranging in duration from 80 to 1,280 ms and L_{AE} (from Kuwano et al. 2007, with permission from the Acoustical Society of Japan)

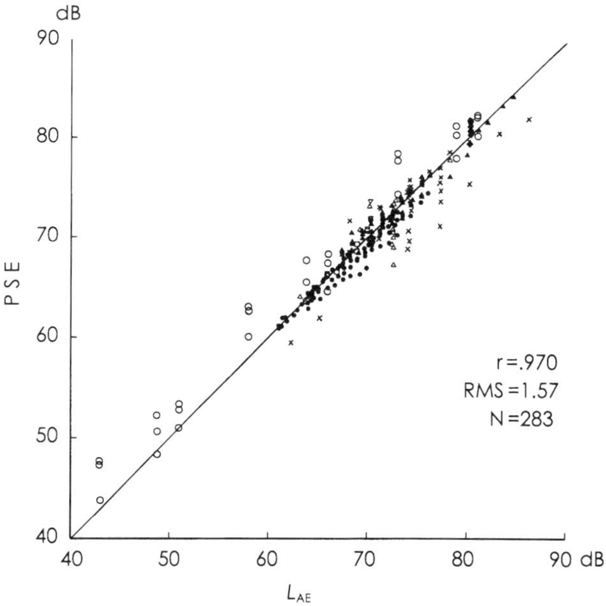

Fig. 6.5 Relation between the loudness of impulsive sounds and L_{AE}. The *symbols* in the figure indicate the impulsive sounds with various temporal structures. *Filled triangles* – impulsive sounds with reverberation; *open circles* – sounds with rise time; *open squares* – sounds with intensity increment (1); *filled circle* – sounds with intensity increment (2); *filled diamond* – impulsive sounds with amplitude-modulated decaying; *inverted open triangles* – sounds with rise and fall time; *open triangles* –sounds with intensity increment (3); X sounds with intensity increment (4) (from Kuwano and Namba 1983, with permission from the Acoustical Society of Japan)

suggests that L_{AE} is a good descriptor for the evaluation of sounds whose duration are shorter than about 1 s even if the envelope patterns of the stimuli are different from one another.

Although there was no systematic effect of rise time on loudness, rise time has an effect on timbre and seems to be related to emotional aspects such as feelings of fear and startle (Izumi 1977).

The influence of changes in loudness can be observed in daily life. For example, Susini and McAdams (2008) reported that the sound produced by a car accelerating was judged louder than decelerating noise. They explain that the end level influences loudness of an increasing sound. Neuhoff (1998, 2004) reported that the change in loudness of a rising tone is judged larger than that of falling tone, and he interpreted this tendency as a survival technique for detecting an approaching sound source. (For more information on loudness is daily life, see Chap. 8.)

6.3.3 The Effect of Temporal Position of an Intensity Increment of a Sound on Loudness

Although sounds whose rise time simply increased with time were used in the experiments of rising sounds, most sounds are temporally varying in a complicated manner. To examine the effect of temporal pattern on loudness, Namba et al. (1976) systematically shifted the temporal position of an intensity increment, while keeping the total energy of the sound constant, as shown in Fig. 6.6. The loudness of these seven sounds was matched to the loudness of a steady-state sound. The carrier was white noise.

The results are shown in Fig. 6.7. The abscissa indicates the position of the intensity increment and the ordinate shows the difference in L_{AE} between PSE and each stimulus. Pattern 1 was judged significantly louder than the other patterns (t-test, $p < 0.001$). A series of similar experiments were conducted with different durations and sound levels of the intensity increment. Though there were slight differences among stimulus conditions, similar tendencies were found. That is, non-steady-state sounds were overestimated compared with L_{AE} when the intensity increment was located at the beginning of the stimulus, whereas loudness was underestimated compared with L_{AE} when the intensity increment was located in the middle of the stimulus. The degrees of overestimation and underestimation were small but statistically significant.

As shown in Fig. 6.5, the loudness of impulsive sounds whose duration is less than 1 s can be approximately evaluated by L_{AE} regardless of the envelope patterns. However, in the case of some specific non-steady-state sounds, as shown in Fig. 6.6, systematic deviation between the loudness and L_{AE} was found. Namba et al. (1976) have proposed a model of dynamic characteristics of hearing, shown in Fig. 6.8a, based on the results in Fig. 6.7. The straight line in Fig. 6.8a indicates the physical level pattern of a sound and the dotted line indicates the sensation to the sound. This model suggests that there are an overshoot at the beginning of a sound, suppression in the middle, and an after-effect after the cessation of the sound. This nonlinear dynamic characteristic of hearing may contribute to emphasizing non-steady-state sounds.

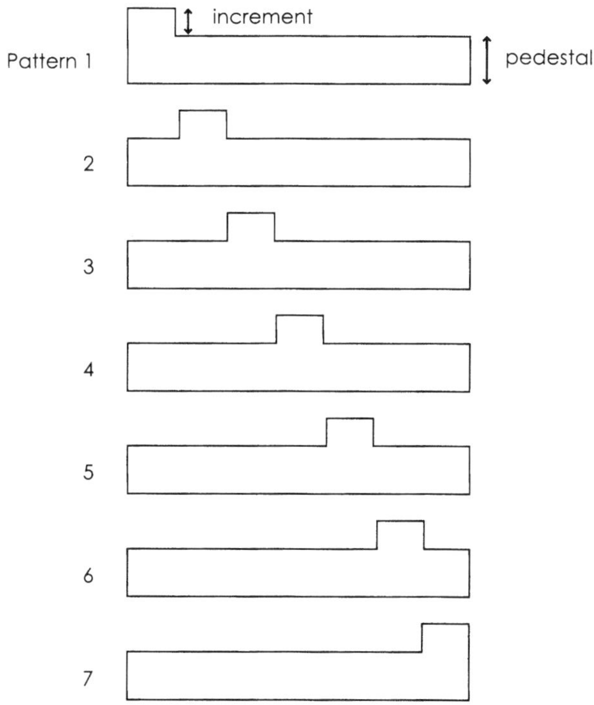

Fig. 6.6 Pattern of non-steady-state sound (from Namba et al. 1976, with permission from the Japanese Psychological Association)

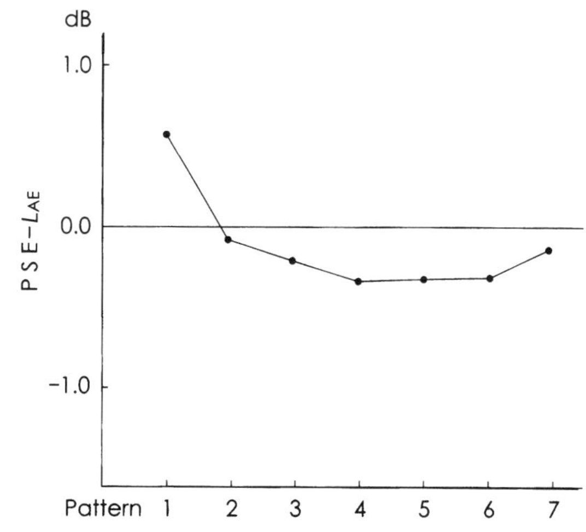

Fig. 6.7 Results of an experiment on loudness of non-steady-state sounds. The abscissa indicates the position of the intensity increment and the ordinate shows the difference in L_{AE} between PSE and each stimulus (from Namba et al. 1976, with permission from the Japanese Psychological Association)

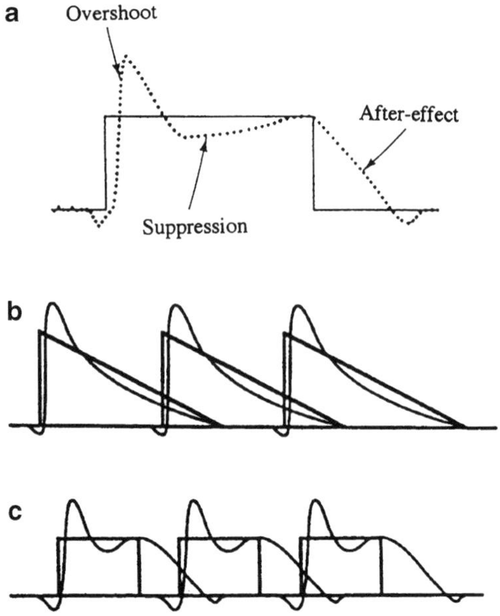

Fig. 6.8 A model of dynamic characteristics of hearing (**a**) Model for a single steady-state sound. (from Namba et al. 1976, with permission from the Japanese Psychological Association) (**b**) The model applied to pulse trains of decaying sounds. (**c**) The model applied to pulse trains of steady-state sounds

To examine the after-effect in the model in Fig. 6.8a, another experiment was conducted wherein the effects of temporal masking were examined using steady-state sounds and sounds with long decay (i.e., decaying sounds), which often exist in daily life, as a masker (Namba et al. 1987). When a steady-state noise was used as a masker, the results were consistent with previous findings (Namba et al. 1976). On the other hand, there was no after-effect except for sounds with a short decay time when decaying sounds were used as a masker.

This dynamic model of hearing was originally applied to a single event of impulsive noise. However, in daily-life situations, pulses often occur successively. Figure 6.8b, c show the application of the model to successive pulse trains. Musical performance is a typical example of this case.

When steady-state sounds are successively presented with short intervals, they are perceived as being continuous without intervals. On the other hand, there is no after-effect in decaying sounds when they overlap each other; the overshoot portion of the following sound masks the tail of the preceding sound. This makes the listener insensitive to the overlapping of the sounds. In the case of a piano performance, sounds should be overlapping to be perceived as "legato" (smoothly connected) because the piano's sound is a typical example of a decaying sound. The smooth impression of legato may be perceived as flowing in a continuous manner (Kuwano et al. 1994).

6.4 Loudness of Irregular Time-Varying Sounds

Sounds in daily life show complicated temporal variations. These sounds are usually treated as sounds with random fluctuation. When some events are randomly fluctuating, there are two methods for calculating the representative values, one of which is based on statistics. The representative values, such as arithmetic mean, geometric mean, and median, are used with an index of variation such as standard deviation and interquartile range. The other is based on the average of physical values such as the mean energy level.

When considering the subjective evaluation of noise, the representative value is required to show good correspondence with the subjective impression in addition to being a good descriptor from a physical viewpoint. The descriptor that should be used in the noise evaluation can be examined by psychophysical experiments using sounds in daily life as stimuli. In fact, many experiments and social surveys have been conducted on this topic. However, because many factors are involved and correlate with one another in daily situations, it is difficult to find orthogonal stimulus conditions. This task continues to be difficult, even now when sounds in daily life can be modified easily using computers.

Experimental factors can be controlled according to the intention of the experimenter when synthetic level-fluctuating sounds are used instead of sounds in daily life. For example, sound level distribution and level fluctuating patterns can be designed freely. This makes it possible to realize orthogonal conditions among different possible descriptors.

It was not easy to generate level-fluctuating sounds in the 1960s, and a system was developed with the ability to alter sound parameters to generate various level-fluctuating sounds (Namba et al. 1973). This system made it possible to generate randomly fluctuating sounds where possible descriptors, such as the mean energy level and median (L_{50}), are orthogonal.

In one of the experiments using level fluctuating sounds, the loudness of level fluctuating sounds was obtained with the method of adjustment (Namba et al. 1972). A good correlation was found between loudness and mean energy level regardless of the distribution pattern of the sound level. In the stimulus condition where the mean energy level and L_{50} were distributed orthogonally (independently) with each other, the loudness clearly shows good correlation with the mean energy level and no correlation with L_{50}. Similar results were found with sounds simulating and actual road traffic noise (Namba et al. 1978a).

The mean energy level shows a strong correlation with the loudness of sounds in daily life, as well as synthesized sounds (Namba et al. 1978b). The results of 11 experiments conducted by the authors are plotted together in Fig. 6.9. The coefficient of correlation is high ($r = 0.979$), and the RMS (root mean square; root of the square sum of the difference between the mean energy level and PSE) is small (RMS = 1.45), indicating that the values are close to each other.

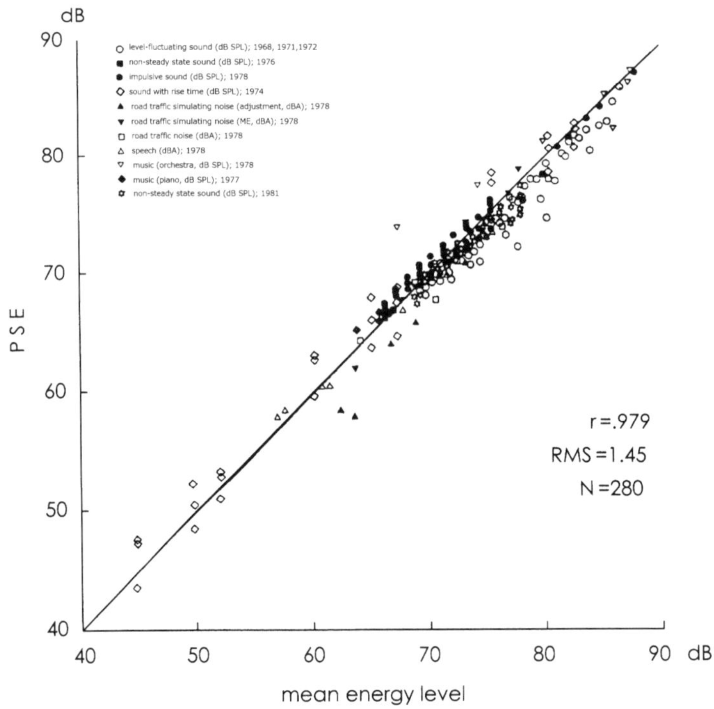

Fig. 6.9 Relation between the mean energy level and the loudness in PSE. The results of 11 experiments are plotted together

As shown in the preceding text, a first approximation to the loudness of various non-steady-state sounds can be determined by the mean energy level. However, the role of patterns in loudness is not clear. Rapid changes in sound level and frequency components may not be perceived as changes in loudness and pitch, but contribute to the formation of a pattern. The difference in pattern may be perceived as the difference in timbre and contributes to the identification of sound sources (Handel 1989).

6.5 Evaluation of Non-Steady-State Sound and Duration

For short-duration sounds, loudness increases as duration becomes longer. This phenomenon shows that the ear integrates energy over time. If this is true, then the total energy of a sound corresponds to the loudness of temporally varying sounds. According to former studies (Scharf 1978; Fastl 1984; Florentine et al. 1988, 1996; Buus et al. 1997), the upper limit of the additive effect is less than 1 s; this limit is called the critical duration of loudness.

Scharf (1978) summarized the results of various studies and concluded that the time constant of hearing is 80 ms and the critical duration is 180 ms. Because this analysis was based on steady-state sounds, it is necessary to examine whether the same values can be applied to the loudness of non-steady-state sounds with various envelope patterns. With non-steady-state sounds, the critical duration may vary depending on the envelope patterns of sounds. Usually, impulsive sounds in daily life have rapid temporal changes and short durations, but human hearing may not follow the individual fluctuations. In this case, loudness may be determined simply by an overall value of the stimulus variation.

L_{AE} seems to be a good descriptor of loudness of impulsive sounds based on the psychophysical evidence shown in Fig. 6.4. In this experiment, the sound energy was integrated up to 1,280 ms, which was the longest duration of the stimuli used, and the limit of loudness summation could not be determined. To find the critical duration of non-steady-state sounds, further investigations using sounds with longer durations are needed.

Namba and Kuwano (2007) conducted an experiment using road traffic noise whose duration varied from 1 to 16 s. Figure 6.10 shows the relation between L_{AE} and loudness. When the total energy level is the same, loudness increases with the mean energy level ($L_{Aeq,T}$). Variations along the ordinate in Fig. 6.10 show this tendency. Fig. 6.11 shows the relation between $L_{Aeq,T}$ and loudness. A good correspondence can be seen, with a correlation coefficient of 0.938. This suggests that the energy of a sound is not integrated, but instead, averaged in the case of

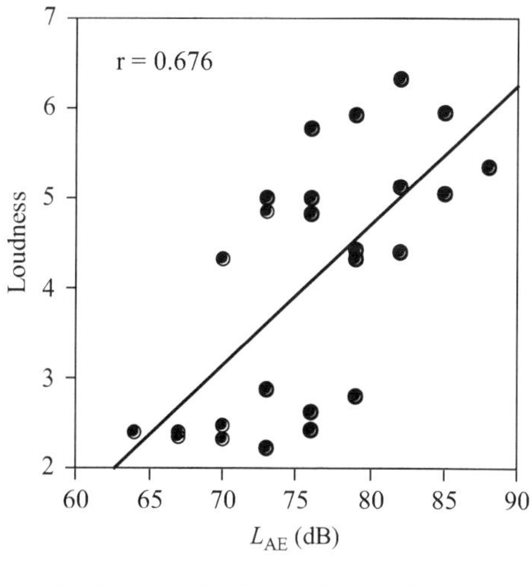

Fig. 6.10 Relation between loudness and L_{AE}. Even when total energy is the same, loudness is not the same (from Namba and Kuwano 2007, with permission from the Acoustical Society of Japan)

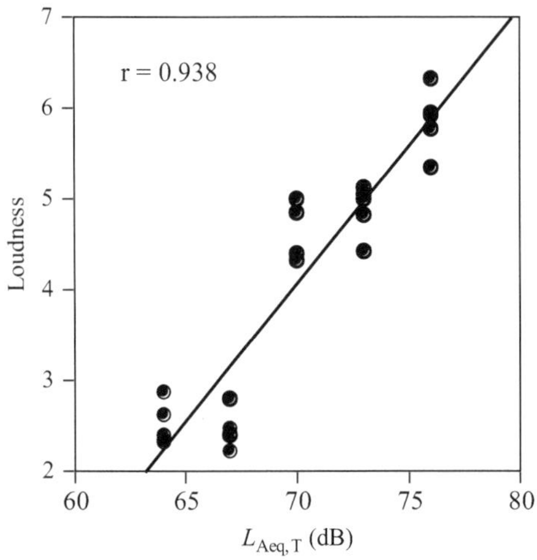

Fig. 6.11 Relation between loudness and the mean energy level ($L_{Aeq,T}$). A Good correlation was found between them when the duration was longer than 1 s (from Namba and Kuwano 2007, with permission from the Acoustical Society of Japan)

sounds longer than the critical duration. Another experiment was conducted using white noise as stimuli (Namba et al. 2006). Similar tendencies were found as in the case of road traffic noise. The loudness showed better correlation with $L_{Aeq,T}$ (r= 0.970) than L_{AE} (r= 0.720) when the duration was longer than 1 s. There is no relation between duration and loudness in either stimulus condition (r= 0.008 in the case of road traffic noise and r= 0.055 in the case of white noise). Even when the value of L_{AE} is the same, loudness is not the same in the case of sounds longer than a critical duration; rather, it changes depending on the value of $L_{Aeq,T}$. $L_{Aeq,T}$ is a good descriptor of the loudness of sounds longer than the critical duration.

These results suggest that the critical duration of the loudness of non-steady-state sounds is about 1 s. According to the classification of ISO 2204, the duration of impulsive sound is less than 1 s. The experimental results mentioned in Sect. 6.3.2 of this chapter show a good correspondence between L_{AE} and the loudness of sounds within 1 s duration. This is a practical advantage in evaluating impulsive sounds in the environment using a sound level meter. In this sense, L_{AE} with a reference duration of 1 s is an appropriate descriptor of the loudness of impulsive sounds.

The sound exposure level of a sound increases as its duration increases. However, owing to the existence of the critical duration, loudness does not increase when the duration becomes longer than the critical duration. It may be possible that the auditory system integrates the sound energy up to the critical duration of loudness, and that the ear has an averaging mechanism of loudness when the duration of a sound becomes longer than the critical duration. Therefore, the loudness of non-steady-state sounds shorter than the critical duration is determined by L_{AE} and the loudness of non-steady-state sounds longer than the critical duration is determined by $L_{Aeq,T}$.

6.6 Continuous Judgments of Loudness Along a Temporal Stream

6.6.1 The Method of Continuous Judgment by Category

Auditory impressions change over time and are different from visual impressions. A still picture can be observed in detail from several distances and directions, and an observer can focus on a specific part of the picture. It can be easily understood that this experience is perception. In hearing, each sound appears and disappears, one after the other. Time cannot be stopped and a sound that has passed cannot be heard again. When a sound stream is presented, however, people do no usually think that they are observing the past sound retrospectively. Actually, sounds pass instantaneously, but the perception of the sound clearly remains. That is to say, this experience along the temporal stream is the world of hearing perception. Does the temporal stream continuously flow away?

It is assumed that while listening to a sound along a temporal stream, people may grasp the information of the sound within each time window (i.e., psychological present) and perceive the loudness of the sound by averaging the energy within the time window. How can the length of psychological present be measured?

Because conventional psychophysical methods, such as the method of constant stimuli, cannot measure subjective impressions along a sound stream, a method called "the method of continuous judgment by category" has been developed (Namba and Kuwano 1980; Kuwano and Namba 1985; Kuwano 1996). In an experiment using this method, participants are asked to judge the instantaneous impression of the sound using seven loudness categories and to press a key on a response box or a computer key board corresponding to their impression at that time. An example of this method is shown in Fig. 6.12.

Participants need not respond if their impressions do not change. When their impressions change, they are required to press a response button corresponding to the number of category. The smooth line in Fig. 6.12 shows the sound level and each step shows the subjective response. Good correspondence can be seen between them, though there is a little temporal gap between the two lines caused by reaction time.

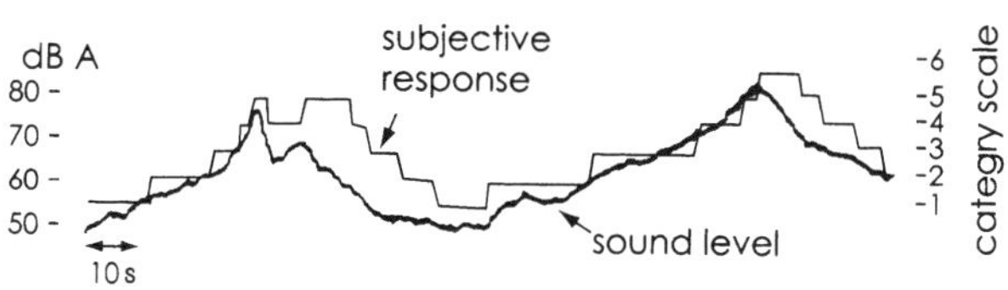

Fig. 6.12 An example of responses obtained by the method of continuous judgment by category (from Kuwano and Namba 1985, with permission from Springer). The *smooth line* shows the sound level and each step shows a subjective response

By systematically changing the time lag (TL) between physical values and participants' responses, the coefficients of correlation between them are calculated. The time lag when the highest correlation can be obtained is presumed to be the reaction time.

A series of calculation of the method of continuous judgment by category is very laborious. When this method was developed in the 1970s, personal computers were not available. Now personal computers are easily available, and the authors have developed software, which runs on MS-WINDOWS OS (Namba et al. 2004). This software can easily analyze the relation between physical parameters of stimuli and categorical judgment or line-length matching for a specific portion in a sound stream. One of the advantages of this method is that various sound sources with different durations can be judged in the same context along a temporal stream. This method and its modifications have been applied to the study of loudness (Namba et al. 1977; Kuwano and Namba 1985, 1990; Kato et al. 1994, 1996, 2000, 2001; Arakawa et al. 1995; Fastl et al. 1996; Kuwano et al. 1997), as well as noisiness, sound quality, and interaction between auditory and visual information (Namba and Kuwano 1980, 1988a, b, 1990; Namba et al. 1982, 1991, 1993, 1997; Kuwano and Namba 1992, 1996).

Methods using continuous judgments of loudness have many varieties. Springer et al. (1997) investigated the instantaneous and overall loudness along a temporal stream and Hedberg (1998) examined the reliability of this method. Susini et al. (2002) developed cross-modal matching of loudness with the feedback of muscular forces, and Kuwano and Namba (1990) proposed cross-modal matching with line length for continuous judgments. Some examples of studies using the method of continuous judgment by category are provided in the following sections.

6.6.2 *Estimation of Psychological Present*

The instantaneous impression of loudness may be affected by preceding portions of a stimulus. That is, it is assumed that the preceding portions may be grouped and contribute to the impression of the loudness of the successive portion. To find the duration of preceding portions of a sound that may affect loudness judgments at each moment, the correlation coefficient was calculated between the sound energy averaged changing integration time interval (TI) and the instantaneous subjective response by sliding along temporal stream (Kuwano and Namba 1985). The purpose of this analysis was to estimate the "psychological present." When TI is 2.5 s and TL is 0, the highest correlation can be seen. An average of 2.5 s may reflect psychological present. This duration is a value close to those proposed as psychological or perceptual present by other researchers using other approaches (e.g., Fraisse 1957; Poeppel 1978).

6.6.3 *Relation Between Overall Judgment and Instantaneous Judgment*

For the assessment of sound environments, long-term effects should be examined. For the evaluation of long-term fluctuating sounds, it is helpful to measure both the instantaneous impression and overall impression. In the experiment using the method of continuous judgment by category, participants are asked to fill in a questionnaire after the continuous judgments, in which various questions on the overall impression are included so that information for interpreting the experimental data can be obtained. The simplest assumption is that the overall or long-term loudness corresponds to an average of instantaneous judgments or loudness of every perceptual present.

Examples of results obtained using this method are shown in Figs. 6.13 and 6.14 (Kuwano and Namba 1985). The abscissa in Fig. 6.13 indicates the average of instantaneous judgments and the ordinate indicates the overall judgments. It can be seen that the overall judgment is louder than the average of the instantaneous judgment. Which portions of the instantaneous loudness determine the overall loudness? Instantaneous loudness was averaged using a cut-off point of 10, 20, or 30 dB lower than the maximum level, omitting the lower-level portions. It was found that the average with a 30-dB cutoff point had values close to the overall loudness as shown in Fig. 6.14. Moreover, the average with a 30-dB cutoff point showed the highest correlation with $L_{\mathrm{Aeq,T}}$.

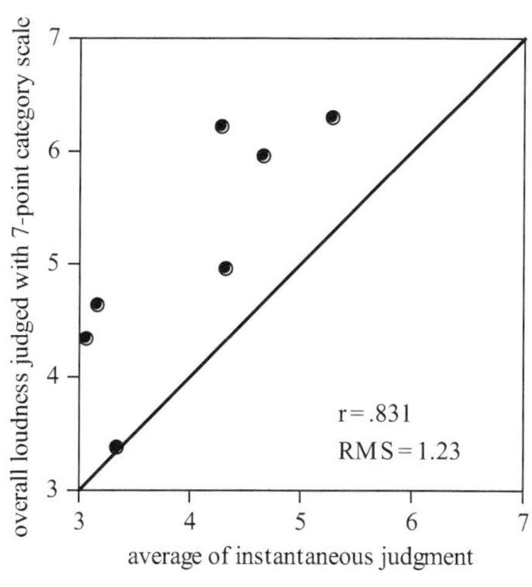

Fig. 6.13 Relation between overall loudness and the average of instantaneous judgment (from Kuwano and Namba 1985, with permission from Springer)

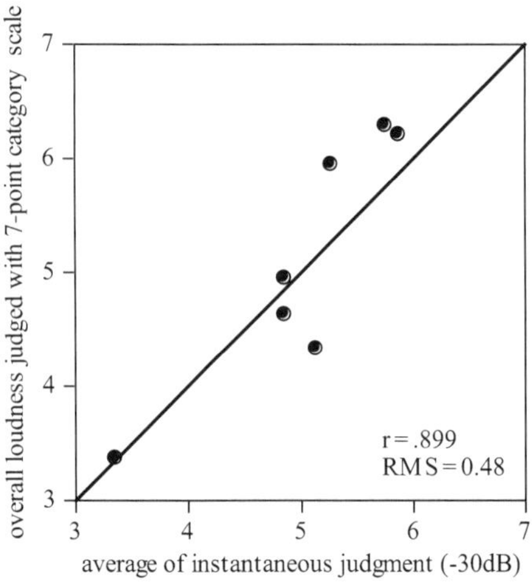

Fig. 6.14 Relation between overall loudness and the instantaneous judgment averaged at cutoff point 30 dB down from the maximum point. A good correlation can be seen between them (from Kuwano and Namba 1985, with permission from Springer)

As time passes, the impression of the prominent portion becomes stronger and that of the less prominent portion becomes weaker (Namba and Kuwano 1980). It is important to find the factors that contribute to the overall impression.

6.7 Application of Laboratory Experiments and Field Studies

Information about the relationship between physical values and subjective impression of sounds is required not only in the laboratory, but also in many situations in daily life. The question of what descriptor is appropriate for the evaluation of the loudness of level-fluctuating sounds is closely related to applied problems, such as the evaluation of road traffic noise and machinery noise. It is also related to legal matters, such as standards for environmental quality.

L_{50} has been used as a descriptor of level-fluctuating sounds, such as road traffic noise in the environmental quality standard in Japan. Maximum level (L_{Asmax}) has been used in Japan for the Shinkansen train (aka "bullet train") noise and aircraft noise.

The Japanese Environmental Quality Standard for general environmental noise including road traffic noise was revised in 1998 and that for aircraft noise in 2007 by adopting rating methods based on $L_{\mathrm{Aeq,T}}$ on the basis of a report by the Central Environmental Council of Japanese Government.

JIS Z8731 "Acoustics – Description and Measurement of Environmental Noise" was revised in 1999. Some papers showing good correlation between $L_{\mathrm{Aeq,T}}$ and loudness of non-steady-state sounds are referred to in JIS Z8731 (Namba et al. 1978a; Namba and Kuwano 1982).

In the evaluation of sounds in daily life, participants can identify the sound sources from their auditory signals. The subjective impression of sound sources may affect the results. That is, cognitive effects inevitably occur when the sound sources are identified. Because cognitive effects influence judgments of sounds in real-life situations, $L_{Aeq,T}$ is used as a basic descriptor, and either different values are used or some correction is added for determining the permissible levels depending on the sound sources.

6.8 Summary

Sounds convey information in terms of temporal variation. The ability to hear temporally varying sounds was discussed by introducing the relation between physical properties of sounds and loudness in relation to the temporal pattern and duration of sounds. The model of dynamic characteristics of hearing was also introduced.

Hearing can trace the instantaneous loudness change of sounds, and an overall impression of loudness can be judged at the end of the sound. The mechanism of how various auditory events are recognized in the environment was examined by analyzing the relation between instantaneous judgments to each portion of an event and the overall judgment of the event as a whole.

The data obtained in laboratory experiments give us clues to the evaluation of sound environment including music, speech and noise. The relation between basic psychoacoustics and its application was also discussed.

Cognitive effects are considerable, especially in the judgments based on memory, and experimental research is being conducted to examine the relation between $L_{Aeq,T}$ and cognitive factors affecting loudness (Kuwano et al. 2003; Kuwano 2007). Further work is ongoing.

References

Arakawa K, Mizunami T, Kuwano S, Namba S (1995) Factors determining the optimum listening level of music performance. J Music Percept Cogn 1:33–42.

Buus S, Florentine, M, Poulsen T (1997) Temporal integration of loudness, loudness discrimination, and the form of the loudness function. J Acoust Soc Am 101:669–680.

Carter NL (1972) Effects of rise time and repetition rate on the loudness of acoustic transient. J Sound Vib 21:227–239.

Davis H, Zerlin S (1966) Acoustic relations of the human vertex potential. J Acoust Soc Am 39:109–116.

Fastl H (1984) An instrument for measuring temporal integration in hearing. Audiol Acoustics 23:164–170.

Fastl H, Zwicker E (2007) Psychoacoustics – Facts and Models, 3rd ed. Berlin: Springer.

Fastl H, Kuwano S, Namba S (1996) Assessing the railway bonus in laboratory studies. J Acoust Soc Jpn (E) 17:139–148.

Florentine M, Fastl H, Buus S (1988) Temporal integration in normal hearing, cochlear impairment, and impairment simulated by masking. J Acoust Soc Am 84:195–203.

Florentine M, Buus S, Poulsen T (1996) Temporal integration of loudness as a function of level. J Acous Soc Am 99:1633–1644.

Fraisse P (1957) Psychology du Temps. Paris: Press University de France.

Fraisse P (1978) Time and rhythm perception. In: Carterette EC, Friedman MP (eds), Handbook of Perception, Vol. V: Perceptual Coding. New York: Academic, pp. 203–254.

Garner WR (1974) The Processing of Information and Structure. Hoboken, NJ: Wiley.

Gjaevenes K, Rimstad ER (1972) The influence of rise time on loudness. J Acoust Soc Am 51:1233–1239.

Handel S (1989) Listening – An Introduction to the Perception of Auditory Events. Cambridge, MA: MIT Press.

Hedberg D (1998) Continuous rating of sound quality. Rep Tech Audiol Karolinska Institute, Report TA 134.

ISO 2204 (1973) Acoustics – Guide to the Measurement of Airborne Acoustical Noise and Evaluation of Its Effects on Man. Geneva, Switzerland: International Organization for Standardization.

Izumi K (1977) Two experiments on the perceived noisiness of periodically intermittent sounds. Noise Control Eng 9:16–23.

Jones MR (1978) Auditory patterns: studies in the perception of structure. In: Carterette EC, Friedman MP (eds), Handbook of Perception, Vol. V: Perceptual Coding. New York: Academic, pp. 255–288.

Kato T, Namba S, Kuwano S (1994) Continuous judgment of loudness by cross-modality matching using line length. Fac Lett Rev Otemon Gakuin Univ 29:15–31.

Kato T, Namba S, Kuwano S (1996) Continuous judgment of loudness by cross-modality matching using line length – a method for describing matching results. Fac Human Rev Otemon Gakuin Univ 2:47–60.

Kato T, Kaku J, Kuwano S, Namba S (2000) Psychological evaluation of environmental noise in field using the method of continuous judgment by category. Fac Human Rev Otemon Gakuin Univ 9:13–26.

Kato T, Namba S, Kuwano S (2001) Continuous judgment of loudness by cross-modality matching using line length–loudness evaluation of the sounds that contain big and small level fluctuation. Fac Human Rev Otemon Gakuin Univ 12:21–28.

Kuwano S (1996) Continuous judgment of temporally fluctuating sounds. In: Fastl H, Kuwano S, Schick A (eds), Recent Trends in Hearing Research. Oldenburg: BIS, pp. 193–214.

Kuwano S (2007) Psychological evaluation of sound environment along temporal stream. Proceedings of the International Congress on Acoustics CD-ROM (http://www.icacommission.org/).

Kuwano S, Namba S (1983) On the dynamic characteristics of hearing and the loudness of impulsive sounds. Trans Tech Com Noise, Acoust Soc Jpn N-8303-13:79–84.

Kuwano S, Namba S (1985) Continuous judgment of level-fluctuating sounds and the relationship between overall loudness and instantaneous loudness. Psychol Res 47:27–37.

Kuwano S, Namba S (1990) Continuous judgment of loudness and annoyance. In: Müller, F (ed), Fechner Day 90. Proceedings of the Sixth Annual Meeting of the International Society of Psychophysics, Würzburg, ISP, pp. 129–134.

Kuwano S, Namba S (1992) Temporal stream of hearing. Stud Hum Soc Sci Col Gen Edu Osaka Univ 40:1–15.

Kuwano S, Namba S (1996) Evaluation of aircraft noise: effects of number of flyovers. Environ Int 22:131–144.

Kuwano S, Namba S, Yamasaki T, Nishiyama K (1994) Impression of smoothness of a sound stream in relation to legato in musical performance. Percept Psychophys 56:173–182.

Kuwano S, Namba S, Hayakawa Y (1997) Comparison of the loudness of inside car noises from various sound sources in the same context. J Acoust Soc Jpn (E) 18:189–193.

Kuwano S, Namba S, Kato T, Hellbrueck J (2003) Memory of the loudness of sounds in relation to overall impression. Acoust Sci Tech 24:194–196.

Kuwano S, Fastl H, Namba S (2007) Relation between loudness of sounds with under critical duration and L_{Aeq} and L_{AE} (2) in the case of impulsive noise. Proc Spring Meeting Acoust Soc Jpn 757–758.

Molino JA (1974) Measuring human aversion to sound without verbal descriptors. Percept Psychophys 16:303–308.

Molino JA, Zerdy GA, Lerner ND, Harwood DL (1979) Use of the "acoustic menu" in assessing human response to audible (corona) noise from electric transmission lines. J Acoust Soc Am 66:1435–1445.

Namba S (1984) Perception of non-steady state sounds. In: Namba S (ed), Handbook of Hearing, Kyoto: Nakanishiya Shuppan, pp. 234–275.

Namba S, Kuwano S (1980) The relation between overall noisiness and instantaneous judgment of noise and the effect of background noise level on noisiness. J Acoust Soc Jpn (E) 1:99–106.

Namba S, Kuwano S (1982) Psychological study on L_{eq} as a measure of loudness of various kinds of noises. J Acoust Soc Jpn 38:774–785.

Namba S, Kuwano S (1988a) Measurement of habituation to noise using the method of continuous judgment by category. J Sound Vib 127:507–511.

Namba S, Kuwano S (1988b) Psychological evaluation of music performance using the method of continuous judgment by selected description. Harmonia Kyoto City Univ Arts 18:6–33.

Namba S, Kuwano S (1990) Continuous multi-dimensional assessment of musical performance. J Acoust Soc Jpn (E) 11:43–52.

Namba S, Kuwano S (2007) Relation between loudness of sounds with over critical duration and L_{Aeq} and $L_{AE.}$ In the case of road traffic noise. Proceedings of the Spring Meeting Acoustical Society of Japan, Tokyo: The Acoustical Society of Japan, pp. 755–756.

Namba S, Nakamura T, Kuwano S (1971) The loudness of level-fluctuating noises. Jpn J Psychol 42:93–103.

Namba S, Nakamura T, Kuwano S (1972) The relation between the loudness and the mean of energy of level-fluctuating noises. Jpn J Psychol 43:251–260.

Namba S, Nakamura T, Kuwano S (1973) Programmable sound control system. Jpn J Psychol 43:309–311.

Namba S, Kuwano S, Kato T (1974) The relation between loudness and rise time as a function of energy. J Acoust Soc Jpn 30:144–150.

Namba S, Kuwano S, Kato T (1976) The loudness of sound with intensity increment. Jpn Psychol Res 18:63–72.

Namba S, Nakamura T, Kuwano S (1977) An analysis of piano performance. Stud Hum Soc Sci Col Gen Edu Osaka Univ 25:25–43.

Namba S, Kuwano S, Kato T (1978a) An investigation of *L*eq and *L*a in relation to loudness. J Acoust Soc Jpn 34:301–307.

Namba S, Kuwano S, Kato T (1978b) On the investigation of *L*eq, L_{10} and L_{50} in relation to loudness. J Acoust Soc Am 64:S58.

Namba S, Kuwano S, Nikaido S (1982) Estimation of tone quality of broadcasting sounds using the method of continuous judgment by category. J Acoust Soc Jpn 38:199–210.

Namba S, Rice C G, Hashimoto T (1987) The loudness of decaying impulsive sounds. J Sound Vib 116:491–507.

Namba S, Kuwano S, Hato T, Kato M (1991) Assessment of musical performance by using the method of continuous judgment by selected description. Music Percept 8:251–276.

Namba S, Kuwano S, Koyasu M (1993) The measurement of temporal stream of hearing by continuous judgments – in the case of the evaluation of helicopter noise. J Acoust Soc Jpn (E) 14:341–352.

Namba S, Kuwano S, Kinoshita A, Hayakawa Y (1997) Psychological evaluation of noise in passenger cars – the effect of visual monitoring and the measurement of habituation. J Sound Vib 205:427–434.

Namba S, Kuwano S, Fastl H, Kato T, Kaku J, Nomachi K (2004) Estimation of reaction time in continuous judgment. Proc Int Cong Acoust 1093–1096.

Namba S, Kuwano S, Kato T (2006) Subjective evaluation of sounds – effects of duration and interval between listening to sound and the judgment to them. Proc Autumn Meeting Inst Noise Control Eng Jpn 117–120.

Neisser U (1976) Cognition and Reality. Principles and Implications of Cognitive Psychology. San Francisco, CA: WH Freeman.
Neuhoff JG (1998) Perceptual bias for rising tones. Nature 395:123–124.
Neuhoff JG (2004) Ecological Psychoacoustics. San Diego, CA: Elsevier, Academic.
Plomp R (2002) The Intelligent Ear: On the Nature of Sound Perception. Mahwah, NJ: Lawrence Erlbaum.
Poeppel E (1978) Time perception. In: Held R, Leibowitz HW, Teuber H-L (eds), Handbook of Sensory Physiology, Vol. III: Perception. Heidelberg: Springer, pp. 713–729.
Rice CG, Zepler EE (1967) Loudness and pitch sensations of an impulsive sound of very short duration. J Sound Vib 5:285–289.
Riesz RR (1928) Differential intensity of the ear for pure tone. Phys Rev 31:867–875.
Scharf B (1978) Loudness. In: Carterette EC, Friedman MP (eds), Handbook of Perception, Vol. IV: Hearing. New York: Academic, pp. 187–242.
Springer N, Weber R, Schick A (1997) Instantaneous and overall loudness of temporally variable pink noise. In: Schick A, Klatte M (eds), Contributions to Psychological Acoustics. Oldenburg: BIS, pp. 91–98.
Susini P, McAdams S (2008) Loudness asymmetry ratings between accelerating decelerating car sounds. J Acoust Soc Am 123:3307.
Susini P, McAdams S, Smith Benett K (2002) Global and continuous estimation of sounds with time-varying intensity. Acta Acoustica 88:536–548.
Symmes D, Chapman LF, Halstead WC (1955) Fusion of intermittent white noise. J Acoust Soc Am 27:470–473.
Terhardt E (1974) On the perception of periodic sound fluctuation (roughness). Acustica 30:201–213.
Vigran E, Gjaevenes K, Arnesen G (1964) Two experiments concerning rise time and loudness. J Acoust Soc Am 36:1468–1470.
Zepler EE, Harel JRP (1965) The loudness of sonic booms and other impulse sounds. J Sound Vib 2:249–256.
Zwicker E (1952) Die Grenzen der Hörbarkeit der Amplitudenmodulation und der Frequenzmodulation eines Tones. Acustica, Akustische Beihefte 3:125–133.

Chapter 7
Binaural Loudness

Ville Pekka Sivonen and Wolfgang Ellermeier

7.1 Introduction

The human auditory system is *binaural*, that is, it consists of two ears that are positioned on the two sides of the head. In a normally functioning system, acoustical pressure waves are picked up by the eardrums at the end of each ear canal. The pressure variations cause the eardrums to vibrate, and the bones of the middle ear transmit the vibrations to the liquid-filled cochlea in the inner ear. The cochlea converts the vibrations into electrical signals that are sent via the auditory nerve fibers to the cochlear nuclei. The nerve signals from the two ears are combined in the superior olivary complex and higher up in the auditory pathways. The brain then utilizes information at the auditory cortices on both sides in forming auditory percepts, such as loudness. Any malfunctioning or individual differences on either side may affect the transduction of pressure waves in the surrounding medium to auditory percepts in the brain, be it peculiarities in the shape of the torso, head, and pinnae, or differences in the peripheral or central auditory pathways (for a review of the physiology of hearing, see Pickles 2008).

Hearing with two functioning ears has several benefits, particularly when gathering spatial information about the listening environment (Blauert 1997a): The benefits include a better performance in localization of sound sources and improved spatial separation of signals from multiple sources. Further binaural benefits are the ability to enhance signals emanating from a chosen source and to bolster the unreflected signal components in a reverberant environment. Binaural hearing thus allows for accurate auditory information of our acoustic surroundings.

Binaural hearing is also beneficial in terms of loudness. It is well established that a sound presented to both ears is perceived as *louder* than the same sound presented to one ear only. This psychophysical effect is termed *binaural loudness summation*. Further, the threshold of hearing is also lower when listening with both ears

V.P. Sivonen (✉)
Department of Signal Processing and Acoustics, Aalto University School of Science and Technology, Otakaari 5 A, 02150 Espoo, Finland
e-mail: ville.sivonen@tkk.fi

M. Florentine et al. (eds.), *Loudness*, Springer Handbook of Auditory Research 37,
DOI 10.1007/978-1-4419-6712-1_7,

(Pollack 1948), bearing evidence that the hearing system is more effective in the transduction of acoustic waves with two, than just one ear. On the other hand, there are inhibitory processes in hearing, such as the suppression of reflected sound in reverberant environments. As opposed to mere summation, binaural inhibition has also been suggested to affect loudness (Hirsh and Pollack 1948; Gigerenzer and Strube 1983).

In auditory experiments in the laboratory, the aim is to observe how various combinations of sound stimuli presented to the two ears affect loudness, as reported by the listener. In such experiments, various modes of stimulation can occur: If a sound is presented to one ear only, stimulation of the auditory system is referred to as *monaural*, or sometimes, as *monotic*. When sounds are presented to both ears, the stimulation is *binaural*. Binaural stimulation, on the other hand, can be either *diotic*, defined as the same sound is presented to both ears, or *dichotic*, defined as different sounds are presented to each ear. Although sounds can be heard and their loudness can be judged using one ear only, loudness is essentially binaural: In real environments, percepts evoked by sounds to the two ears are fused into a single binaural loudness percept.

This chapter begins with a summary of methodological issues in measuring binaural loudness, followed by a review of classical studies on binaural loudness utilizing headphone playback. Then, binaural loudness of free, diffuse, and directional sound fields is discussed, and finally, possible applications for binaural loudness are outlined.

7.2 Measuring Binaural Loudness

The question of whether, and how, the auditory system integrates inputs from the two ears to form a unitary binaural loudness percept has been studied for almost 100 years, not only to address a substantial psychoacoustical question, but also, because it provides methodological opportunities to test the feasibility of scaling the loudness sensation in the first place. A major concept for the measurement of sensation is the additivity of the entities to be measured (see Luce and Tukey 1964; Luce 2002; Chap. 2), and the binaural summation of loudness appeared to be a very straightforward instance of such additivity.

Several methodological approaches have been used to study binaural summation, the most important ones being: (1) intramodal matching, (2) direct psychophysical scaling, and (3) measurement-theory based strategies of investigating additivity. Each of these approaches is briefly inspected for its merits and limitations.

7.2.1 Intramodal Matching

Concerned primarily with the observation of binaural beats, Seebeck (1846) was the first to report evidence for binaural loudness summation. He used a "double siren" of two rotating discs with holes drilled at regular intervals from which tubes led to the observers' left and right ears. Seebeck found that when both tubes were open, the "tone appeared to be stronger" (p. 451) than when one tube was shut off. This amplification appeared to occur independently of the phase relation (manipulated by changing the length of the tubes) between the air streams exciting the two ears.

It took almost another century until binaural loudness was systematically studied for its own sake: In their seminal article on loudness, Fletcher and Munson (1933) measured the sound pressure levels (SPLs) of sinusoids (0.125, 1, or 4 kHz) presented to one ear only that matched the loudness of the same tones when presented to both ears simultaneously, and in phase. They found the monotic presentation to require higher levels, by approximately 5 dB at a loudness level of 20 phons, and by up to 10 dB from approximately 50 phons upwards. These measurements clearly demonstrate binaural summation to occur, and they suggest a straightforward method of measuring what in the literature is often called the *binaural gain* (g; in dB), or sometimes the *binaural level difference for equal loudness* (BLDEL), that is, the decibel difference between equally loud monotic and diotic presentations. Fletcher and Munson (1933) further *assumed* that binaural (diotic) loudness is twice monaural loudness, and used this assumption to construct a loudness scale that bears remarkable similarity to the sone scale of loudness as it is used today (see Chap. 2).

7.2.2 Psychophysical Scaling

The majority of investigations of binaural summation, however, has employed subjective (numerical) estimates of loudness based on the methodologies developed by Stevens (1956b, 1975) and his students (see Chap. 2). A paradigmatic study addressing the issue of binaural loudness summation was performed by Marks (1978). He had listeners judge all 81 combinations of 9 SPLs (including a subthreshold level) delivered to the right ear with the same 9 levels delivered to the left ear. Thus monaural, binaural, and various dichotic exposures were used, allowing for a more general test of additivity. The scaling method employed was (free) magnitude estimation. Figure 7.1 shows a subset of these data, namely the average monaural and binaural loudness estimates.

It is evident that the average monaural loudness functions (circles and squares) for this sample of normally hearing listeners largely coincide. Binaural loudness, by contrast, received noticeably higher estimates, and the mean binaural loudness data (marked by triangles in Fig. 7.1) come quite close to the sum of the monaural loudness values (indicated by the dashed lines in Fig. 7.1). Marks' (1978) study, therefore, is one of a few providing evidence for "perfect" binaural loudness summation, that is, a binaural-to-monaural loudness ratio of 2.

Other studies, however, have not found perfect loudness summation, but rather considerably lower summation ratios such as 1.5 (Algom et al. 1989b; Zwicker and Zwicker 1991) or 1.3 (Marozeau et al. 2006). How can these discrepancies be reconciled? Clearly, one may speculate that the "raw" estimates made by human observers are not the "true" loudness readings on their inner psychological scale, and

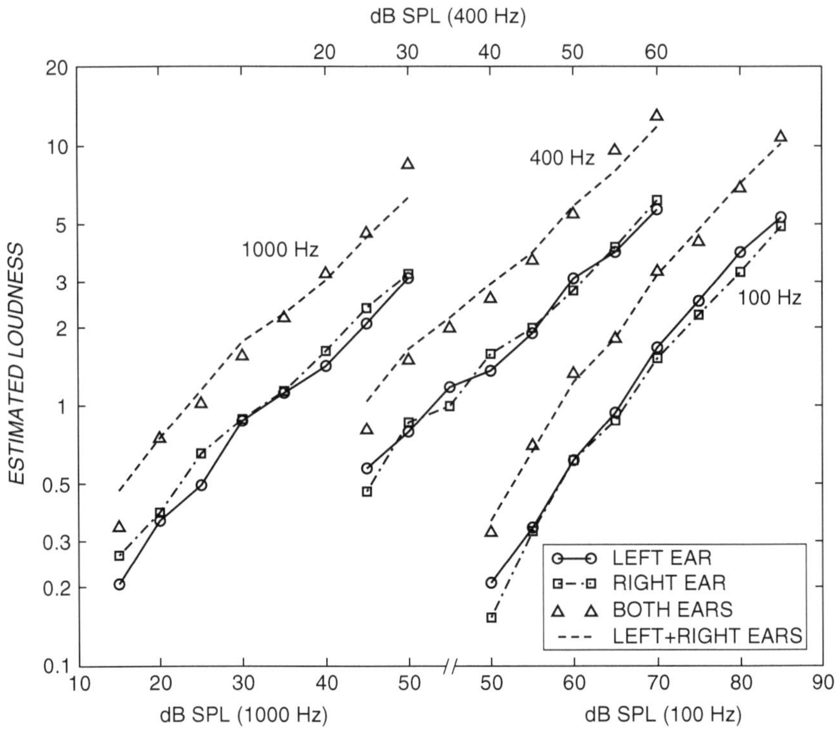

Fig. 7.1 Direct scaling of binaural loudness: Monaural and binaural loudness functions (at 0.1, 0.4, and 1 kHz) determined by magnitude estimation. Monaural estimates denoted by circles (*left ear*) and squares (*right ear*); binaural data by triangles. The sum of the left-ear and right-ear estimates is marked by the dashed lines. The figure is redrawn after Marks (1978, Fig 7.5). Numerical values provided by Larry Marks (personal communication, September 2008)

that some sort of distortion has biased the overt responses. That assumption was in fact made by Marks (1987) and by Algom et al. (1989a), who rescaled the raw data obtained in their experiments on binaural summation to comply with the sone scale of loudness. In these two studies, the rescaled loudness functions turned out to be consistent with perfect summation, but that is not always the case (e.g., Algom et al. 1989b).

It may be instructive to analyze the relationship among (a) the binaural gain, (b) the monaural-to-binaural loudness ratio, and (c) the psychophysical function more closely. Assume, loudness (N) grows as a power function of sound pressure (p), then (assuming identical functions for the left and right ear) the monaural and binaural loudness functions may be written as

$$N_m = k_m p_m^{\beta_m} \tag{7.1}$$

$$N_b = k_b p_b^{\beta_b} \tag{7.2}$$

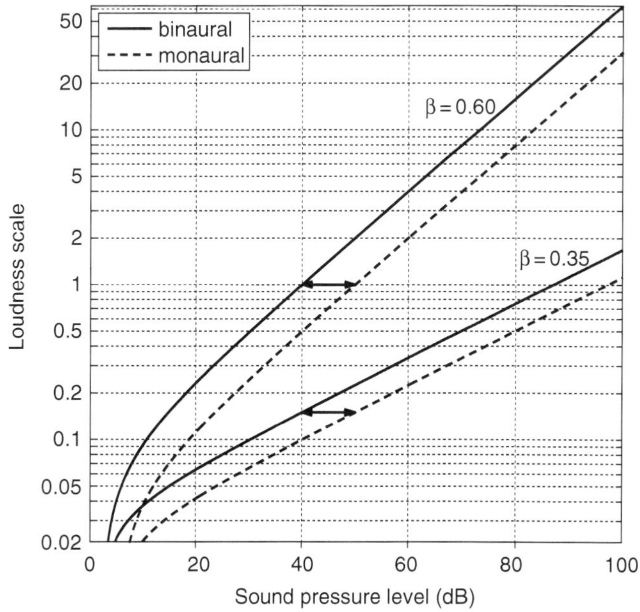

Fig. 7.2 Hypothetical binaural and monaural loudness functions. The upper two curves represent monaural (*dashed line*) and binaural (*solid line*) loudness as described by the sone scale, i.e., loudness grows as a power function of sound pressure ($\beta = 0.6$), and binaural loudness is twice monaural loudness. The lower two curves preserve the "binaural gain" (vertical distance between monaural and binaural curves: 10 dB, as indicated by the *arrows*), but represent different psychophysical functions (with $\beta = 0.35$). As a consequence, the binaural-to-monaural loudness ratio is only 1.5 for the lower curves. For both sets of curves, the growth of loudness below 20 dB SPL is steeper than what is defined by the exponent β

Perfect loudness summation (as observed by Marks 1978) holds, if $\beta_m = \beta_b$ and $k_b/k_m = 2$. That situation is depicted in the two upper curves of Fig. 7.2, with the added feature that the loudness functions get steeper at low SPLs, as has been often observed (for a summary, see Scharf 1978 or Chap. 5). Note that above 20 dB SPL the slopes of the monaural and binaural functions are identical, and that the two functions are (vertically) offset by a ratio of 2. Further, the upper two curves in Fig.7.2 are drawn so that the monaural and binaural functions are horizontally displaced by 10 dB (at least above 20 dB SPL), implying (a) that the binaural gain is constant and (b) that for both functions, a ratio of 2 on the ordinate corresponds to a 10-dB increase on the abscissa, as specified by the standard loudness exponent of $\beta = 0.6$, that is, Stevens' power law.

Note, however, that various other scaling outcomes are equally consistent with a given binaural gain (in decibels). The two lower curves in Fig. 7.2 show imperfect loudness summation, that is, a monaural-to-binaural loudness ratio of just 1.5. They were generated preserving the same binaural gain as the two lower curves (i.e., the same horizontal distances in decibels), and – to achieve that – require shallower loudness functions, in this case having exponents of approximately 0.35. Assuming power functions for loudness, and a level-independent binaural gain g in decibels, the exponent β may be predicted from the binaural-to-monaural loudness ratio N_b/N_m and the value of the binaural gain:

$$\beta = \frac{20\log_{10}(N_b / N_m)}{g} \tag{7.3}$$

Thus there are three parameters involved in a simple comparison of binaural and monaural loudness functions for headphone-delivered narrow-band sounds: the binaural gain (in dB), the binaural-to-monaural loudness ratio, and the shape of the loudness function (e.g., its exponent). As will be seen, when the relevant empirical studies are reviewed below, depending on which of these parameters is measured (and which one is fixed based on assumptions), various, seemingly contradictory outcomes may be observed. Owing to the interdependencies of these parameters, results may often be reconciled by rescaling the data for compliance with a certain binaural-to-monaural loudness ratio or with a certain power function exponent. Thus the two top curves in Fig. 7.2 may be considered rescaled versions of the two bottom curves, where the rescaling consists of arriving at a binaural-to-monaural loudness ratio of 2:1, and compliance with the sone scale of loudness.

It thus appears that making a loudness match, or determining the level difference required to make monotic and diotic stimulation with the same sound to appear equally loud is the most fundamental type of fact to be established in this realm. Obtaining quantitative loudness judgments takes one a step further and requires additional assumptions, for example, on the relationship between the numbers uttered by the participants to the underlying (latent) loudness scale (Narens 1996; Ellermeier and Faulhammer 2000). Given that direct loudness scaling may contain some such undeterminacies, and that several types of scaling outcomes are consistent with the same kind of binaural gain, it appears safe to consider intramodal matches (e.g., determinations of binaural gain) as the most fundamental type of data to be gathered on binaural summation. They are what remains invariant with respect to various transformations of the response scale.

7.2.3 Axiomatic Measurement and Nonmetric Scaling

A handful of studies have addressed the issue of binaural additivity without relying on quantitative judgments or matches. Based on concepts of axiomatic measurement theory, they investigated preconditions (axioms) for additivity to hold. One such precondition requires combinations of different levels of left-ear and right-ear inputs to exhibit some form of transitivity (technically: double cancellation, see Luce and Tukey 1964; Krantz et al. 1971). Formally, if a, b, c are levels presented to the left ear, and q, r, s are levels presented to the right ear, then

$$(a,q) \geq (b,r) \tag{7.4}$$

and

$$(b,s) \geq (c,q) \tag{7.5}$$

$$\Rightarrow (a,s) \geq (c,r) \tag{7.6}$$

with the notation (a, q) referring to a simultaneously presented (dichotic) pair of levels, and the symbol ≥ meaning "is judged to have greater or equal loudness as." Obviously, it requires all but *ordinal* comparisons to test whether the conclusion holds given the two premises. If the orderings of dichotic level combinations satisfy the double-cancellation axiom, additivity may be assumed to hold. Levelt et al. (1972) investigated double cancellation with 1-kHz tones in three different ranges between 10 and 80 dB SPL and found the axiom to hold in more than 97% of the several hundred tests performed on two listeners.

Schneider and Cohen (1997), using a similar experimental design but broadband noise, confirmed this finding with much larger samples (45 children and 40 adults), finding violations of the double-cancellation axiom in a mere 1–3% of all tests performed. As Levelt et al. (1972) had done, they went on to estimate left-ear, right-ear, and binaural loudness scales consistent with the paired comparisons made, and found these scales to be power functions of sound pressure and to exhibit binaural summation (albeit with a slightly stronger weight for the right-ear input).

As part of a larger study investigating issues fundamental to the measurement of loudness, in their experiment 3, Steingrimsson and Luce (2005) tested the weaker Thomsen condition (Krantz et al. 1971), which is obtained when the greater-or-equal (≥) operation in (7.4)–(7.6) is replaced with equivalence (~) and thus may be implemented in an intramodal matching paradigm. Steingrimsson and Luce also found support for additivity in the large majority of their tests, combined with significant ear asymmetries in individual listeners. The main conclusion to be derived from the axiomatic studies of binaural summation is that the validity of an additive summation rule for loudness may be determined without having to subjectively scale the monaural and binaural inputs, and that the weight of the evidence suggests that additivity may be assumed to hold.

7.3 Headphone Investigations

A straightforward way of investigating binaural loudness is by means of headphone playback. Assuming there is no crosstalk between the ears (i.e., leakage of sound pressure from one earphone to the contralateral ear), monaural stimulation can be achieved simply by switching off the other earphone, and the experimenter can manipulate the sound signals at the two ears independently of one another. Headphone playback may, however, result in interaural signal combinations far exceeding what would naturally occur. Note that in a real sound field, there is a relation between the two at-ear signals defined by the size and shape of the head, torso, and pinnae of the listener (see Sect. 7.4). Moreover, without spatial synthesis, more or less lateralized, "inside-the-head" percepts will be evoked by headphone playback, as opposed to percepts localized outside the head in sound field listening. Thus the benefit of rigorous experimental control over what reaches the eardrums comes at the expense of disregarding what is natural, ecologically valid stimulation.

Below, main outcomes of studies investigating binaural loudness via matching and scaling methodologies are summarized. In these studies, results have typically been obtained in headphone playback by varying the mode of stimulation (monaural, diotic, and dichotic), as well as the interaural properties of the signals delivered to the two ears.

7.3.1 Outcome of Matching Studies: Binaural Gain

Though, based on a-priori considerations (see Sect. 7.2.1), intramodal matches determining which binaural, monaural, or dichotic sound-pressure combinations sound equally loud appear to constitute the most straightforward subjective measurement to be made, the data collected on the binaural gain are scarce, and so variable across studies that they do not warrant unequivocal conclusions. Early studies using the method of adjustment to obtain loudness matches, and employing pure tones (Fletcher and Munson 1933) or band-limited noise (Reynolds and Stevens 1960) found the binaural gain to vary between 6 and 10 dB, with a tendency for it to increase with level. Matching data collected by Irwin (1965) and Marks (1978) are consistent with these findings, and further extend them to dichotic level combinations.

Further, Scharf (1969) used tones of different frequencies in each ear and found the degree of summation (6–8 dB) to be largely independent of the frequency separation. Mulligan et al. (1985) observed much larger binaural gains (of 9–17 dB), but they used low-frequency tones (250 Hz), short stimuli (40 ms) and a protocol delivering standard and comparison in immediate succession, which makes their results hard to compare with those of other reports.

A comprehensive study of the binaural gain as measured with monaural vs. binaural stimulation under headphones was recently reported by Whilby et al. (2006). This study used a finer grain of levels (from 10 to 90 dB SL) than most of its precursors, and a modern adaptive forced-choice procedure to obtain loudness matches. The overall outcome – as well as the pattern for individual listeners – showed clear evidence for a level dependence of the binaural gain: It increased from an average of 4 dB just above threshold to a maximum of almost 10 dB in the 50–60 dB SPL range to drop down to 5–6 dB at the highest levels (100–105 dB SPL) again. Average data, both for the long (200 ms) and short (5 ms) 1-kHz tones used are depicted in Fig. 7.3.

Note that this feature of a variable (in this case: nonmonotonic) binaural gain is not consistent with the hypothetical functions sketched in Fig. 7.2, where a constant horizontal displacement (i.e., binaural gain) of the monaural and binaural loudness

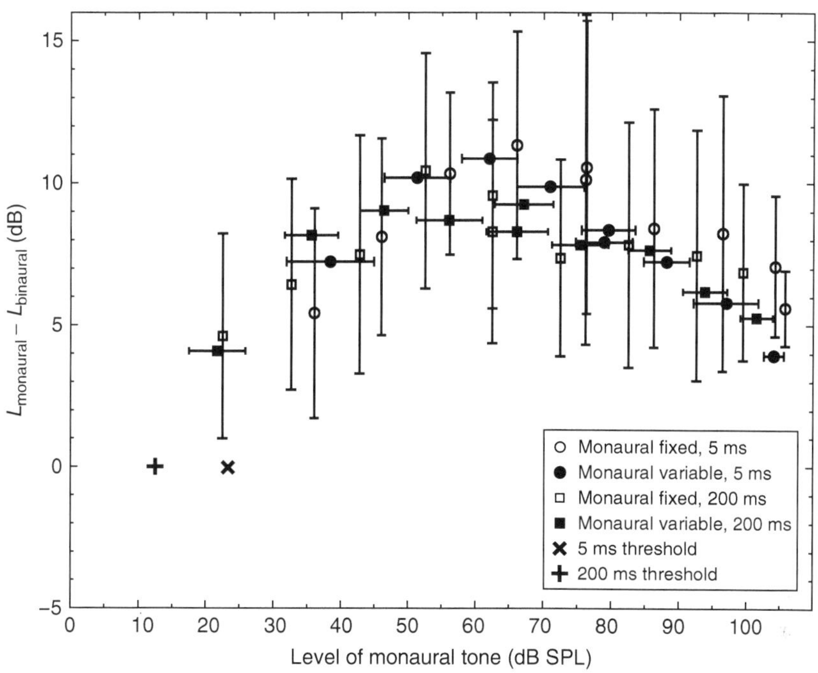

Fig. 7.3 Binaural-to-monaural loudness matches determined by Whilby et al. (2006) for 5-ms (*circles*) and 200-ms (*squares*) 1-kHz tones. The level difference required to obtain equal loudness is plotted as a function of the level of the monaural tone. Either the monaural tone (*closed symbols*) or the binaural tone (*open symbols*) was matched by the adaptive procedure. Error bars represent standard deviations for data averaged across listeners

functions was assumed. As will be seen when the scaling results are taken into account, this nonmonotonicity is either indicative of level-dependent changes in the monaural-to-binaural loudness ratio, or it suggests systematic deviations from loudness being a simple power function of sound pressure.

7.3.2 Outcome of Scaling Studies

Reynolds and Stevens (1960) were the first to measure binaural and monaural loudness directly, using the subjective scaling methods developed just a few years earlier, namely magnitude estimation, magnitude production, and cross-modality matching (see Chap. 2). Using bandpass noise (100–500 Hz), they found both monaural and binaural loudness to grow as power functions of sound pressure, but with a slightly smaller exponent for the former (0.54 vs. 0.60). As a consequence, the binaural-to-monaural loudness ratio appeared to increase from an average of 1.5 at 50 dB SPL to 2.2 at 100 dB SPL.

Hellman and Zwislocki (1963) later refuted that finding by comparing new data on the monaural loudness of 1-kHz tones with their binaural data collected earlier on a different group of listeners. They found nearly perfect loudness summation, that is, a binaural-to-monaural loudness ratio of 2, and rejected the idea of a level dependence of that ratio. Further support for perfect summation came from the experiment by Marks (1978) using factorial combinations of levels presented to the right and left ears. Subsets of these data, reproduced in Fig. 7.1, agree with the hypothetical functions depicted in Fig. 7.2 (top curves) showing: (a) perfect summation, that is, a ratio of 2:1, (b) power functions with exponents near 0.6, and (c) no level dependency of the binaural-to-monaural loudness ratio.

But there is evidence for less-than-perfect summation as well. Scharf and Fishken (1970), for example, found loudness ratios between 1.7 and 1.4 with very little evidence for a level dependency (if anything, it ran counter to the trend observed in the earlier studies). Marks (1987) found a binaural-to-monaural ratio of 1.6 in his raw data, but rescaled the response measure for compliance with the sone scale ($\beta = 0.60$), thus arriving at a ratio of 2. This is quite like transforming the lower set of functions in Fig. 7.2 to produce the upper set of curves. Zwicker and Zwicker (1991) reported also less-than-perfect summation, the binaural-to-monaural loudness ratio being about 1.5 and independent of level.

The most recent investigation of the issue (Marozeau et al. 2006) explicitly addresses the "binaural equal-loudness-ratio hypothesis," that is, the hypothesis that this ratio is constant across levels. That was found to hold, albeit with evidence for a very low binaural-to-monaural loudness ratio of 1.3 in the average data. It should be noted though, that the response was measured based on cross-modality matching of string length (cut by the listener on each trial) to perceived magnitude, thus requiring the additional assumption that this measure constitutes a ratio scale of loudness. Marozeau et al. (2006) further found the loudness functions thus derived to exhibit sinusoidal departures from a power function – a finding recently reported in other investigations as well (e.g., Buus et al. 1997; see Chap. 5) – thus leading them to propose a fourth-order binomial model to describe them (Marozeau and Florentine 2009). Taken together with the loudness matches (Whilby et al. 2006) showing a nonuniform binaural gain in decibels, these data suggest the monaural and binaural loudness functions to be parallel (constant ratio, as in Fig. 7.2) but, due to their S-shape (nonmonotonic horizontal displacement, unlike in Fig. 7.2) to provide evidence for deviations from a plain power law.

7.3.3 Stimulus Variables Affecting Binaural Summation

The stimulus properties affecting binaural summation in headphone investigations are quickly summarized, because relatively little was varied in these studies that largely confined themselves to using sinusoids or random noise, typically in nonspatial monotic or diotic configurations. The few exceptions are summarized here.

In some of the early scaling studies (e.g., Scharf and Fishken 1970) binaural summation for tones was compared with (white) noise. While differences in the shape of loudness functions emerged, the same kinds of binaural-to-monaural loudness ratios were obtained. Keen (1972) obtained matches between diotic and dichotic stimulation to investigate the effect of at-ear level asymmetry on binaural loudness. A different contour for preserving binaural loudness was observed for a 1-kHz tone than for a noise or triangular transient signal. No significant dependency on the tone frequency, however, was found. Likewise, Zwicker and Zwicker (1991) found no significant influence of the center frequency of third-octave-band noises, or of the overall spectral shape of broadband noise. Even presenting two different tone frequencies to the left and right ear did not seem to matter: Scharf (1969) found the degree of summation – measured in a matching paradigm – to be the same no matter how large the frequency separation (from zero to several thousand Hertz).

In some studies, the interaural configuration was systematically varied. Hirsh and Pollack (1948) obtained loudness matches between binaural tones varying in phase and a monaural reference tone, both presented against the same diotic background noise. For 0.25- and 1-kHz tones at low signal-to-noise ratios, less level for interaurally out-of-phase than for in-phase tones was required for equal loudness with the reference tone. Similarly, Marks (1987) presented 1-kHz tones in narrow bands of noise and found greater degrees of binaural summation when the noise was presented to one ear only, or when dichotic configurations were used. These effects are reminiscent of the binaural unmasking occurring in experiments on the "masking level difference." Mulligan et al. (1985) observed the greatest degrees of summation when two sinusoids were interaurally shifted by phase angles that might realistically occur due to head size, and less summation at phase angles beyond. Since the conditions of this study have not been replicated, it might just serve to point to the potential importance of "natural" interaural parameters.

The effect of interaural correlation on binaural loudness has also been under investigation, with no effects being found for wideband sounds (Dubrovskii and Chernyak 1969; Culling and Edmonds 2007). With low-frequency, narrow-band sounds, however, Culling and Edmonds (2007) found an interaurally uncorrelated stimulus to require approximately 2 dB less to match the loudness of a diotic stimulus, or one for which the phase of the signal to the other ear was merely inverted. The results of Culling and Edmonds (2007), at least at low frequencies, thus suggest a greater amount of binaural summation when the signals at the two ears are uncorrelated than when they are correlated or in antiphase.

In one study (Whilby et al. 2006) the duration of 1-kHz sinusoids was varied (5 vs. 200 ms) in a matching paradigm measuring the binaural level difference for equal loudness. In the range studied, however, no evidence for duration effects was found.

To summarize, headphone investigations of binaural summation provide relatively little evidence for the effects of such stimulus properties as spectrum, level, or interaural disparity. This may be largely due to the scarcity of auditory cues provided by the synthetic stimuli employed. In the following, therefore, the review will focus on the few studies that included spatial information in the stimuli used when studying binaural summation.

7.4 Loudness of Sound Fields

Acoustical waves in a sound field reach both ears of a listener, and even in the simplest fields, the two at-ear signals differ at least to some extent from one another. In fact, monaural stimulation in a sound field occurs only with severe hearing impairment or deafness in one ear, or it may be approximated by occluding one of the ears as efficiently as possible. The listening in a sound field is thus *binaural* and typically *dichotic*. While listening over headphones gives the experimenter more control over the exact stimulation at the ears, sound field listening better resembles the real-life situation of being immersed in an acoustic environment. Psychoacoustic experiments in sound fields are typically carried out in environments which are simplistic enough in acoustical terms, but yet allow for generalization to as many real-life situations as possible. Before going into the issue of loudness, the basic properties of sound fields are elaborated below.

Acoustic environments surrounding us vary greatly in their temporal and spatial properties. Due to the finite speed of sound in a medium (air), sound traveling for a shorter distance will reach the listener earlier than sound from a longer distance. *Indirect* reflections from obstacles and boundaries of the surrounding environment will travel for a longer distance than the *direct* sound. Therefore, the direct sound from the source arrives at the listener's ears first, then early reflections will arrive, and lastly, the reverberant energy of the acoustic space composed of sound reflected a large number of times. All of these time signals are superimposed at the listener's ears.

In addition to temporal properties, sound sources and reflecting surfaces may be distributed or positioned in a number of spatial locations with reference to the listener. While the direct sound and early reflections reach the listener from distinct directions, reverberation is typically fairly *diffuse*, meaning that it not restricted to any specific direction. Further, the acoustic environment may be *static*, when the properties remain unchanged over time, or *dynamic*, when there is movement of sources, of sound-reflecting surfaces, or of the listener.

The distance to the source also influences the composition of the sound field. In the near field, very close to the source, the reactive field dominates. This field is associated with the local fluctuating flow of sound energy, and it does not result in energy propagated into the far field of the source (e.g., Crocker 1991). Moving further away from the source, into the far field, sound pressure decreases linearly along a radial line connecting with the source (Beranek 1986), or in terms of sound pressure level, by 6 dB for each doubling of the distance from the source (Harris 1991). That is, sound pressure is inversely proportional to distance $(p \propto 1/r)$ and sound intensity to distance squared $(I \propto 1/r^2)$. This decrease with distance is called the *inverse-square law*, and it holds for a point source of constant power in *anechoic*, reflection-free conditions. In reverberant surroundings, the decrease is less than 6 dB for each doubling of distance, due to reflected sound. When reflections are much weaker than the direct sound, the inverse-square law can be approximated over a limited range, making the acoustic conditions *quasi-anechoic*. In a *reverberant* field, on the other hand, reflections are much stronger than the direct sound, and the resulting distribution of sound is characterized by reflections, rather than distance to the source. Be it anechoic or reverberant, psychoacoustical experiments are typically carried out in the far field of the source.

In a sound field, the signals reaching the listener's two ears depend greatly on the direction and distance of sources relative to the listener, as well as on the amount of reflected sound in the acoustic environment. The effect of these dependencies on binaural loudness are discussed in the text that follows.

7.4.1 Free Field

The simplest example of a sound field used in listening experiments is one in which sound arrives only from the direction of the source, without reflections (i.e., echoes) from the boundaries of the surrounding environment. This type of an acoustic environment is called an anechoic or a *free* sound field. In everyday situations, sound travels via a number of reflections when reaching the listener's ears, and thus, anechoic conditions can only be approximated in the open air, or in laboratory test chambers specifically designed to absorb most of the sound energy by the walls.

In loudness research, free-field listening implies that sound reaches the listener from the *frontal* direction only, that is, the listener is facing the sound source at a large enough distance to be in the far field of the source. In such a situation, sound signals reaching the two ears are fairly similar due to the apparent symmetry of the human head, torso and pinnae. In fact, this mode of listening comes fairly close to a diotic situation.

Stimulus frequency has been a primary variable in investigating the loudness of binaural, free-field listening. By matching the loudness of a test sound with a given frequency to a reference sound at 1 kHz, *equal-loudness-level contours* have been derived (see Chap. 5). A classical study by Fletcher and Munson (1933) derived such contours for headphone presentation, while having the headphones electrically compensated at each test frequency to produce the same loudness as a source in front of a listener in the free field. These contours were revised by Robinson and Dadson (1956), this time obtained via loudspeaker listening in an anechoic chamber. The most recent set of equal-loudness-level contours for the free field can be found in an international standard ISO 226 (2003), based on the best account for a number of investigations.

In Fig. 7.4, 40-phon loudness-level and absolute-threshold contours are plotted both for the free field (solid lines) and the diffuse field (dashed lines, see Sect. 7.4.2). The contours show a considerable frequency dependency, for example, in the free field at 20 Hz, 100 dB SPL is required for a loudness match with the 1-kHz reference at 40 dB SPL. Sound pressure level on the ordinate of Fig. 7.4 is measured in the absence of a listener in a reference position, where the center of the listener's head would be.

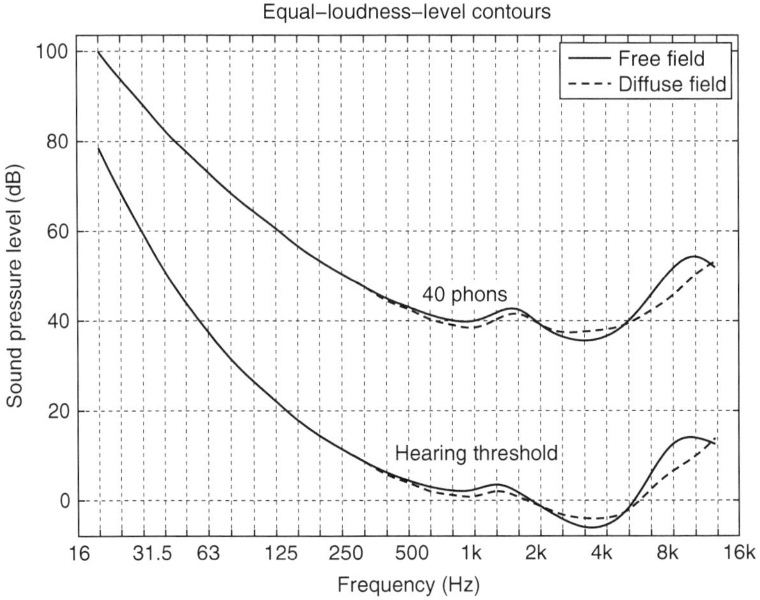

Fig. 7.4 A 40-phon equal-loudness-level contour and the threshold of hearing for binaural listening in the free and the diffuse field (ISO 226 2003; ISO 389 2005)

Although equal-loudness-level contours are level dependent (Chap. 5), the hearing threshold contour obtained in a detection task shows remarkable similarity in shape to that of the 40-phon contour. Equal loudness-level contours thus demonstrate that the sensitivity of the binaural human auditory system (including the head, torso, and pinnae immersed in a sound field) varies greatly with frequency. These contours form the basis for auditory modeling such as when loudness models are used to predict loudness from measurements of sound pressure, as perceived by an average, normally hearing listener (see Sect. 7.5 and Chap. 10).

7.4.2 *Diffuse Field*

A more complex example of the loudness of sound fields is the diffuse field. In such a field, sound travels via a large number of reflections before reaching the listener. In an ideal diffuse field, sound energy is not concentrated in any specific location, but rather, it arrives with equal acoustic intensity from all directions. Diffuse field conditions can be approximated by exciting a highly reverberant room with an omnidirectional sound source, and by making measurements at some distance from the source and the boundaries of the room. Rooms designed for such purposes are called reverberation chambers and they have long reverberation times, on the order of many seconds. Reverberation chambers are useful in sound power measurements of directional sources, but their use in psychoacoustical experiments may be problematic: long reverberation time (i.e., the slow decay rate of sound) may make paired loudness comparisons of successive sounds difficult to obtain in practice.

The loudness of diffuse fields has been of interest, since free-field conditions do not apply to normal rooms. To investigate loudness in conditions differing from those of the free-field case, equal-loudness-level contours were obtained in reverberant rooms with added diffusion (Stevens 1956a; Kuhl and Westphal 1959), or using an artificial diffuse field generated by multiple incoherent sound sources in an anechoic chamber (Robinson et al. 1961). The aim of all of these studies was to maximize the amount of diffuse sound in the listening conditions. The results showed that equal-loudness-level contours thus obtained differ from those in the free field. This difference is illustrated in Fig. 7.4 by the solid and the dashed lines for the free and the diffuse fields, both at 40 phons and at threshold. As shown in Fig. 7.4, the difference between the two sound fields is negligible at low frequencies and it increases with frequency.

From a listener's viewpoint, the main difference between the sound fields is that the free field sound arrives from the frontal direction only, whereas in the diffuse field sound arrives equally from all directions. Even though the same sound pressure might be present in both sound fields in the absence of a listener, the difference in the angle of incidence would cause the sound pressure at the ears to vary considerably. This direction-dependent variation can be determined by the use of head-related transfer functions (HRTFs; Shaw 1974; Mehrgardt and Mellert 1977; Wightman and Kistler 1989a; Møller et al. 1995; Blauert 1997b), as elaborated below. HRTFs represent the transfer of sound from an unobstructed field to the two ears, for a given incidence angle.

Figure 7.5 depicts the attenuation needed for a narrow-band sound in a diffuse field to be perceived as equally loud as the same sound in a free field (Robinson et al. 1961), see the "actual" third-octave-band levels of Fig. 7.5. On the other hand, the "predicted" data of Fig. 7.5. represent the difference in at-ear third-octave-band sound pressure level between the free and the diffuse fields, as derived from artificial-head HRTFs (ISO 11904-2 2004). Note that both in the free-field and the diffuse-field case,

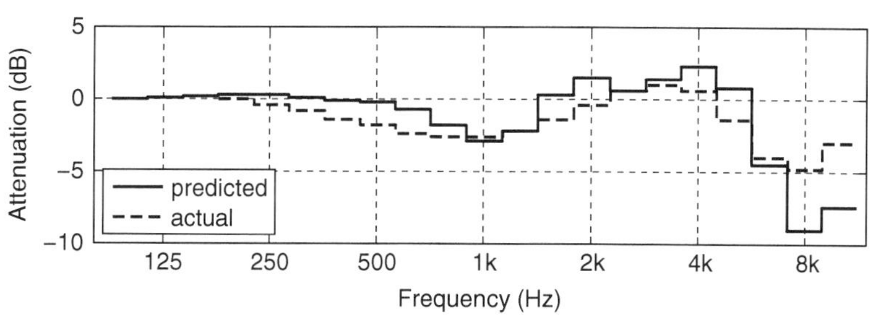

Fig. 7.5 Attenuation for a sound in diffuse field to be equally loud as the same sound in free field. "Predicted" data are based on the physics of a dummy-head (ISO 11904-2 2004), "actual" data on psychophysical loudness measurements (Robinson et al. 1961)

there are only small differences in the long-term spectra between the ears, due to symmetry of the human body along the interaural axis. Therefore, the attenuation in the predicted curve of Fig. 7.5 has the same effect at both ears and the effect can be described by a single curve. Comparison of the predicted curve with the actual curve reveals a fair agreement between objective measurements of at-ear level changes and subjective measurements of loudness. At 1 kHz, for example, the level arriving at the listener's ears in a free field is approximately 3 dB lower than in a diffuse field, and thus, a diffuse field sound needs to be attenuated by an equal amount to appear as loud as the same sound in a free field. It follows that equating the at-ear level of a narrow-band sound results in equal loudness in both fields (Robinson et al. 1961).

The fact that the loudness difference between the free and the diffuse fields can be explained by a simple linear filtering effect of the body suggests that there are no other differences in loudness processing between the two sound fields. Thus spectral summation of loudness across critical bands (Zwicker and Fastl 1999; Moore 2003; Chap. 5) and binaural summation of loudness across ears (Reynolds and Stevens 1960; Hellman and Zwislocki 1963; Marks 1978; Zwicker and Zwicker 1991; Sect. 7.3) are not affected. Further support for the simple filtering concept comes from the average hearing thresholds in the two sound fields, namely, it is generally accepted that the filtering has identical effects on hearing thresholds.

7.4.3 Directional Sound Fields

Sound may impinge on the listener from various locations in space, and the angle of incidence in real sound fields often differs from that of the free and the diffuse field case. As is evident from the research on HRTFs, the physics of the human body cause the inputs to the two ears to vary greatly with direction. In free and diffuse fields, changes in at-ear sound pressure show a fair agreement with loudness; therefore, effects on loudness may also be expected in sound fields where the direction, from which a sound reaches the listener, is varied. Spatial dependencies in HRTFs are often investigated in hearing research as sound localization cues (e.g., Blauert 1997b), and fairly few studies have investigated the effect of these dependencies on loudness.

Two investigations have explored the effect of direction on loudness for narrow-band stimuli (Robinson and Whittle 1960; Sivonen and Ellermeier 2006). The listeners' task in both studies was to match the loudness of a comparison sound from a given direction to that of the same sound from a reference direction. The sound pressure levels matched for equal loudness were then measured in a position where the center of the listener's head would be. The matched levels were reported as directional loudness contours relative to the frontal direction: If a sound coming from a given direction had to be attenuated by, for example, 3 dB to achieve a loudness match with the reference source ahead, a data point at +3 dB would be plotted, to indicate that the direction (and center frequency) in question was perceived as being *louder* than the reference direction by an amount corresponding to this decibel difference.

Directional loudness contours thus defined are plotted in three polar diagrams, for three center frequencies each, in Fig. 7.6. Even though Robinson and Whittle (1960) obtained directional loudness data for sources in both hemispheres of the horizontal plane, they combined data from the two sides in their report due to symmetry. Here, the left-hand side of each diagram shows contours in the horizontal plane from ahead to the (left or right) side and to behind a listener, and the right-hand side shows contours in the median plane from ahead to above and to behind a listener. The contours are based on mean data averaged across listeners. None of the contours trace the 0-dB line, which would indicate omnidirectionality for loudness. Rather, the direction from which a sound reaches the listener clearly has an effect on loudness for all center frequencies. The direction to the side is generally perceived as louder and the direction behind as softer than the frontal direction. Further, the effect of direction increases with center frequency, as indicated by the larger changes in the radii of the contours: the matches for narrow-band sounds vary up to 8 dB between a comparison and the reference direction at higher center frequencies.

Unlike the case of a free or diffuse field, the physical effects of the human body may be asymmetrical as a function of sound incidence angle, resulting in time, level and spectrum differences between the two ears. For example, in the horizontal plane with a source to the side of the head, sound arrives earlier at the ipsilateral ear closer to the source than at the contralateral ear further away from the source. This delay is called the interaural time difference (ITD). Reflections and diffractions from the body also result in interaural level differences (ILD), which depend heavily on frequency. Figure 7.7 shows HRTF magnitude spectra, measured for both ears of an artificial head at directions ahead and to the side of the head. For the frontal direction, the spectra at the ears are fairly similar (see the solid lines in Fig. 7.7), whereas for the direction to the side of the head the spectra at the two ears differ considerably. For that particular direction, and head, the ILD is on the order of few dB at low frequencies and up to 30 dB at high frequencies.

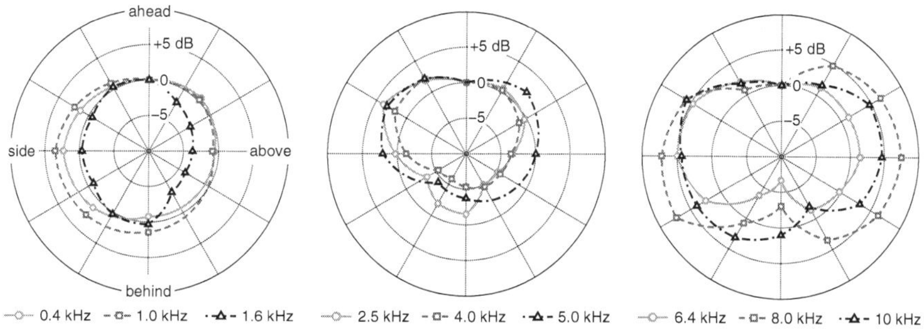

Fig. 7.6 Directional loudness contours for narrow-band sounds in an anechoic sound field. Loudness reference is located ahead the listener, and the symbols mark matched directions at each center frequency. The greater the radius the louder the sound produced in that direction by a source of constant output. Contours with center frequencies 0.4, 1.0, and 5.0 kHz are based on Sivonen and Ellermeier (2006), and the rest of the contours on Robinson and Whittle (1960). Left-hand and right-hand sides of each graph show data in the horizontal and the median planes, respectively. See text for details

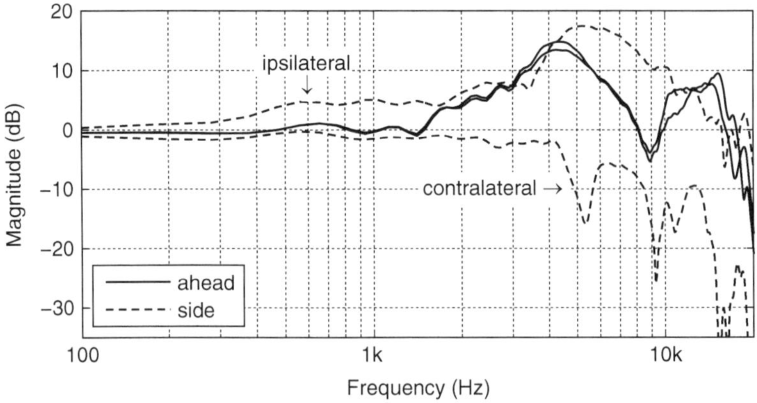

Fig. 7.7 Magnitude spectra of head-related transfer functions for directions ahead and side to the head (based on artificial-head measurements by Sivonen and Ellermeier; 2008)

The physical filtering of sound by the human head, torso, and pinnae may be inspected in an attempt to explain the directional loudness dependencies of Fig. 7.6. In fact, up to about 1.5 kHz, the HRTF of the ipsilateral ear has higher levels than the HRTFs for the frontal direction (see the upper dashed line and the solid lines in Fig. 7.7). This agrees well with the loudness contours of Fig. 7.6; the direction to the side is perceived as louder than the frontal reference both at 0.4 and 1.0 kHz (see the data points to the side in the left polar diagram of Fig. 7.6). The same agreement holds for frequencies between 5 and 10 kHz. Between 1.5 and 4 kHz the ipsilateral level is close to the frontal level, and the effect of direction on loudness is small (see the data points to the side at 1.6, 2.5, and 4.0 kHz in the left and middle diagram of Fig. 7.6). Qualitatively, there appears to be fair agreement between at-ear levels derived from HRTFs and directional loudness contours for narrow-band sounds, although the physics of the artificial head producing the HRTFs in Fig. 7.7 most likely differed to some extent from the average physical dimensions of the listeners of Robinson and Whittle (1960) and Sivonen and Ellermeier (2006).

Bearing in mind that loudness is essentially a binaural percept, the contralateral signal may also contribute to the loudness in sound fields. Thus, in addition to the physical filtering of sound from an unobstructed field to the ears, the (psychophysical) binaural summation of loudness must also be considered. Determining at-ear sound pressure levels for each center frequency and direction, both Robinson and Whittle (1960) and Sivonen and Ellermeier (2006) showed that changes in left- and right-ear levels largely account for the effects of direction on loudness. Namely, for narrow-band sounds, a relatively simple formula was suggested for combining sound pressure levels at the two ears (L_{left} and L_{right}) to an equivalent monaural level:

$$L_{mon} = g \times \log_2 \left(2^{L_{left}/g} + 2^{L_{right}/g}\right) \tag{7.7}$$

where g is the binaural gain discussed in Sect. 7.2.1. Robinson and Whittle (1960) reported a binaural gain of 6 dB to underlie directional loudness matches, while in Sivonen and Ellermeier (2006) a gain of 3 dB was reported to fit their mean data best. Note that despite the same unit (dB) with the directional loudness contours of Fig. 7.6, all of which were obtained binaurally, the gain constants here are theoretical, and merely express by how many decibels a monaural sound would have to be increased to be perceived as equally loud as the same sound presented diotically.

Further, assuming the standard loudness function to hold ($\beta = 0.6$), the corresponding loudness ratios of the binaural gains reported by Robinson and Whittle (1960) and Sivonen and Ellermeier (2006) would be approximately 1.5 and 1.2, respectively, both indicating less than perfect summation of loudness. Similar attempts to investigate the binaural summation in directional sound fields have been made by Remmers and Prante (1991) and Sørensen et al. (1995), however, based on fewer data and less systematic modeling. Nevertheless, all aforementioned studies explain directional effects on loudness based solely on magnitude spectra at the ears, assuming the effect of other binaural parameters, such as the interaural time difference or correlation to be negligible.

Quantifiable, although smaller directional effects (than those illustrated in Fig. 7.6) on loudness matches have also been reported for anechoic wideband (Remmers and Prante 1991; Sivonen 2007, up to 3 dB), for reverberant narrowband (Sivonen 2007, less than 2 dB), as well as for reverberant wideband sounds (Bech 1998, approximately 1 dB). The use of a wider bandwidth in combination with diffuse reverberation appears to wash out directional effects on loudness. In an extreme case of a diffuse sound field, the orientation of the listener supposedly has no effect on loudness because sound arrives at the listening position equally from all directions. The spatial distribution or multiplicity of sound sources has also been under investigation, reporting no (Scharf 1974) or only a small effect (Song et al. 2005) on loudness.

7.4.4 Distance and Loudness Constancy

In the open air, the amount of sound energy arriving at the listener's ears decreases with distance. Given large enough changes in distance, loudness is bound to be affected by the distance between the listener and a source of a constant output: A person speaking at a great distance is inaudible, if the sound waves propagating in the air are attenuated below the listener's threshold of hearing. When the distance is continuously decreased, the speaker will become audible at some point, and loudness will increase. Speaking with the same vocal effort directly to a listener's ears, a speaker is likely to be perceived as louder than when speaking from a distance.

Because spatial dependencies of at-the-ear signals form the basis for determining the direction of a sound source, similar cues have been proposed for auditory distance perception. These include the sound pressure level (or sound intensity), direct-to-reverberant energy ratio, spectrum changes by absorption, and binaural, as well as, dynamic cues (Zahorik et al. 2005).

In anechoic surroundings, the primary cue for judging auditory distance is sound pressure level. In such conditions, a dependence of loudness on distance has been observed (Stevens and Guirao 1962; Petersen 1990), showing a lawful relationship with the changes in sound pressure level according to the inverse-square law. These findings are in agreement with studies on the loudness of free, diffuse and directional sound fields (see Sects. 7.4.1–7.4.3), where differences in perceived loudness can be attributed to changes in the signal levels arriving at the two ears.

In reverberant surroundings, direct sound is similarly affected by the inverse-square law, while reverberant energy is fairly diffuse over varying source positions. This provides the listener with an additional cue for judging distance, that is, the changing ratio between direct and reverberant sound. Reverberation appears to enable listeners, when asked to do so, to judge the loudness at the *source*, despite profound changes in the at-ear signals. This type of *loudness constancy* with varying source distance has been reported for sounds in reverberant rooms (Mohrmann 1939; Zahorik and Wightman 2001). The terminology is analogous to the one used in visual perception where lightness, color, and size constancy have been classical topics of study (see Goldstein 2010), the latter meaning that perceived size, for example, the height of a person, is judged to remain the same, irrespective of the distance between the person and the viewer, and of the changes in the size of the person's image on the viewer's retina.

In the study by Zahorik and Wightman (2001), individual binaural room impulse responses (BRIRs) for 12 source distances were measured in a lecture hall, thus capturing the transfer of sound from the source to each listener's ears, including the characteristics of the room environment. The BRIRs were then convolved with noise bursts of different levels, and presented to the same listeners over headphones. Even though at-the-ear sound pressure levels varied with distance over a range of 20 dB, source loudness judgments, obtained via a magnitude estimation task, remained largely constant. Zahorik and Wightman's (2001) finding does not appear to require accurate distance estimates of the sources, nor can it be explained by abnormal growth of loudness in their subjects: loudness estimates of diotic (anechoic) noise bursts obtained from the listeners prior to the main experiment were fairly congruent with the standard loudness function ($\beta = 0.27$ for sound intensity, corresponding to $\beta = 0.54$ for pressure).

Zahorik and Wightman (2001) suggested loudness constancy to be related to reverberant sound energy, which remains relatively constant in rooms with varying source distance. In the absence of reverberation, they further suggested loudness estimates to be based on the direct sound, as was observed for their control condition of diotically presented, anechoic noises. Similar parsing of direct and reverberant sound has been suggested by Stecker and Hafter (2000), who reported stimuli with slow attack and fast decay in the temporal envelope to be perceived as louder than when the envelope was reversed. Stecker and Hafter (2000) argued stimuli with fast attack and slow decay (similar to sounds observed in rooms) to be compensated for the effects of the acoustic environment, and thus, to be subject to perceptual constancy due to a general mechanism of (higher-level) decay suppression. The suppression of reverberance is suggested to be a virtue of binaural hearing (Blauert 2005), although evidence for the inferior performance of monaural listening in this respect is largely anecdotal. Note that in the above-mentioned studies reporting loudness constancy (Mohrmann 1939; Stecker and Hafter 2000; Zahorik and Wightman 2001), sounds were presented to the listeners binaurally.

Another form of loudness constancy has also been recently proposed: Epstein and Florentine (2009) obtained monaural and binaural loudness estimates from listeners, who were presented with a 1-kHz tone, recorded spondees (two-syllable words) and monitored live-voice spondees either via headphones or loudspeakers. In all of their experimental conditions, Epstein and Florentine reported binaural loudness summation to occur, but to be less than perfect. More interestingly, for their live-voice condition, where the listeners were watching an audiologist uttering the spondees to a microphone and this signal was transduced in real-time to the listening booth, the mean binaural-to-monaural loudness ratio was only about 1.1. This was significantly less than the corresponding ratios for the tone or the recorded spondees, or what has been observed in classical studies on binaural loudness summation in the laboratory (see Sect. 7.3.2). The outcome suggests that the loudness of a familiar talker is little affected by switching between binaural and monaural listening, and similarly to loudness constancy with varying source distance, it may reveal an effect of higher-level processing on the loudness judgments of everyday sounds. Epstein and Florentine (2009) termed the phenomenon "binaural loudness constancy." An analogous visual phenomenon may be found in binocular brightness judgments, where at suprathreshold stimulus levels, closing one eye does not make the world to appear markedly less bright, when monocular brightnesses are nearly equal (Arditi 1986).

In summary, the finding of loudness constancy argues against loudness being solely based on the signal at the sensory receptors, as reported in most laboratory studies using typically anechoic stimuli delivered via headphones or loudspeakers. It still needs to be uncovered, to what extent constancy plays a role in real-life listening, outside the laboratory (for further information, see Chap. 8).

7.4.5 Headphone vs. Loudspeaker Presentation

Early research on loudness, and on hearing thresholds, has suggested a systematic difference in the eardrum level required for a match when data are collected using headphones or loudspeakers ("the missing 6 dB"; for a review, see Rudmose 1982). Namely, for equal loudness, at-ear levels of a 100 Hz tone had to be approximately 6 dB higher when headphones were used compared to when a loudspeaker provided the source. This effect had been observed both for monaural and for binaural listening. On the basis of his review, Rudmose (1982) refuted this effect by pointing out several factors causing such a difference, including masking caused by physiological noise transferred from the head to the ear canal when obtaining thresholds by headphones, as well as issues with mechanical coupling, distortion, listening test procedure and efficient ear occlusion in monaural listening when balancing loudness between headphone and loudspeaker playback.

The most compelling evidence against any such difference comes from outside of loudness research. Langendijk and Bronkhorst (2000) used HRTFs in binaural synthesis to reproduce the same at-ear signals over headphones, as would reach the listeners' ears from real sources from various spatial locations (for fundamentals of this synthesis technique, see Wightman and Kistler 1989a). Using three different psychoacoustical procedures and wideband (0.5–16 kHz) noise bursts as stimuli, Langendijk and Bronkhorst (2000) showed their listeners to be unable to discriminate whether the playback mode was "virtual" over headphones or "real" over loudspeakers. Had there been a systematic loudness difference requiring a 6-dB higher level from headphones, their listeners could have used this difference to discriminate between the playback modes. Note that the just-noticeable difference in level is well below 6 dB, on the order of, or less than 1 dB (Jesteadt et al. 1977; Florentine et al. 1987).

Similar data have been presented by Sivonen et al. (2005), who obtained binaural, directional loudness matches between virtual sources at various angles of incidence using binaural synthesis over headphones and three narrow bands of noise as stimuli (center frequencies at 0.4, 1.0 and 5.0 kHz). When comparing these data to data obtained from the same listeners using real sources (Sivonen and Ellermeier 2006), no significant differences between the mode of playback were observed. Although not directly comparing headphone and loudspeaker playback, and hence not being able to refute the systematic difference discussed by Rudmose (1982), the data of Sivonen et al. (2005) show that the directional effects on loudness (and the binaural judgments underlying these effects, see Sect. 7.4.3) are independent of the mode of playback.

The results of Langendijk and Bronkhorst (2000) and Sivonen et al. (2005) thus lead to the suggestion that when headphones are used in binaural synthesis (to reproduce spatial auditory percepts at a given distance and direction) no loudness difference is to be expected, compared to a situation where the percept is evoked by a loudspeaker at that same location. Note that binaural synthesis implies reproducing the same at-ear signals (including spectra and interaural characteristics) in both situations on an individual basis, as opposed to a mere calibration of eardrum sound pressure level. This is especially important when stimuli other than pure tones or narrow-band noises are used, as loudness depends heavily on the spectrum of sound (see Chap. 5).

What remains unexplained in Rudmose's (1982) data, however, is an observation that when matching a 100-Hz tone for loudness, a lower sound pressure level at the eardrum was required for a distant than for a nearby loudspeaker for three out of four listeners (see his Fig. 9). No physical explanation for this observation is given. Rather, it is suggested that the perceived "acoustic size" may have affected the match, namely, some subjects perceive the near source as much "smaller" and consequently, it must produce more sound pressure to be equally loud as the distant source. This observation may be related to the issue of loudness constancy discussed in Sect. 7.4.4. In line with the observation of Rudmose (1982), distant and near sources were given equal loudness estimates (Zahorik and Wightman 2001), despite the reduced amount of at-ear sound pressure from distant sources.

This location effect may, however, be somewhat plastic, and Rudmose (1982) noted that listeners can be trained to eliminate the effect.

Furthermore, the loudness scaling data by Epstein and Florentine (2009) showed differences in binaural-to-monaural loudness ratios between headphone and loudspeaker playback: Namely, the amount of binaural summation was significantly less when listening to a loudspeaker than when listening over headphones. This is in agreement with the binaural gain results when comparing classical headphones studies (gain of 4 to well over 10 dB, see Sect. 7.3.1) with the directional loudness studies in sound fields (3–6 dB, see Sect. 7.4.3). The study of Epstein and Florentine (2009) thus provides the first evidence that a direct comparison of results between the two playback modes may not be warranted. Note that Epstein and Florentine (2009) fed the same electrical signal to the headphones as was fed to the loudspeaker in a (somewhat reverberant) listening cabin. Therefore, there may be several reasons for the difference in binaural summation, such as the location effect (Rudmose 1982, "inside" vs. "outside-the-head" percepts) or the fact that room reverberation was present in the loudspeaker condition.

7.5 Applications

In addition to providing methodological and substantial insight into the functioning of auditory perception, knowledge about binaural loudness may be utilized in a number of applied settings. One such application is given by the increasing number of bilateral hearing-aids being fitted. There is evidence for individual differences in binaural loudness summation being larger in hearing-impaired listeners, and thus, the issue of binaural summation bears upon how gain is applied in the aid of each ear, on an individual basis (see Whilby et al. 2006).

Another application of binaural loudness is the formulation of auditory models, specifically with respect to improving measurement procedures simulating properties of human hearing (see Chap. 10). Here, by contrast, the emphasis is on instrumental models to *predict* loudness from acoustical measurement of sound pressure, as perceived by an average (normally hearing) listener. The importance of loudness in instrumental analysis of sound is prominent: Loudness forms the basis of instrumental metrics developed for sound-quality evaluation, such as roughness and sharpness (Zwicker and Fastl 1999), and it strongly correlates with other psychoacoustic attributes, such as the perceived quality of sound-reproduction systems (Gabrielsson and Sjögren 1979; Illényi and Korpassy 1981), or annoyance caused by noise (Hellman 1985; Berglund et al. 1990).

The most common way of measuring sound pressure is by using a single microphone, that is, *monophonically*. Based on the vast amount of psychoacoustical research, models for predicting loudness from monophonic measurements have been developed (Zwicker 1960; Stevens 1961). These pioneering studies have served as a basis for the first standardized loudness model for the free and the diffuse sound fields (ISO 532 1975). In such monophonic modeling, sound pressure is measured with an omnidirectional microphone in a position where the center of the listener's head would be, and the input to the model is the long-term spectrum of the sound signal, in octave-band or third-octave-band levels.

When computing loudness according to ISO 532 (1975), it must be specified, whether the sound field under measurement is "free" or "diffuse." This is due to equal-loudness-level contours, which describe the frequency-dependent sensitivity of the human auditory system, being dependent on the sound field type, see Figs. 7.4 and 7.5. The long-term spectrum is then used in the computations of the model to yield a prediction of loudness level in phons and loudness in sones. It is worth noting that even though a single monophonic sound signal is needed as input to ISO 532 (1975), the model is based on binaural listening in the free and the diffuse fields. Thus, it should not be referred to as a "monaural loudness," where a sound is presented to one ear only.

The drawback of the standardized monophonic model (ISO 532 1975) is that it is based on two specific sound field types, namely, the free and the diffuse field. Due to symmetry of the human head, torso and pinnae, the long-term spectra are the same at both ears in these two sound fields, and thus, the model cannot be applied to situations resulting in spectral differences between the ears. Nowadays, as part of so-called binaural technology (Møller 1992), artificial heads with a microphone placed in each ear, or miniature microphones placed in real ears, are used more frequently in acoustical measurements. As the signals at the ears are by and large different from another, binaural loudness models are needed for the analysis of such measurements.

The first model to facilitate the use of binaural ear-drum sound pressures for loudness computations was published by Moore et al. (1997). This model uses knowledge on HRTFs to separate the acoustical transformation from an unobstructed field to the ears, which was an integral part in the earlier model discussed above. To integrate the two at-ear signals, the model further assumes perfect summation of loudness between the ears (a binaural-to-monaural loudness ratio of 2), that is, monaural loudness being half of the corresponding binaural (diotic) loudness. Thus, any combinations of binaural signals can be processed by the model to arrive at a single binaural loudness prediction.

Sivonen and Ellermeier (2006) showed that the model of Moore et al. (1997) did not accurately predict directional loudness matches of anechoic, narrow-band sounds reaching the listener from various spatial locations. Rather, a binaural "power-summation" model, the use of which for artificial-head measurements is delineated in Sivonen and Ellermeier (2008), gave better predictions of binaural loudness for spatial sounds. This was corroborated for wide-band and reverberant sounds by Sivonen (2007). To better account for directional loudness data and other data showing less-than-perfect summation across ears, a revised loudness model was proposed by Moore and Glasberg (2007). This model includes inhibitory interactions to binaural loudness processing, namely, a signal at the left ear reduces the loudness evoked by a signal at the right ear, and vice versa. According to the revised model, the ratio between binaural and monaural loudness ratio is 1.5.

Note that all prediction models described above utilize the long-term spectrum measured in the absence of a listener, or at the ear of a listener or an artificial head. These models therefore neglect possible effects of interaural characteristics (such as correlation; Culling and Edmonds 2007) other than what is defined by at-ear spectra. Furthermore, the higher-level effects discussed above (loudness constancy; Zahorik and Wightman 2001; Epstein and Florentine 2009) are not covered by the present instrumental loudness models.

7.6 Summary

Binaural loudness continues to be an active field in auditory research that has considerably gained in breadth recently. While classical studies using headphone stimulation suggested "perfect" binaural loudness summation to hold, i.e. binaural loudness to be twice monaural loudness, more recent studies employing a greater variety of methodologies and sounds tend to obtain smaller binaural-to-monaural loudness ratios in the range between 1.2 and 1.5, remaining constant with changes in absolute level, and a binaural gain of 3–6 dB.

To broaden the scope of research on binaural loudness, a number of investigators have studied spatial conditions, approximating natural listening in situations ranging from free and diffuse sound fields to listening to directional sounds both in anechoic and reverberant environments. In these situations, binaural loudness is determined (a) by the (measurable) physical filtering imposed by the head, torso and pinnae of the listener and (b) by the (unobservable) binaural summation of the at-ear signals. This research has shown both considerable, frequency-dependent directional effects, and evidence for binaural summation largely converging with the outcome of recent headphone studies. An interesting exception occurs in situations – typically listening in reverberant sound fields – in which loudness "constancy" is preserved despite profound changes in at-ear stimulation. Future research may concentrate on investigating the binaural loudness of natural, time-varying sounds, instead of stationary tones and noises traditionally used, on uncovering the role of direct and reverberant energy in the perception of loudness in sound fields, as well as on looking into the physiological processing of binaural loudness in the central auditory pathways.

Findings on binaural loudness have only recently been incorporated into general models of hearing and initial attempts to provide guidelines for computing binaural loudness, e.g. based on artificial-head measurements, have been made. All of these developments suggest that the research on binaural loudness has evolved from being a special topic in psychophysical scaling to making rich connections to other fields of hearing research such as spatial hearing, binaural technology, and ecological (psycho-)acoustics.

Acknowledgments Ville Sivonen received funding from NIH DC008168, the Academy of Finland, and Emil Aaltonen's Foundation.

References

Algom D, Ben-Aharon B, Cohen-Raz, L (1989a) Dichotic, diotic, and monaural summation of loudness: a comprehensive analysis of composition and psychophysical functions. Percept Psychophys 46:567–578.

Algom D, Rubin A, Cohen-Raz, L (1989b) Binaural loudness and temporal integration of the loudness of tones and noises. Percept Psychophys 46:155–166.

Arditi A (1986) Binocular vision. In: Boff KR, Kaufman L, Thomas JP (eds), Handbook of Perception and Human Performance Vol. I – Sensory Processes and Perception. New York: John Wiley & Sons, pp. 23.1–23.41.

Bech S (1998) Calibration of relative level differences of a domestic multichannel sound reproduction system. J Audio Eng Soc 46:304–313.

Beranek LL (1986) Acoustics. New York: American Institute of Physics.

Berglund B, Preis A, Rankin K (1990) Relationship between loudness and annoyance for ten community sounds. Environ Int 16:523–531.

Blauert J (1997a) An introduction to binaural technology. In: Gilkey RH, Anderson TR (eds), Binaural and Spatial Hearing in Real and Virtual Environments. Mahwah, NJ: Lawrence Erlbaum Associates, pp. 593–609.

Blauert J (1997b) Spatial Hearing – The Psychophysics of Human Sound Localization. Cambridge, MA: MIT Press.

Blauert J (2005) Communication Acoustics. Berlin: Springer-Verlag.

Buus S, Florentine M, Poulsen T (1997) Temporal integration of loudness, loudness discrimination, and the form of the loudness function. J Acoust Soc Am 101:669–680.

Crocker MJ (1991) Measurement of sound intensity. In: Harris CM (ed), Handbook of Acoustical Measurements and Noise Control. New York: McGraw-Hill, pp. 14.1–14.7.

Culling JF, Edmonds BA (2007) Interaural correlation and loudness. In: Kollmeier B, Klump G, Hohmann V, Langemann U, Mauermann M, Uppenkamp S, Verhey J (eds), Hearing – from Sensory Processing to Perception. Heidelberg: Springer Verlag, pp. 359–368.

Dubrovskii NA, Chernyak RI (1969) Binaural summation of differently correlated noises. Sov Phys Acoust 14:326–332.

Ellermeier W, Faulhammer G (2000) Empirical evaluation of axioms fundamental to Stevens' ratio-scaling approach: I. Loudness production. Percept Psychophys 62:1505–1511.

Epstein M, Florentine M (2009) Binaural loudness summation for speech and tones presented via earphones and loudspeakers. Ear Hear 30:234–237.

Fletcher H, Munson WA (1933) Loudness, its definition, measurement and calculation. J Acoust Soc Am 5:82–108.

Florentine M, Buus S, Mason CR (1987) Level discrimination as a function of level for tones from 0.25 to 16 kHz. J Acoust Soc Am 81:1528–1541.

Gabrielsson A, Sjögren H (1979) Perceived sound quality of sound-reproducing systems. J Acoust Soc Am 65:1019–1033.

Gigerenzer G, Strube G (1983) Are there limits to binaural additivity of loudness? J Exp Psychol 9:126–136.

Goldstein EB (2010) Sensation and Perception. Pacific Grove, CA: Wadsworth.

Harris CM (1991) Handbook of Acoustical Measurements and Noise Control. New York: McGraw-Hill.

Hellman RP (1985) Perceived magnitude of two-tone-noise complexes: loudness, annoyance and noisiness. J Acoust Soc Am 77:1497–1504.

Hellman RP, Zwislocki JJ (1963) Monaural loudness function at 1000 cps and interaural summation. J Acoust Soc Am 35:856–865.

Hirsh IJ, Pollack I (1948) The role of interaural phase in loudness. J Acoust Soc Am 20:761–766.

Illényi A, Korpassy P (1981) Correlation of loudness and quality of stereophonic loudspeakers. Acustica 49:334–336.

Irwin RJ (1965) Binaural summation of thermal noises of equal and unequal power in each ear. Am J Psychol 78:57–65.
ISO 226 (2003) Acoustics – Normal equal-loudness-level contours. Geneva, Switzerland: International Organization for Standardization.
ISO 389-7 (2005) Acoustics – Reference zero for the calibration of audiometric equipment – Part 7: Reference threshold of hearing under free-field and diffuse-field listening conditions. Geneva, Switzerland: International Organization for Standardization.
ISO 532 (1975) Acoustics – Method for calculating loudness level. Geneva, Switzerland: International Organization for Standardization.
ISO 11904-2 (2004) Acoustics– Determination of sound emission from sound sources placed close to the ear – Part 2: Technique using a manikin. Geneva, Switzerland: International Organization for Standardization.
Jesteadt W, Wier GC, Green DM (1977) Intensity discrimination as a function of frequency and sensation level. J Acoust Soc Am 61:169–177.
Keen K (1972) Preservation of constant loudness with interaural amplitude asymmetry. J Acoust Soc Am 52:1193–1195.
Krantz DH, Luce RD, Suppes P, Tversky A (1971) Foundations of measurement Vol. I. New York: Academic Press.
Kuhl W, Westphal W (1959) Unterschiede der Lautstärken in der ebenen Welle und im diffusen Schallfeld. Acustica 9:407–408.
Langendijk EHO, Bronkorst AW (2000) Fidelity of three-dimensional-sound reproduction using a virtual auditory display. J Acoust Soc Am 107:528–537.
Levelt WJM, Riemersma JB, Bunt AA (1972) Binaural additivity of loudness. Br J Math Statist Psychol 25:51–68.
Luce RD (2002) A psychophysical theory of intensity proportions, joint presentations, and matches. Psych Rev 109:520–532.
Luce RD, Tukey JW (1964) Simultaneous conjoint measurement: a new type of fundamental measurement. J Math Psych 1:1–27.
Marks LE (1978) Binaural summation of the loudness of pure tones. J Acoust Soc Am 64:107–113.
Marks LE (1987) Binaural versus monaural loudness: supersummation of tone partially masked by noise. J Acoust Soc Am 81:122–128.
Marozeau J, Florentine M (2009) Testing the binaural equal-loudness-ratio hypothesis with hearing-impaired listeners. J Acoust Soc Am 126:310–317.
Marozeau J, Epstein M, Florentine M, Daley B (2006) A test of the Binaural Equal-Loudness-Ratio hypothesis for tones. J Acoust Soc Am 120:3870–3877.
Mehrgardt S, Mellert V (1977) Transformation characteristics of the external human ear. J Acoust Soc Am 61:1567–1576.
Mohrmann K (1939) Lautheitskonstanz im Eutfernungswechsel. Zeitschr. f. Psychologie 145:145–199.
Møller H (1992) Fundamentals of binaural technology. Appl Acoust 36:171–218.
Møller H, Sørensen MF, Hammershøi D, Jensen CB (1995) Head-related transfer functions of human subjects. J Audio Eng Soc 43:300–321.
Moore BCJ (2003) An Introduction to the Psychology of Hearing. London: Academic Press.
Moore BCJ, Glasberg BR (2007) Modeling binaural loudness. J Acoust Soc Am 121:1604–1612.
Moore BCJ, Glasberg BR, Baer T (1997) A model for the prediction of thresholds, loudness and partial loudness. J Audio Eng Soc 45:224–240.
Mulligan BE, Goodman LS, Gleisner DP, Faupel ML (1985) Steps in loudness summation. J Acoust Soc Am 77:1141–1154.
Narens L (1996) A theory of ratio magnitude estimation. J Math Psychol 40:109–129.
Petersen J (1990) Estimation of loudness and apparent distance of pure tones in a free-field. Acustica 70:61–65.
Pickles JO (2008) An Introduction to the Physiology of Hearing. London: Academic Press.

Pollack I (1948) Monaural and binaural threshold sensitivity for tones and for white noise. J Acoust Soc Am 20: 52–57.
Remmers H, Prante H (1991) Untersuchung zur Richtungsabhängigkeit der Lautstärkeempfindung von breitbandigen Schallen, Fortschritte der Akustik–Deutsche Arbeitsgemeinschaft für Akustik (DAGA), pp. 537–540.
Reynolds GS, Stevens SS (1960) Binaural summation of loudness. J Acoust Soc Am 32:1337–1344.
Robinson DW, Dadson RS (1956) A re-determination of the equal-loudness relations for pure tones. Br J Appl Phys 7:166–181.
Robinson DW, Whittle LS (1960) The loudness of directional sound fields. Acustica 10:74–80.
Robinson DW, Whittle LS, Bowsher JM (1961) The loudness of diffuse sound fields. Acustica 11:397–404.
Rudmose W (1982) The case of the missing 6 dB. J Acoust Soc Am 71:650–659.
Scharf B (1969) Dichotic summation of loudness. J Acoust Soc Am 45:1193–1205.
Scharf B (1974) Loudness summation between tones from two loudspeakers. J Acoust Soc Am 56:589–593.
Scharf B (1978) Loudness. In: Carterette EC, Friedman MP (eds), Handbook of Perception Vol. IV – Hearing. New York: Academic Press, pp. 187–242.
Scharf B, Fishken D (1970) Binaural summation of loudness: reconsidered. J Exp Psychol 86:374–379.
Schneider BA, Cohen AJ (1997) Binaural additivity of loudness in children and adults. Percept Psychophys 59:655–664.
Seebeck A (1846) Beiträge zur Psychologie des Gehör-und Gesichtssinnes. Poggendorffs Annalen 68:450–465.
Shaw EAG (1974) Transformation of sound pressure level from the free field to the eardrum in the horizontal plane. J Acoust Soc Am 56:1848–1861.
Sivonen VP (2007) Directional loudness and binaural summation for wideband and reverberant sounds. J Acoust Soc Am 121:2852–2861.
Sivonen VP, Ellermeier W (2006) Directional loudness in an anechoic sound field, head-related transfer functions and binaural summation. J Acoust Soc Am 119:2965–2980.
Sivonen VP, Ellermeier W (2008) Binaural loudness for artificial-head measurements in directional sound fields. J Audio Eng Soc 56:452–461.
Sivonen VP, Minnaar P, Ellermeier W (2005) Effect of direction on loudness in individual binaural synthesis. In Proceedings of the Audio Engineering Society 118th Conference, Barcelona, Spain, Paper No. 6512.
Song W, Ellermeier W, Minnaar P (2005) Perceived loudness of spatially distributed sound sources. Proceedings of Forum Acusticum, Budapest, Hungary, pp. 1665–1670.
Sørensen MF, Lydolf M, Frandsen PC, Møller H (1995) Directional dependence of loudness and binaural summation. In 15th International Congress on Acoustics, pp 293–296.
Stecker GC, Hafter ER (2000) An effect of temporal asymmetry on loudness. J Acoust Soc Am 107:3358–3368.
Steingrimsson R, Luce RD (2005) Evaluating a model of global psychophysical judgments – I: Behavioral properties of summations and productions. J Math Psych 49:290–307.
Stevens SS (1956a) Calculation of the loudness of complex noise. J Acoust Soc Am 28:807–832.
Stevens SS (1956b) The direct estimation of sensory magnitude – loudness. Am J Psychol 69:1–15.
Stevens SS (1961) Procedure for Calculating Loudness: Mark VI. J Acoust Soc Am 33:1577–1585.
Stevens SS (1975) Psychophysics. Introduction to its perceptual, neural, and social prospects. New York: John Wiley & Sons.
Stevens SS, Guirao M (1962) Loudness, reciprocality, and partition scales. J Acoust Soc Am 34:1466–1471.

Whilby S, Florentine M, Wagner E, Marozeau J (2006) Monaural and binaural loudness of 5- and 200-ms tones in normal and impaired hearing. J Acoust Soc Am 119:3931–3939.
Wightman FL, Kistler DJ (1989) Headphone simulation of free field listening. I: Stimulus synthesis. J Acoust Soc Am 85:858–867.
Zahorik P, Brungart DS, Bronkhorst AW (2005) Auditory distance perception in humans: a summary of past and present research. Acta Acustica united with Acustica 91:409–420.
Zahorik P, Wightman FL (2001) Loudness constancy with varying sound source distance. Nat Neurosci 4:78–83.
Zwicker E (1960) Ein Verfahren zur Berechnung der Lautstärke. Acustica 10:304–308.
Zwicker E, Fastl H (1999) Psychoacoustics – Facts and Models. Berlin: Springer-Verlag.
Zwicker E, Zwicker UT (1991). Dependence of binaural loudness summation on interaural level differences, spectral distribution, and temporal distribution. J Acoust Soc Am 89:756–764.

Chapter 8
Loudness in Daily Environments

Hugo Fastl and Mary Florentine

8.1 Introduction

Acquiring a comprehensive understanding of how people perceive loudness in daily environments requires us to take the study of loudness out of artificial controlled laboratories and observe the perception of loudness under more ecologically valid conditions. In doing so, we find that some assumptions made in the laboratories do not apply to daily environments.

The purpose of this chapter is to review work that attempts to bridge the gap between laboratory studies of loudness and our perception of loudness in daily environments. This is not an easy task. In addition to those of language and culture described earlier (see Chap. 1), "real-world" studies of loudness are fraught with problems. One of these problems is that the study of loudness can be confounded by annoyance. Therefore, the first section deals with loudness and annoyance. The second section discusses loudness as it relates to music. Loudness gives music its dynamics, and musicians have understood much about loudness long before it was studied in a laboratory. As early as the eighteenth century, musicians commonly used a notation scale of relative loudness, and they understood that increasing the duration of a brief sound could increase loudness. The third section covers audio–visual and audio–tactile interactions in ratings of loudness. It shows that the same sound can have different loudness ratings depending on its visual and/or tactile context. In the fourth section, cognitive effects in loudness ratings are discussed, including the context in which sounds are heard. Finally, levels required for optimal loudness for groups of people in public spaces are discussed. This chapter concludes with an introduction to some current apparatus used for setting sounds to optimal loudness.

H. Fastl (✉)
Department of Technical Acoustics, AG Technische Akustik, MMK,
Technische Universität München, 80333 München, Germany
e-mail: fastl@mmk.ei.tum.de

M. Florentine et al. (eds.), *Loudness*, Springer Handbook of Auditory Research 37,
DOI 10.1007/978-1-4419-6712-1_8,

8.2 Loudness and Annoyance

8.2.1 Annoyance

Loudness can be an important factor in annoyance, but loudness and annoyance are separate attributes (Berglund et al. 1976; Hellman 1982). Whereas annoyance often correlates with loudness, not all annoying sounds are loud. Imagine a dripping faucet. The sound is soft, but annoying. It is important to understand factors that relate to annoyance because judgments of loudness can be confounded by annoyance. The acoustic parameters that are related to annoyance are different from those that are related to loudness. Those related to loudness can be found in Chaps. 5 and 6.

Annoyance depends much more on the psychological state of the individual listener than loudness. For example, a loud distorted sound is usually perceived as annoying. However, distortion can be perceived as enjoyable when heard in the context of modern electric guitar music, such as rock, blues, jazz, and pop (Zwicker and Buus 1998). Likewise, tonal components are enjoyed in a musical context, whereas they are judged as annoying in an industrial noise context. For this reason, many countries have instituted "tone penalties" of up to 5 dB for tonal industrial noise (e.g., ANSI S1.13 2005; TA Lärm 1998). However, most people judge sounds with high-frequency components – like sounds from a circular saw or a dentist's drill – to be annoying. This annoying aspect of sound can be examined by measurements of the percept of sharpness (Bismarck 1974; Fastl and Zwicker 2007; DIN 45 692 2009).

The annoying aspects of repetitive sounds can be examined by measuring the percept of fluctuation strength (Terhardt 1968; Fastl and Zwicker 2007). Whereas some repetitive sounds like dripping faucets or pile drivers can be annoying and undesirable, others can be very useful. For example, fluctuation strength is an indispensable asset for a warning signal. A warbling police siren with large fluctuation strength is essentially a frequency-modulated tone at low (4 Hz) modulation frequency (Fastl 2006b).

There is a common view that annoyance is more subject dependent than loudness, and that the intersubject variation in judgments is greater for annoyance than for loudness. In at least one laboratory study, this was not found to be the case (Berglund and Preis 1997), although the reason for this finding is not clear.

8.2.2 Impact of the Meaning of a Sound Source on Annoyance and Loudness

It is widely accepted that liking or disliking a sound source can influence ratings of annoyance, and possibly loudness. To determine the magnitude of this effect, a procedure was developed that keeps the loudness–time function the same, but largely obscures recognition of the sound source (Fastl 2001). In essence, the procedure is as follows: First, an algorithm called Fourier time transform (FTT, see

Terhardt 1985) is used to process the original sound, such as a church bell. This spectrotemporal analysis is somewhat similar to conventional fast Fourier transform (FFT), but differs in two main features: a sliding time window with essentially exponential weighting and filter bandwidth increases for high frequencies in accord with the auditory system. In this way, an FTT spectrum with about 600 spectral lines is created. The next step, called spectral broadening, squeezes the 600 lines into 24 channels. After resynthesis, a sound is achieved, which produces the same loudness–time function as the original signal, but has lost all spectral detail. The FTT spectrum of the processed sound is a blurred image of the original FTT spectrum. After processing, the sound of the church bell is changed to be reminiscent of a sound produced when starting an aircraft. Therefore, two sounds with the same loudness–time function can be created that differ substantially in their meaning. For details and acoustic demonstrations see Fastl and Zwicker (2007, p. 331).

Figure 8.1 gives an example of FTT processing of a sound produced by a trumpet playing a scale. In the left panel, the harmonic structure of the unprocessed sound is clearly visible and the increase in critical bandrate indicates that a scale is being played. In the right panel, some harmonic structure and an increase in critical bandrate are still visible; however, the whole image is blurred.

The FTT method has been shown to be quite effective. Whereas the sound sources of more than 90% of the original sounds can be named correctly, only 15% of processed sounds can be recognized correctly (Ellermeier et al. 2004). Some categories of processed sounds are recognized more easily than others. For example, human speech can be recognized whether it is processed or not. On the other hand, musical instruments are never correctly identified when the processed version of the sound is presented.

As a rule, the loudness of the original sound and the processed sound is the same (Zeitler et al. 2003). However, subjects rate the annoyance of the original sound and the processed sound differently (Ellermeier et al. 2004). For example, the sound of clinking wine glasses or a coffee machine brewing fresh coffee is rated much less

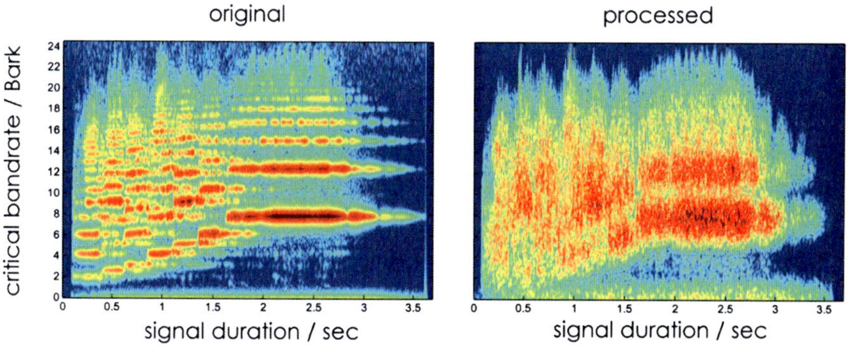

Fig. 8.1 An example of FFT processing (see text) of a sound produced by a trumpet playing a scale. Critical bandrate is plotted as a function of signal duration for the unprocessed sound (*left*) and the processed sound (*right*)

annoying as the original (recognized) sound than as processed (unrecognized) sound. Therefore, laboratory studies indicate that the meaning of a sound clearly affects judgments of annoyance, but may have little, if any, affect on judgments of loudness. The concept of "acceptable loudness" in our daily environment is intertwined with annoyance as is further illustrated in the next section.

8.2.3 Complex Nature of Acceptable Loudness Levels in Communities

According to William J. Cavanaugh (personal communication) and other noise consultants, "loudness of community noise" problems usually have at least one or more nonacoustical factors that influence judgments of acceptable loudness levels. Nonacoustic factors encompass a wide range of types (e.g., neighborhood problems, personal relationships, financial compensation, etc.). One of his stories from more than 50 years of experience illustrates the complex nature of acceptable loudness levels:

About 25 years ago, residents of a home development complex about 1,200–1,500 ft. from the performing stage house of a 15,000-capacity outdoor music pavilion in Northeastern United States complained about certain performing groups during the first summer season of operations. Three nonacoustic factors proved to be dominant during the initial concert season. First, many residents found the concert music from certain popular performers, like Willy Nelson, quite acceptable; whereas the same levels of concert music from less popular artists were judged to be unacceptable. This can be understood in light of some psychoacoustic data that indicate the more listeners like a particular type of music the louder they prefer it (see Sect. 8.3.5 of this chapter). Second, even when many residents judged the loudness of the concert as acceptable, sounds from the adjacent parking facility both before and after a concert were considered unacceptable. This may be understood in light of data indicating that the type or meaning of the sound influences its acceptable loudness. As described in the previous section, laboratory studies indicate that the meaning of a sound clearly affects judgments of annoyance – and hence, "acceptable loudness." Finally, it became known that the "leader" of the complaining residents had a daughter who had not been hired as a summer employee at the facility. It was thought that lack of financial compensation could have influenced subjective judgments of acceptable loudness. In subsequent years, the facility adopted a policy of preferential hiring of local residents and implemented a community relations program, including many nonacoustical features. To date, this outdoor concert facility has been in successful operation for nearly 25 seasons and, at the annual renewal of "permit to operate" hearings, the former leader of the home community complainers has become one of the concert facilities staunchest supporters. Lessons learned about taking nonacoustic factors into account have been successfully applied to countless other outdoor concert facilities throughout the United States.

The preceding story illustrates the nature of the complexity of addressing acceptable loudness of community noises. Some of the general practices of noise consultants are based on observations of what seem to generate successful outcomes. Cognitive factors – as they relate specifically to loudness – are covered in Sect. 8.4 of this chapter.

8.3 Loudness and Music

8.3.1 Music Can Be Loud and Not Annoying

Any person who has enjoyed music knows that music can be loud but not annoying. The human response to music seems to be different from that of all other sounds; in general, very loud sounds are annoying (Kuwano et al. 1992). In fact, loud music creates a unique human response that appears to transcend cultural differences (Kuwano et al. 1992). This discovery was made by an international team of scientists, who were brought together to study cross-cultural attitudes toward noise and annoyance under the leadership of Sonoko Kuwano and Seiichiro Namba from Osaka University in Japan. In one of their studies, recordings were made of various sounds, including aircraft noise, train noise, road traffic noise, construction noise, music, and speech. These recordings were played at different levels to subjects in research laboratories at Academia Sinica in China, Osaka University in Japan, and Northeastern University in the United States. Data from a subset of their numerous studies are shown in Fig. 8.2, which show a correlation between judgments of annoyance and loudness. The data indicate that people in China, Japan, and the United States respond differently to some music than the other sounds; music can become quite loud without becoming annoying – unlike aircraft noise, train noise, traffic noise, construction noise, and even speech. The exact reason why music elicits a response to loud sounds that is different from other sounds is unclear.

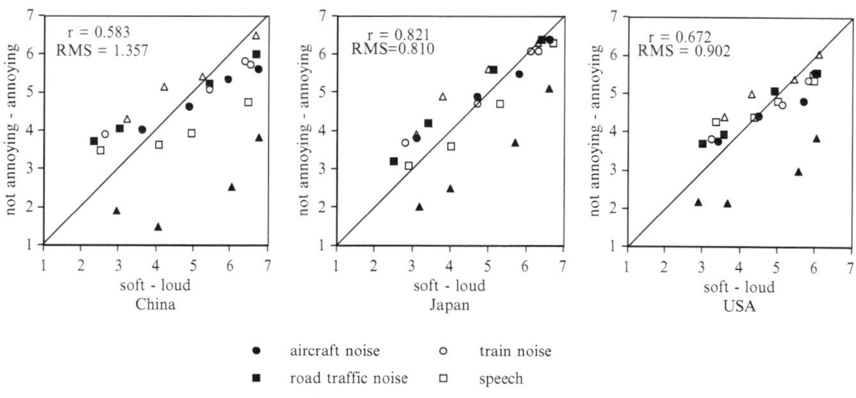

Fig. 8.2 Correlations between annoyance and loudness of six types of sounds (see *insert*) obtained in three countries: China (*left*), Japan (*middle*), and the United States (*right*). These graphs were replotted and provided courtesy of Sonoko Kuwano

8.3.2 Loud Music and Hearing Loss

Because most loud sounds are annoying, people tend to protect their hearing by reducing the level of potentially damaging sounds reaching their ears. For example, they cover their ears and move away from a sound if they cannot simply reduce its level. This protective mechanism may be absent in the case of well-liked loud music and there may be greater potential for noise-induced hearing loss caused by damaging levels of music.

Recording engineers are faced with a dilemma: If the loudness of a music reproduction is too soft, they cannot check for accurate intonations of instruments, appropriate tonal balance, the typical acoustic "scene" of a symphony orchestra or a group of artists, etc. On the other hand, recording engineers should avoid exposure to sounds that are too loud to prevent hearing loss, tinnitus, and auditory distortions – such as diplacusis, an abnormal perception of pitch. Auditory damage depends on *both* level and duration of sound exposure, as well as individual susceptibility to hearing loss. It seems reasonable to listen at levels below about 85 dBA and allow rest periods, although research compiled over the past two decades is somewhat ambiguous regarding the specific limits of music exposure levels (see Chap. 1, Sect. 1.1). The 85 dBA sound exposure limit is recommended on the basis of data that were obtained primarily from adult men who were exposed to industrial noise, not music. These data were used to develop standards to limit noise exposure to "acceptable" levels based on audiometric data and economic factors; they were not designed to prevent hearing loss in all noise-exposed workers. A maximum exposure of 85 dBA for 8 h per day over 40 years in an industrial-noise setting has been estimated to result in about 3–8% of workers having a hearing handicap (Prince et al. 1997). Hearing handicap is usually defined as a deviation from normal average thresholds of greater than 25 dB for both ears at selected frequencies. Most recommended standards also include a time-weighted average (TWA), which is a trade off (or "exchange rate") in decibels between level and exposure duration that is estimated to keep the risk of hearing loss constant. For example, a 3-dB exchange rate means that each 3-dB increase in sound level is exchanged against a factor of two in duration within a workday. Most developed countries limit worker sound exposure to 85 dBA, 8-h TWA with a 3-dB exchange rate (e.g., ISO 1999 and NIOSH 1998).

News reports have warned of the dangers of the high levels produced by portable media players. Data indicate that hearing loss can result from the abusive use of these devices. Portnuff and Fligor (2006) estimate that a typical person could safely listen to an iPod for 4.6 h at 70% of the listening range using the supplied earphones without greatly increasing the risk of hearing loss; one should never listen at the highest levels produced by the device. For a summary of evidence on the risk of hearing loss from the use of portable media players, see Fligor (2009).

When precautions are taken, music can be enjoyed at reasonably loud levels for limited amounts of time without risk of hearing loss. Young people seem to be able to set the level of music to a reasonable "preferred listening level." In fact, Laumann and colleagues (2006) asked 88 Japanese students with a mean age of about 19 years to sit in a lecture hall and listen to 60 pieces of music played at different levels through loudspeakers. They were asked to rate each musical piece on a five-step line in terms of loudness (too soft, slightly too soft, appropriate loudness, slightly too loud, and too loud) and preference (I like it very much, I like it, I neither like or dislike it, I dislike it, I dislike it very much). In this experiment, the subjects responded that a moderate loudness (i.e., L(Aeq) = 65 dB) was their preferred listening level. Because the experimenters were very surprised by these results, other experiments have been performed in Japan and Germany, and these results have been confirmed. Research is ongoing to determine why reasonable levels are preferred in a research setting, whereas potentially damaging levels have been reported at rock concerts. One possible reason is that crowd noise typical for rock concerts often exceeds levels of 85 dBA; for the music to be heard (and felt) over the crowd noise the rock bands must increase their performance levels to values that may be potentially dangerous. Another reason may be social control due to the research environment and the presence of authority figures (i.e., the researchers themselves).

One of the puzzling questions about loud music is why some people listen to music at levels high enough to damage their hearing. There have been numerous warnings in the public media on the risk of hearing loss caused by excessive music listening, but not all people heed these warnings. Two hypothesis have been put forth to explain why some people continue to listen to loud music, despite knowledge of harm caused by loud music. They are not mutually exclusive. One is that loud music excites the vestibular system and is associated with thrill seeking (Todd and Cody 2000). The other is that some people may have a behavioral dependency disorder related to loud music listening. To provide insight into the behavioral characteristics of people who listen excessively to loud music, a survey was developed and administered to groups of listeners encompassing a range of music-listening behaviors (Florentine et al. 1998). Results indicated that some listeners scored within a range that would suggest the presence of a maladaptive pattern of music-listening behavior similar to that exhibited by substance abusers. This behavior was not associated with the age of the listener or the type of music. For example, one 56-year-old man reported blasting classical music while his college-age daughter complained that his music was too loud. Like all experiences around which behavioral dependencies develop, music alters moods, reduces pain, and has the tendency to elicit the experience of craving (i.e., the need to hear a piece of music again and again).

8.3.3 Language of Loudness for Music (Musical Notation)

Since the late eighteenth century, it is common for composers to use written notation to indicate the level at which a piece of music should be played to produce a corresponding loudness. These are called dynamic markings and are usually in Italian. These dynamic markings refer to relative levels, as opposed to specific levels, so that a passage of music may be a little louder or softer than another. Typical dynamic markings include the ones shown in Table 8.1. Therefore, musicians used a

Table 8.1 Typical dynamic markings

fff	*fortississimo* or *forte fortissimo*	extremely loud
ff	*fortissimo*	very loud
f	*forte*	loud
mf	*mezzo forte*	medium loud
mp	*mezzo piano*	medium soft
p	*piano*	soft
pp	*pianissimo*	very soft
ppp	*Pianississimo* or *piano pianissimo*	extremely soft

Likert-like category scale before it was used in the laboratory. (For discussion of measuring loudness using category scaling, see Chap. 2.)

The language of loudness notation for music also includes *crescendo* to indicate an increase in loudness, and *decrescendo* or *diminuendo* to indicate a decrease in loudness. In addition, a composer can indicate that one note should be louder than the other notes by using *accents*. Such notations help to give music the proper loudness dynamics and indicate the beat of the music.

It is noteworthy that musically trained listeners can usually identify the level at which music was recorded independent of the level at which the music is played back. This effect is quite obvious with whispered speech; it can be played back at high levels and perceived as loud, but it is obvious that the originally recorded sound was soft whispered speech. Musical instruments produce audible changes in the spectral envelope at different production levels that may be used to cue the listener (Luce 1975). Miskiewicz and Rakowski (1994) used Zwicker's and Steven's procedures for loudness calculation to predict the loudness level of short scale segments played pianissimo and fortissimo on various orchestral instruments. Results demonstrated that due to spectral loudness summation, the changes in frequency bandwidth with playing level enhance the dynamic loudness level range.

8.3.4 Increasing Loudness Without Increasing Level

Long before the first laboratory experiments on loudness, musicians discovered how physical dimensions – other than level – influence loudness. Musicians use various methods to increase the loudness of an instrument other than simply increasing the level of an amplifier. For example, increasing the intensity of a pluck on a string will increase loudness, as will increasing the duration of a short sound (i.e., temporal integration of loudness; see Chap. 6).

Ways to increase loudness without increasing the level limit of sounds have been used to circumvent broadcasting regulations. For example, broadcasting stations frequently process their program material by a device called a "loudness maximizer." The brochures for these devices maintain that the loudness of program material can be increased without introducing distortions or affecting spaciousness or tone color. Results of psychoacoustic experiments (Chalupper 2000) reveal that indeed the loudness of pop music can be increased significantly without audible negative effects. However, the increase in loudness is much smaller than predicted for synthetic sounds like broadband noise or pure tones, and pure-tone distortions are clearly audible. Nevertheless, for pop music an increase of loudness by a factor of 2 is possible. For classical music somewhat less than double loudness can be obtained, but only with the presence of annoying distortions and an increased sharpness of the music. Its effects can be explained simply on the basis of classic loudness models as described in DIN 45 631/A1 (2008; see also Chap. 10). The only "trick" is that an instantaneous limiter follows a slow compression. Simply speaking, a loudness maximizer acts most of the time like an amplifier, producing higher critical band levels and, therefore, greater total loudness. Therefore, loudness increases while the monitored peak level, required by broadcasting regulations, remains unchanged (Chalupper 2000).

A countermeasure for the use of a loudness maximizer, or similar device, is the use of a sonemeter, which approximates the human response to sounds. (For information on sonemeters, see Sect. 8.6.2 of this chapter.) About 50 years ago, such devices were proposed by several scientists (Zwicker 1959; Pfeiffer 1964; Benjamin Bauer of CBS (see Torick et al. 1968)) as a better device for broadcast monitoring of radio and television, especially for commercials (for references, see Zwicker and Zwicker 1991; Moore et al. 2003).

8.3.5 Loudness Rating for Different Types/Styles of Music

There is an interaction between the type/style of music and the preferred listening level for individual listeners. For example, well-liked music may be perceived as less loud than music that is not liked when both samples of music are played at the same intensity (Fucci et al. 1993). In a recent study, Laumann et al. (2007) asked students about their preferred type of music (e.g., classic, jazz, hip hop, hard rock, reggae, and so forth). Then, they were presented segments of music at different reproduction levels and they rated preferred loudness. The preferred listening level was higher – corresponding to greater loudness – for preferred musical styles. It is clear from this study and others that the more listeners like a particular type of music, the louder they prefer to listen to it (e.g., Cullari and Semanchick 1989). This explains the common behavior of turning up the level of a favorite song to increase the enjoyment of it.

Laumann et al. (2007) studied loudness judgments and preferred types of music using another method, which was taken from a method used to study industrial noise immissions. (Immissions are sounds received at the ears of a listener that are a composite of all sounds in the vicinity of the listener.) It is well known that overall loudness ratings of noise immissions are larger than the average of instantaneous ratings of immissions (e.g., Kuwano and Namba 1985). The two rating tasks are different. In overall loudness ratings, listeners are asked to listen to a reasonably long-duration sound (e.g., 10 min) and judge the loudness of it at the end of the sound. In instantaneous loudness ratings, listeners are presented the same sound and asked to track their perceived loudness continuously by varying line length (at various time intervals) while the sound is being heard (Kuwano 1996; Chap. 6). Results show that for a given noise the overall ratings of loudness are larger than the average of the instantaneous ratings. The authors point out that this may indicate that loud events are better remembered and contribute more to the overall loudness. For example, a very loud sound from a motorcycle may overpower the memory of an otherwise moderately loud background traffic sound. Other data indicate that memory can influence judgments of loudness (Ward 1987; Chap. 2). This phenomenon has been observed for road noise, railway noise, air traffic noise, and industrial noise.

Laumann et al. (2007) asked if this same relationship observed with industrial noise immissions is found for an excerpt of music when it is disliked and considered as noise. They hypothesized that the difference between an overall loudness rating and a rating based on the average of instantaneous loudness ratings should be larger for disliked music (noise) than preferred music. As a rule (one exception) the hypothesis was borne out by the data. This indicates that the preference of musical style may influence loudness ratings. In short, one person's music is another person's noise!

8.4 Multisensory Interactions in Ratings of Loudness

Some cross-modality interactions with loudness in laboratory settings have been introduced under the topic of context effects (see Chap. 3, Sect. 3.7). For example, low-level noise bursts are rated louder when they are heard in the presence of lights than in their absence (Odgaard et al. 2004).

8.4.1 Audio–Visual Interactions

When visual stimuli are presented together with acoustic stimuli in experiments that are designed to start to bridge the gap between laboratory experiments and our perceptions in daily environments, audio–visual interactions can be observed. In general, these effects are small, but significant. For example, loudness ratings of a white noise depend on the type of picture shown together with the noise (Suzuki et al. 2000). If white noise is combined with a picture of a waterfall, the white noise is judged to be softer than when the sound is heard without the picture, and it is also judged to sound more pleasant. Similarly, ratings of loudness and pleasantness of car interior noises can be influenced significantly when a video filmed through a front car window is presented together with noise (e.g., Namba et al. 1997). Nice landscapes like mountain areas have a positive impact on the rating of car sounds, whereas videos of a car stuck in a traffic jam have a negative impact.

Audio–visual interactions have also been observed under more ecologically valid conditions (e.g., Fujiwara et al. 2006). Fujiwara et al. (2006) took participants in an experiment to a newly developed residential town in Japan where they were asked to judge the auditory and visual environment. Their judgments were made using a semantic differential method (Osgood et al. 1957, see also Chap. 1) in which participants rated their impressions on adjective scales to obtain information on the meaning of the perception. They were tested in three conditions: auditory environment alone, visual environment alone, and overall (auditory and visual) environment. Results indicate that for the adjective pair "quiet–noisy" the addition of a visual environment to the auditory environment results in judgments of a quieter overall environment than the auditory environment alone.

The loudness of sounds presented at the same sound pressure level can be judged to be different depending on the picture presented with the sound. For example, the loudness of a sound of a train passing by decreases when auditory and visual inputs are presented together (Patsouras et al. 2003). This effect even holds for Japanese subjects rating the loudness of German Super Express Trains (ICE) and Japanese Super Express Trains (Shinkansen) (Rader et al. 2004). When non-sound-related pictures are presented with sounds, the results are not clear. Some studies show an effect (e.g., Fastl 2004), whereas other studies do not (e.g., Menzel et al. 2008a). There seems to be no consistent statistically significant effect on non-sound related pictures on loudness.

If sounds are presented together with moving pictures instead of still pictures, the situation becomes somewhat more realistic. However, the decrease in perceived loudness is about the same magnitude for still vs. moving pictures (Fastl 2004). An even more realistic situation for the listener exists when sounds and moving pictures recorded at the passenger seat of a driving car are reproduced in a car simulator (Fastl 2004). In these situations, the perceived loudness is decreased by about 15% compared to judgments of the sound without the visual stimuli. However, for some individual listeners, the audio–visual interaction can lead to a loudness reduction of even more than 50%, whereas for others the effect is very small and only a few percent.

8.4.2 *Influence of Color on Loudness*

Small but significant audio–visual interactions have been observed by simply changing the color of a visual stimulus while keeping a sound constant. For example, Fastl and Patsouras (2004) showed that when the same sound of a train passing by is combined with a picture of a train in different colors, the perceived loudness of the train depends on the color of the train. At the same sound pressure level, red trains are considered to be louder than trains in a light green color. Figure 8.3 shows data from German and Japanese subjects, who rated loudness using a magnitude estimation procedure. When the relative loudness for the green train is set to 100%, German subjects rate the red train at 115% (i.e., the red train is considered as being 15% louder). Japanese subjects

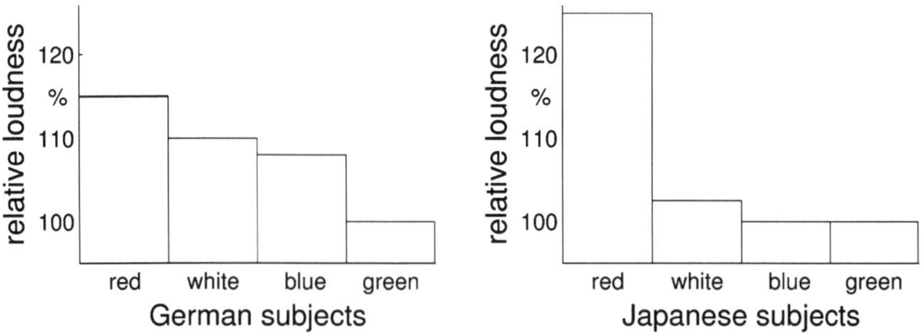

Fig. 8.3 Relative loudness ratings for trains of different colors by German subjects (*left*) and Japanese subjects (*right*)

rate the red train at about 125%. This effect also holds true whether sounds and pictures from Japanese high-speed trains (Shinkansen) or Japanese commuter lines (Chuo line) are used.

The 10% difference between the ratings of the German and Japanese subjects in Fig. 8.3 could be due to variability in the measurements. The difference could also be due to cross-cultural differences. An experiment was designed to address this issue using another group of subjects from these rather different cultural backgrounds (Fastl et al. 2008). The method of semantic differential (Osgood et al. 1957) was used in which subjects rated different colors on adjective scales to obtain information on the meaning of the perception. Results show generally similar ratings from both groups. In particular, both German and Japanese subjects rated "red" as a "loud" color and "green" as a "soft" color.

The same effect has been shown for cars. For sports cars, it is somewhat expected that "red" is a "loud" color because for many sports cars (e.g., Ferrari) red is the preferred color. However, British sports cars frequently come in dark green. To assess whether dark green is an "appropriate" color for a sports car, the sound of a sports car was combined with a picture of an Aston Martin in red, light blue, light green, and dark green (Menzel et al. 2008b). As expected, not only "red" but also "dark green" qualified as "loud" colors (i.e., when sounds of a sports car were combined with pictures of a car in these colors, the perceived loudness was greater than for pictures with cars in light blue or light green).

8.4.3 Audio–Tactile Interactions and Loudness

Although there are relatively few studies on the interaction between loudness and the sense of touch, the data are fascinating. Jousmäki and Hari (1998) reported an interaction that they called the "parchment-skin illusion." They demonstrated that sound presented while subjects rubbed their hands together can strongly modify tactile sensation. In their experiment, recorded sounds of subjects rubbing their hands together were played back to the subjects at different levels while they engaged in rubbing their hands. When the level of the hand-rubbing sound was decreased from a comfortable loudness to soft, the skin started to feel less paper-like (i.e., the skin was reported to feel more smooth and moist).

Studies by Schürmann et al. (2004) and Gillmeister and Eimer (2007) indicate that the loudness of low-level sounds increases in the presence of vibrotactile stimulation. Current thinking on this facilitatory effect can be found in Chap. 3, Sect. 3.7.

It is likely that other audio–tactile interactions with loudness exist, but the authors are unaware of any controlled studies that measured loudness, not annoyance or perceived quality. In the absence of such data, it seems possible that vibrations could change the perception of loudness.

8.5 Cognitive Effects in Loudness Ratings

8.5.1 Complex Influences of Context on Loudness

In the laboratory, loudness changes in the context of other sounds (see Chap. 3). In daily environments, context may play an even greater role than in laboratory experiments. People have expectations about what they hear that are based on their perceived listening situation. This occurs in loudness, as well as in other perceptual modalities. Often the nature of these cognitive factors can be quite complex. An example of this type of complexity comes from studies on visual shielding of the noise source. Visual shielding is used as part of community noise abatement strategies.

Although visual barriers – such as fences – can reduce the sound level transmitted through them, barriers that provide no measurable reduction in sound level may be effective in reducing neighborhood noise for some residents (DeFrain 1973). Aylor and Marks (1976) designed an outdoor experiment to measure the loudness of noise transmitted through three barriers (acoustic tile, a slat fence, and hemlock) and no barrier. When listeners were blindfolded, they judged the loudness of a narrowband noise stimulus to be the same under all four conditions. When the listeners could see the barriers, there were sizable differences in loudness estimates. The presence of a barrier that partially obscured the sound source reduced loudness judgments to less than those obtained in the absence of a barrier. On the other hand, the presence of a barrier that *totally* obscured the sound source increased loudness judgments to *greater* than when the sound source could be seen completely or partially. In other words, the barriers that reduced the visibility of the sound source – without entirely eliminating it – reduced the apparent loudness. However, the barrier that totally obscured the sight of the sound source increased the apparent loudness. In an attempt to explain this unexpected result, the authors propose that the listeners' expectations about the effectiveness of the solid barrier influences loudness. It is well known that expectation can influence perception. This demonstrates the complex nature of loudness judgments in ecologically valid context.

Because of the complex nature of loudness judgments in the real world, it is recommended that loudness experiments be performed under the most realistic conditions possible. For example, the loudness of car noise could be evaluated using an apparatus similar to that used in the evaluation of acoustical comfort of vehicles. A "SoundCar" provides a listening environment that consists of a real vehicle cabin with authentic control instruments and equipped with acoustic and vibration simulators (e.g., Genuit 2008).

8.5.2 Loudness and Distance, Loudness Constancy, and Binaural Loudness Constancy

As distance increases between a listener and a sound source, the intensity of the sound at the ears of the listener decreases. In controlled laboratory conditions of an anechoic environment, it decreases 6 dB for every doubling of distance according to the inverse-square law. Daily environments are not anechoic and there are numerous possible cues that vary with distance from a listener, such as spectral shape, reverberations, binaural cues, and onset and offset cues. In addition, most waveforms of natural sounds are dynamically changing and have gradual onsets and offsets. The impact of these onsets and offsets on loudness has been described well in Schlauch (2004; see also Chap. 6, Sect. 6.3.2).

The physical correlate theory of perception predicts a link between loudness and distance of a sound based on the physical properties of sound and a listener's past experience (Warren 1981). This interesting theory has its critics and is not universally accepted; a discussion can be found in Schlauch (2004).

Loudness constancy refers to the phenomena by which loudness remains constant in the presence of substantial changes in the physical stimulus caused by varying sound distance. For example, conversational speech can remain constant even when the distance between the speaker and listener changes (Mohrmann 1939; Zahorik and Wightman 2001). It is noteworthy that loudness constancy is absent when level is the only cue available to the listener (Stevens and Guirao 1962). It is highly likely that the brain combines multiple cues, although how this is accomplished is not yet understood.

Binaural loudness summation refers to the finding that a sound presented binaurally is louder than the same sound presented monaurally. In 1933, Fletcher and Munson made the assumption that the binaural-to-monaural ratio was 2. In other words, a sound presented to two ears was twice as loud as a sound presented to only one ear. From 1960 until today, subsequent laboratory experiments have been performed using tones or noises presented via earphones. Results of these experiments show a binaural-to-monaural ratio ranging from about 1.3 to almost 2.0 (Marozeau et al. 2006; Epstein and Florentine 2009; Chap. 7). The inference was often made to explain how people perceived loudness under daily environmental conditions. It was frequently stated and written that a sound heard with two ears is almost two times as loud as a sound heard only with one ear.

When binaural loudness summation is tested with introspection experiments in daily environments – such as a classroom with a visually present talker – there is little loudness advantage when listening with two ears. An experiment by Epstein and Florentine (2009) indicates that (1) the amount of binaural loudness summation is significantly less for speech from a visually present talker than for recorded speech or tones, (2) the amount of binaural loudness summation is significantly less when sounds are presented via loudspeakers than when sounds are presented via earphones, and (3) the amount of binaural loudness summation is least for speech from a visually present talker presented via loudspeakers than any of their other test conditions. Some recent data suggest that there is less binaural loudness summation in ecologically valid listening situations than listening through commercially available earphones (Florentine and Epstein 2010). Why laboratory measurements of binaural loudness summation obtained using tones presented via earphones do not apply to binaural loudness summation outside the laboratory is an interesting question, and there are a number of possible answers. Cognitive factors may play a role, because a listener has learned that the sound a talker produces should not change whether listening with one or two ears. High-level processing in a listener's brain could account for this perception. Two likely cues for this response may be reverberation and spectral shape. The lack of a binaural loudness advantage in rooms has been called Binaural Loudness Constancy, because of its relation to a similar effect called Loudness Constancy. (For a review of binaural loudness, see Chap. 7.)

8.6 Optimal Loudness for Groups of People in Various Environments

8.6.1 Optimal Loudness for Music Halls

Setting sounds to optimal loudness requires an understanding of the many factors that influence perception. Previous chapters have explained how loudness changes with the type of sound, background sounds, and physical and psychological state of the listener. Loudness also changes with physical environment and architecture, as is so clearly experienced when comparing the same music produced in various concert halls and opera houses. This is because optimal loudness is essential for a full appreciation of the dynamics of music. The listener must be able to experience the full loudness range from piano pianissimo to forte fortissimo. Think "William Tell Overture" or Beethoven's Ninth Symphony! A fortissimo is possible only if the hall is not too large and there is a minimum of carpets, draperies, and a lack of heavily upholstered seats (Beranek 2004). In addition to the magnitude of the loudness range, the relative loudness of bass sounds is important when the full orchestra is playing, because it gives music its "warmth" (Beranek 2004). Warmth is an important subjective attribute of sound that is often described as the "richness in the bass."

There is a consensus among musicians that the loudness of both the early and reverberant sounds are essential for the optimal perceptual experience in a space used for listening to music. The early sound comes directly from the source or from early reflections within the first 80 ms. The reverberant portion of the sound occurs after the first 80 ms. Music sounds somewhat louder in a highly reverberant hall than in a dead hall (Beranek 2004).

Whereas it may be intuitive that the overall loudness of a performance should decrease as a listener moves away from the stage, this does not occur (Zahorik and Wightman 2001; Barron 2007; also see Beranek 2008). Of course, this should not be surprising after learning about binaural loudness constancy and loudness constancy described in Sect. 8.5.2.

The reader is referred to Beranek (2004) for more information on loudness in concert halls and opera houses, which is a comprehensive study of 100 of the world's halls for music. See also Fastl and Zwicker (2007).

8.6.2 Optimal Loudness in Background Noise

It is difficult to set optimal loudness in a quiet environment – such as a music hall, but it is even more challenging to obtain optimal loudness in noisy environments. Background sounds are almost always present in our daily environments (Thompson 2002; see Chap. 1). When background sounds are present, people raise their voices to be heard over the background sounds. As the overall level increases, so does the perceived overall loudness, but speech is soft against the background sounds due to partial masking of loudness (Scharf 1964). When this happens, people need to raise their voices even more. The increase in loudness continues to escalate and usually stops when it becomes too much effort for people to raise their voices over the noise. At this point in time, only young native speakers of the language with normal hearing can still communicate via speech. Background sounds can create communication problems, especially for people with hearing losses (Chap. 9), children (Nelson et al. 2002), older adults (Kim et al. 2006), and non-native speakers of a language (e.g., Mayo et al. 1997; Lecumberri and Cooke 2006). Public places are often too loud, and loudness has psychological and physical impact (Chap. 4).

Whereas problems arise when a sound is too loud, problems also arise when a sound is too soft. We may not be aware of a soft sound that we want to hear, because it may blend into background sounds in the environment. This can be dangerous. For example, sometimes hybrid vehicles are not heard early enough in the urban soundscape to avoid collisions between the hybrid vehicles and pedestrians (Kerber 2006). Simply increasing the level of sounds produced by hybrid vehicles is not a desirable solution, because this would increase the level of an already too loud urban soundscape (Schafer 1977). Therefore, warning signals have been proposed that would only be used when a vehicle is behind a pedestrian, similar to those used to prevent rear-end collisions (Kerber and Fastl 2007).

8.6.3 Sound-Level Meters vs. Sonemeters

Specific sound levels (e.g., 60 dB SPL) have little meaning to the general public with respect to loudness. For this reason, acoustic meters have been designed with numbers that are associated with specific sounds. Level meters give values in weighted sound levels. An example of a level meter can be seen at the left of Fig. 8.4. It shows the approximate sound-level weighting in dB(A) of well-known sounds (Fastl et al. 2006). On this level meter, a trickling faucet with 30 dB(A) is given as an example of a very soft sound and a jackhammer with a level in excess of 100 dB(A) as a very loud sound. The advantage of level meters is that they give quick psychological reference values that correspond to dB(A) values of generally known types of sounds. However, dB(A) values do not always give good estimates that correspond to average perceived loudness.

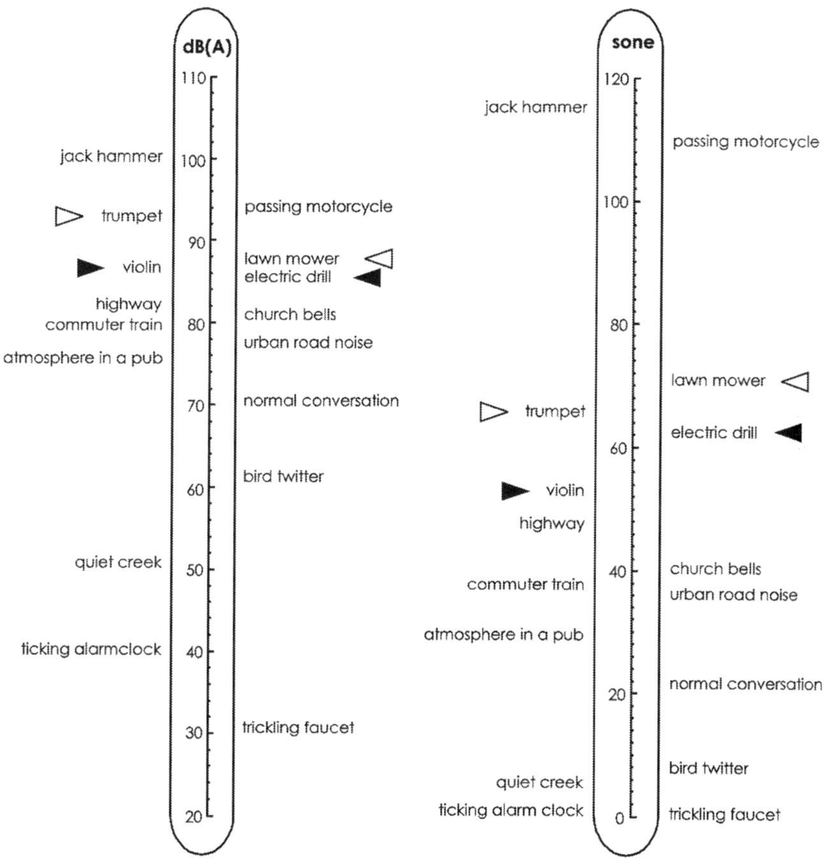

Fig. 8.4 A level meter (*left*) is shown in comparison to a sonemeter (*right*)

A recent improvement to the level meter is the sonemeter (aka loudness meter), which shows values that correspond more closely to perceived loudness for most people than do sound pressure levels. At the right of Fig. 8.4, a sonemeter is shown with the loudness of well-known sounds in sones, which is a unit of loudness described in Chap. 5. Of course such a device does not measure loudness *per se*, but it does give values that more closely correspond to the percept of loudness for most people. A sonemeter has at least two advantages over a level meter. First, loudness is displayed as perceived by the average, normal-hearing listener, taking into account spectral differences of sounds. Second, the sone scale allows for direct comparison of the loudness of different sounds (i.e., three times the sone value corresponds to about three times as loud).

8.6.4 Estimating Loudness for Groups of People

Different environments have different loudness requirements. Because it is not practical to ask a group of human listeners to estimate the loudness of sounds every time a loudness estimate is needed, attempts have been made to design meters that give a good estimate of how an average normal-hearing person perceives loudness. Therefore, sound-level meters were developed in an attempt to estimate the loudness of sounds. In the 1920s, the first sound-level meters with analog circuits took advantage of simple single-channel processing. A problem with single-channel processing is that it squeezes low, middle, and high frequencies into a single channel. This causes errors in the loudness estimations.

Attempts have been made to develop devices that give more accurate estimates of loudness based on psychoacoustic models of loudness. These devices more closely mirror the human auditory system and use multichannel processing that is easily engineered with current technology. These so-called "loudness meters" or "sonemeters" are improvements over sound-level meters, because they take into account differences in spectral composition that impact the perception of loudness (e.g., Fastl 2006a). For example, sounds of different timbre like a flute vs. a pipe organ at full register can differ by as much as 15 dB and have the same perceived loudness. This is because sounds with large bandwidths (e.g., an organ sound) have to be presented about 15 dB SPL lower than sounds with narrow bandwidths (e.g., a flute) to have the same perceived loudness. On the other hand, if sounds of different spectra are presented at the same sound pressure level, the corresponding perceived loudnesses can differ by as much as a factor of three (i.e., the sound with the broadband spectrum is perceived to be about three times as loud as the sound with the narrowband spectrum). Therefore, sonemeters have advantages when trying to assess loudness for groups of people compared to other spectral displays, such as FFT analyzers, 1/3-oct band spectra, wavelets, gammatone filters, etc. (for more, see Zwicker 1988, Fastl and Zwicker 2007).

Of course a sonemeter does not actually measure loudness, because loudness is a psychological percept. However, it does give values that more closely correspond to the percept of loudness in average normal-hearing people than other meters.

It should be noted that these meters might not provide accurate predictions for individual listeners or for all situations, but they are the best available devices for estimating optimal loudness for groups of people.

8.7 Summary

Scientists are beginning to bridge the gap between laboratory studies of loudness and our perception of loudness in the daily environments. In doing this, they face a host of difficulties in addition to issues related to language and culture. In practical situations loudness and annoyance are related, but are clearly different concepts. In the laboratory, the impact of the meaning of a sound can influence its annoyance, but may have little effect on its overall loudness. Anecdotal reports from noise consultants indicate that nonacoustic parameters can influence ratings of loudness and annoyance. The memory of the loudness of a sound may be different from its actual perception at the time the sound was heard; loudness memory is highly likely to be important in how people respond to sounds in daily environments. Music is one sound that can be loud, but not annoying. Perhaps this is why some people enjoy rock music at loudness levels that can be hazardous to hearing. Loudness gives music its dynamics and musicians in daily environments knew much about loudness before it was studied in any laboratory; they had a language notation for loudness and knew how to increase loudness without increasing level. Methods based on psychoacoustical data have been developed to circumvent regulations designed to limit the loudness of media broadcasts. It is suggested that sonemeters be used, instead of level meters, to set regulations. Finally, sensory and cognitive contexts of sounds were reviewed. Whereas multisensory interactions (i.e., audio–visual and audio–tactile) typically have a relatively small effect on loudness, they can play a dominant role with respect to annoyance, pleasantness and perceived quality. Cognitive effects of loudness can be quite complex, especially in daily environments.

References

American National Standards Institute (2005) ANSI S1.13 American National Standard Methods for the Measurement of Sound Pressure Levels in Air. New York: American National Standards Institute.

Aylor DE, Marks LE (1976) Perception of noise transmitted through barriers. J Acoust Soc Am 59:397–400.

Barron M (2007) When is a concert hall too quiet? In: Proceedings of the 19th International Congress on Acoustics, 2–7 September, 2007, Madrid, Paper RBA-06–006.

Beranek L (2004) Concert Halls and Opera Houses: Music, Acoustics, and Architecture, 2nd ed. New York: Springer.

Beranek L (2008) Concert hall acoustics–2008. J Audio Eng Soc 56:532–544.

Berglund B, Preis A (1997) Is perceived annoyance more subject-dependent than perceived loudness? Acustica 83:313–319.

Berglund B, Berglund U, Lindvall T (1976) Scaling loudness, noisiness, and annoyance of community noises. J Acoust Soc Am 60:1119–1125.
Chalupper J (2000) Aural Exciter and Loudness Maximizer: What's psychoacoustic about "psychoacoustic processors"? In: Proceedings of the 109th Audio Engineering Society Convention. New York: Audio Engineering Society, Reprint 5208.
Cullari S, Semanchick O (1989) Music preferences and perception of loudness. Percept Mot Skills 68(1):186.
DeFrain LE (1973) Highway noise abatement by design? Proc Natl Conf Noise Control Eng 15–17:99–104.
DIN 45 692 (2009) Messtechnische Simulation der Hörempfindung Schärfe, Beuth Verlag Berlin.
DIN 45 631/A1 (2008) Berechnung des Lautstärkepegels und der Lautheit aus dem Geräuschspektrum – Verfahren nach E. Zwicker – Änderung 1: Berechnung der Lautheit zeitvarianter Geräusche, Beuth Verlag Berlin.
Ellermeier W, Zeitler A, Fastl H (2004) Impact of source identifiability on perceived loudness. In: Proceedings of ICA 2004, 18. International Congress on Acoustics, Kyoto, Japan, volume II, pp. 1492–1494. Acoustic Science and Technology for Quality of Life, 4–9 April 2004. International Commission on Acoustics.
Epstein M, Florentine M (2009) Binaural loudness summation for speech and tones presented via earphones and loudspeakers. Ear Hear 30:234–237.
Fastl H (2001) Neutralizing the meaning of sound for sound quality evaluations. In: Proceedings of the International Congress on Acoustics ICA 2–7 September 2001, Rome, Italy, volume IV. Commission on Acoustics.
Fastl H (2004) Audio-visual interactions in loudness evaluation. In: Proceedings of the ICA 2004, 18. International Congress on Acoustics, Kyoto, Japan, Vol. II, pp. 1161–1166, 2004. "Acoustic Science and Technology for Quality of Life," 4–9 April 2004. International Commission on Acoustics.
Fastl H (2006a) Advanced procedures for psychoacoustic noise evaluation. In: Proceedings of the 6th European Conference on Noise Control EURONOISE 2006, Tampere, Finland. European Acoustics Association (EAA), 2006. CD-ROM Abstract: Acta Acustica united with Acustica 92(Suppl 1):10.
Fastl H (2006b) Psychoacoustic basis of sound quality evaluation and sound engineering. In: Eberhardtsteiner, Mang HA, Waubke H (eds), Proceedings of the 13th International Congress on Sound and Vibration ICSV13, Vienna, Austria.
Fastl H, Patsouras C (2004) The sound of colour. German Res 26:25–26.
Fastl H, Zwicker E (2007) Psychoacoustics - Facts and Models, 3rd ed. Berlin: Springer.
Fastl H, Menzel D, Krause M (2006) Loudness-thermometer: evidence for cognitive effects? In: Proceedings of the 35th Intern Congress and Exposition on Noise Control Engineering, Inter-Noise 2006, Honolulu, Hawaii. Institute of Noise Control Engineering of the USA, 3–6 December 2006.
Fastl H, Rader T, Van den Boogart G, Kuwano S, Namba S (2008) Cross-cultural comparison of colour evaluation using semantic differential. In: Fortschritte der Akustik - DAGA 2008, DEGA Berlin, pp. 385–386.
Fletcher H, Munson WA (1933) Loudness, its definition, measurement and calculation. J Acoust Soc Am 5:82–108.
Fligor BJ (2009) Risk for noise-induced hearing loss from use of portable media players: a summary of evidence through 2008. Perspect Audiol 5(1):10–20.
Florentine M, Epstein M (2010) Ecological loudness: binaural loudness constancy. In: Proceedings of the International Congress on Acoustics ICA, Sydney, Australia, 23–27 August 2010. Commission on Acoustics.
Florentine M, Hunter W, Robinson M, Ballou M, Buus S (1998) On the behavioral characteristics of loud-music listening. Ear Hear 19:420–427.
Fucci D, Harris D, Petrosino L, Banks M (1993) The effect of preference for rock music on magnitude estimation scaling behavior in young adults. Percept Mot Skills 76:1171–1176.

Fujiwara M, Aono S, Kuwano S (2006) Audio-visual interaction in the image evaluation of the environment – an on-site investigation. Inter-Noise 2006, Honolulu Hawaii, 3–6 December 2006.

Genuit K (2008) Vehicle interior noise – a combination of sound, vibration and interactivity. In: The 37th Congress & Exhibition on Noise Control Engineering, InterNoise 2008, Shanghai China, 26–29 October 2008.

Gillmeister H, Eimer M (2007) Tactile enhancement of auditory detection and perceived loudness. Brain Res 1160:58–68.

Hellman R (1982) Scaling loudness, noisiness, and annoyance of community noises. J Acoust Soc Am 60:1119–1125.

ISO (1999, 1990) Acoustics – Determination of occupational noise exposure and estimation of noise-induced hearing impairment. Geneva, Switzerland: International Organization for Standardization.

Jousmäki V, Hari R (1998) Parchment-skin illusion: sound-biased touch. Curr Biol 8:190.

Kerber S (2006) The importance of vehicle exterior noise levels in urban traffic for pedestrian – vehicle interaction. ATZ Automobiltech Z Worldwide 108:19–21.

Kerber S, Fastl H (2007) Perceptibility of approaching vehicles in urban background noise. In: Proceedings of the 36th International Congress on Noise Control Engineering Inter-Noise 2007, Istanbul, Turkey, 28–31 August 2007.

Kim SH, Frisina RD, Mapes FM, Hickman ED, Frisina DR (2006) Effect of age on binaural speech intelligibility in normal hearing adults. Speech Commun 48:591–597.

Kuwano S (1996) Continuous judgment of temporally fluctuating sounds. In: Fastl H, Kuwano S, Schick A (eds), Recent Trends in Hearing Research. Oldenburg: BIS, pp. 193–214.

Kuwano S, Namba S (1985) Continuous judgment of level-fluctuating sounds and the relationship between overall loudness and instantaneous loudness. Psychol Res 47:27–37.

Kuwano S, Namba S, Florentine M, Zheng DR, Hashimoto T (1992) Factor analysis of the timbre of noise – comparison of the data obtained in three different laboratories. Proc Acoust Soc Jpn N92–4–3:559–560.

Laumann K, Fastl H, Kuwano S, Namba S, Fujiwara M (2006) Loudness evaluation of various musical genres and types of listening behavior. In: Proceedings of the Spring Meeting of ASJ. Tokyo, Japan: Acoustic Society of Japan.

Laumann K, Fastl H, Kuwano S, Namba S (2007) Overall loudness versus average of instantaneous loudness for excerpts of music: effects of musical style. In: Fortschritte der Akustik – DAGA 2007, Berlin: DEGA, pp. 865–866.

Lecumberri MLG, Cooke M (2006) Effect of masker type on native and non-native consonant perception in noise. J Acoust Soc Am 119:2445–2454.

Luce DA (1975) Dynamic spectrum changes of orchestral instruments. J Audio Eng Soc 23:565–568.

Marozeau J, Epstein M, Florentine M, Daley B (2006) A test of the binaural equal-loudness-ratio hypothesis for tones. J Acoust Soc Am 120:3870–3877.

Mayo LFH, Florentine M, Buus S (1997) Age of second-language acquisition and perception of speech in noise. J Speech Lang Hear Res 40:686–693.

Menzel D, Faccinelli E, Fastl H (2008a) Untersuchung von Farbeinflüssen auf die Lautheit mit einem Einregelverfahren. In: Fortschritte der Akustik - DAGA 2008, Berlin: DEGA, pp. 383–384.

Menzel D, Fastl H, Graf R, Hellbrück J (2008b) Influence of vehicle color on loudness judgments. J Acoust Soc Am 123:2477–2479.

Miskiewicz A, Rakowski A (1994) Loudness level versus sound pressure level: a comparison of musical instruments. J Acoust Soc Am 96:3375–3379.

Mohrmann K (1939) Lautheitskonstanz im Entfernungswechsel. Z Psychol 145:146–199.

Moore BCJ, Glasberg BR, Stone MA (2003) Why are commercials so loud? Perception and modelling of the loudness of amplitude compressed speech. J Audio Eng Soc 51:1123–1132.

Namba S, Kuwano S, Kinoshita A, Hayakawa Y (1997) Psychological evaluation of noise in passenger cars – the effect of visual monitoring and the measurement of habituation. J Sound Vib 205:427–434.

Nelson PB, Soli SD, Seltz A (2002) Classroom Acoustics II: Acoustical Barriers to Learning. Melville, NY: Acoustical Society of America.
NIOSH (1998) Criteria for a recommended standard: occupational noise exposure, revised criteria. Publication No. 98–126 of the National Institute for Occupational Safety and Health, Cincinnati, OH.
Odgaard EC, Arieh Y, Marks LE (2004) Brighter noise: sensory enhancement of perceived loudness by concurrent visual stimulation. J Cogn Affect Behav Neuro 4:127–132.
Osgood CE, Suci G, Tannenbaum P (1957) The Measurement of Meaning. Urbana, IL: University of Illinois Press.
Patsouras Ch, Böhm M, Fastl H (2003) Beeinflussung des Lautheitsurteils durch schallfremde, stehende Bilder. In: Fortschritte der Akustik, DAGA 2003, Oldenburg: Dt. Gesell. für Akustik e V, pp. 616–617.
Pfeiffer T (1964) Ein neuer Lautstärkemesser. Acustica 14:162–167.
Portnuff CDF, Fligor BJ (2006) Sound output levels of the iPod and other MP3 players: is there potential risk to hearing? Paper presented at the NIHL in Children Meeting, Cincinnati, OH. Available at http://www.hearingconservation.org/docs/virtualPressRoom/portnuff.htm
Prince MM, Stayner LT, Smith RJ, Gilbert SJ (1997) A re-examination of risk estimates from the NIOSH Occupational Noise and Hearing Survey (ONHS). J Acoust Soc Am 101: 950–963.
Rader T, Morinaga M, Matsui T, Fastl H, Kuwano S, Namba S (2004) Crosscultural effects in audio-visual interactions. Transactions of the TC Noise and Vibration of Acoust Soc Jpn, N-2004–31.
Schafer RM (1977) The Tuning of the World. Toronto: Random House.
Scharf B (1964) Partial masking. Acustica 14:17–23.
Schlauch RS (2004) Loudness. In: Neuhoff JG (ed), Ecological Psychoacoustics. New York: Elsevier, pp. 318–345.
Schürmann M, Caetano G, Jousmäki V, Hari R (2004) Hands help hearing: facilitatory audiotactile interaction at low sound-intensity levels. J Acoust Soc Am 115:830–832.
Stevens SS, Guirao M (1962) Loudness, reciprocality, and partition scales. J Acoust Soc Am 34:1466–1471.
Suzuki Y, Abe K, Ozawa K, Sone T (2000) Factors for perceiving sound environments and the effects of visual and verbal information on these factors. In: Contributions to Psychological Acoustics: Eighth Oldenburg Symposium on Psychological Acoustics, pp. 209–232.
TA Lärm (1998) www.umweltbundesamt.de/laermprobleme/publikationen/talaerm.pdf.
Terhardt E (1968) Über akustische Rauhigkeit und Schwankungsstärke. Acustica 20:215–224.
Terhardt E (1985) Fourier transform of time signals: conceptual revision. Acustica 57:242–256.
Thompson E (2002) The Soundscape of Modernity. Cambridge, MA: MIT Press.
Todd NPM, Cody FW (2000) Vestibular responses to loud dance music: a physiological basis of the "rock and roll threshold"? J Acoust Soc Am 107:496–500.
Torick EL, Allen RG, Bauer BB (1968) Automatic control of loudness level. IEEE Trans Broadcasting BC-14(4):143–146.
von Bismarck G (1974) Sharpness as an attribute of the timbre of steady sounds. Acustica 30:159–172.
Ward LM (1987) Remembrance of sounds past: Memory and psychophysical scaling. J Exp Psych Hum Percept Perform 13:216–227.
Warren RM (1981) Measurement of sensory intensity. Behav Brain Sci 4:175–188.
Zahorik P, Wightman FL (2001) Loudness constancy with varying sound source distance. Nature Neurosci 4:78–83.
Zeitler A, Fastl H, Hellbrück J (2003) Einfluss der Bedeutung auf die Lautstärkebeurteilung von Umweltgeräuschen. In: Fortschritte der Akustik, DAGA 2003, Oldenburg: Dt. Gesell. für Akustik e. V., pp. 602–603.
Zwicker E (1959) Lautstärke und Lautheit. In: Proceedings of the 3rd International Congress on Acoustics, Stuttgart, 14–16 September 1959, pp. 63–78.

Zwicker E (1988) Loudness patterns (ISO 532 B), an excellent guide to noise-reduced design and to expected public reaction. In: Bolton JS (ed), Proceedings of the Noise-Control 88. New York: Noise Control Foundation, pp. 15–26.
Zwicker T, Buus S (1998) When bad amplification is good: distortion as an artistic tool for guitar players. J Acoust Soc Am 103:2797.
Zwicker E, Zwicker UT (1991) Audio engineering and psychoacoustics: matching signals to the final receiver, the human hearing system. J Audio Eng Soc 39:115–126.

Chapter 9
Loudness and Hearing Loss

Karolina Smeds and Arne Leijon

9.1 Introduction

A hearing loss affects many aspects of sound perception. The study of loudness in relation to hearing loss is scientifically important and interesting for several reasons.

- It is clinically interesting to understand the physiological reasons for the abnormal loudness perception that is common in people with hearing losses.
- Knowledge about individual loudness perception is central for the habilitation/ rehabilitation of people with impaired hearing. Hearing aids are designed and individually adjusted to compensate, as much as possible, for abnormal loudness perception.
- General knowledge about loudness perception can be gained by studying the effects of hearing loss on loudness perception.

A major consequence of a cochlear hearing loss, which is the most common type of hearing impairment, is closely related to loudness perception. Low-level sounds that are audible, but quiet, for a normal-hearing listener cannot be perceived at all by a person with a cochlear hearing loss, whereas high-level sounds that are perceived as very loud by a normal-hearing listener can have the same, or only slightly lower, loudness for a person with a cochlear hearing loss. This is illustrated in Fig. 9.1, where the results of a large number of uncomfortable and most comfortable loudness level measurements for pure tones at 0.5, 1, 2, and 4 kHz are plotted against hearing threshold levels (Pascoe 1988). It can be seen that the mean uncomfortable loudness levels are approximately the same for normal-hearing listeners (represented by a hearing threshold of 0 dB HL[1] in the graph) and for a

[1] A tone level expressed as x dB hearing level (HL) means that the level is x dB above an average of normal hearing listeners' hearing threshold at the tone frequency.

K. Smeds (✉)
ORCA Europe, Widex A/S, Maria Bangata 4, 118 63 Stockholm, Sweden
e-mail: karolina.smeds@orca-eu.info

M. Florentine et al. (eds.), *Loudness*, Springer Handbook of Auditory Research 37,
DOI 10.1007/978-1-4419-6712-1_9,

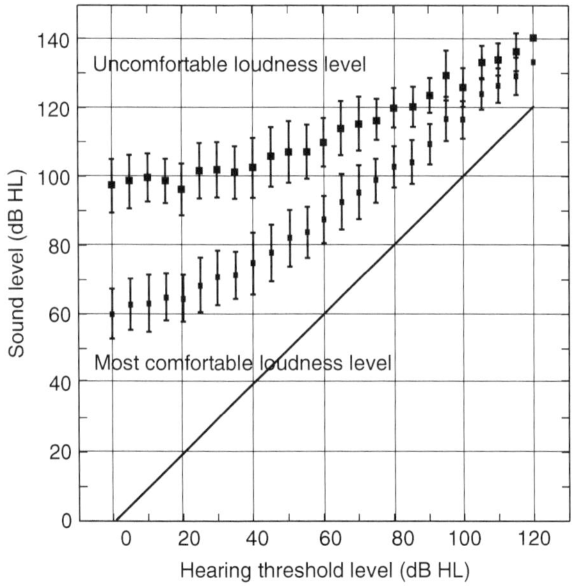

Fig. 9.1 Uncomfortable and most comfortable loudness levels, measured for 508 ears using pure tones at 0.5, 1, 2, and 4 kHz, are plotted against the HTL. Means (*dots*) and ±1 standard deviation (*bars*) are shown (redrawn from Fig. 4 in Pascoe (1988), with permission of GN ReSound)

person with a hearing loss up to about 40 dB HL. For greater hearing losses, the mean UCLs increase about 5 dB for every 10 dB increase in hearing loss. The range from threshold to uncomfortable loudness is compressed into a smaller dynamic range of hearing for people with a hearing loss than for normal-hearing listeners. Section 9.2 in this chapter first reviews the physiological basis of loudness formation, as an approach toward understanding the loudness perception effects of various types of hearing impairment. The clinically important problems with tinnitus and hyperacusis are briefly discussed from a loudness-perception point of view. Clinically useful loudness measurements are covered briefly in Sect. 9.3.

The reduced auditory dynamic range associated with most hearing impairments has consequences for the design and fitting of hearing devices. A hearing aid that provides the same amount of gain for low-level and high-level sounds can provide appropriate loudness only for one input level. If the gain is selected to give normal loudness for average-level speech, the hearing aid will provide too little gain to make low-level sounds audible, and high-level sounds will be amplified to uncomfortable loudness (Fig. 9.2, left panel). Instead, the hearing aid needs to provide more gain for low-level sounds than for high-level sounds to compensate for the hearing loss (Fig. 9.2, right panel). Loudness considerations in relation to hearing aids are covered in Sect. 9.4, and loudness aspects of cochlear implants used by people with severe or profound hearing loss are briefly reviewed in Sect. 9.5. The chapter is summarized in Sect. 9.6.

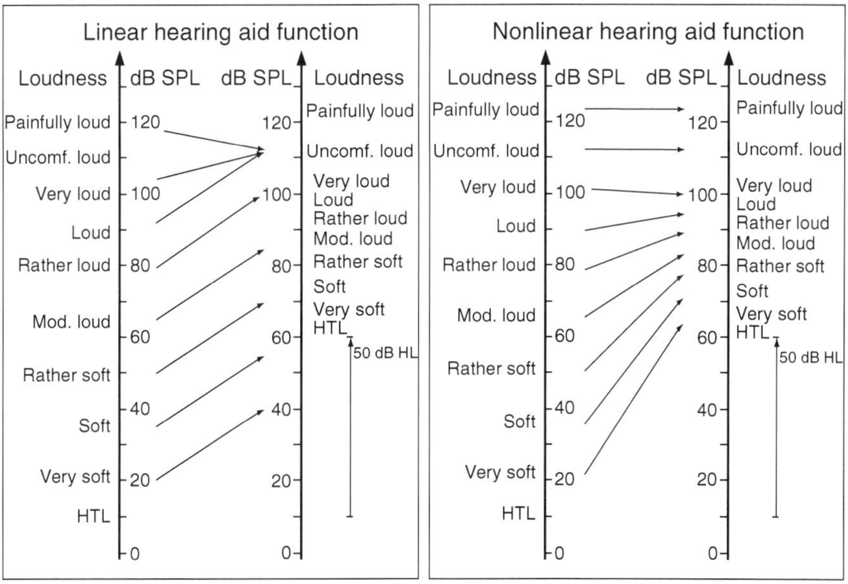

Fig. 9.2 The *left scale* in each panel shows schematically the relation between sound pressure levels and loudness for normal hearing. The *right scale* in each panel shows a hypothetical relation for a person with a 50 dB cochlear hearing loss. The *left panel* shows how a hearing aid with constant gain, a so-called linear hearing aid, works. The hearing aid is adjusted so that sounds that are judged to be moderately loud for the normal-hearing listener are also moderately loud for the hearing-aid user (approximately 20 dB gain). With this gain, sounds that are judged as soft by the normal-hearing listener are not audible for the hearing-aid user, whereas sounds that are judged as loud by the normal-hearing listener are judged as uncomfortably loud by the hearing-aid user. The *right panel* shows the situation for a hearing aid that changes its gain depending on the input level, a so-called nonlinear hearing aid. Here, the gain is set so that the loudness impression of the hearing-aid user is restored to normal; this means low-levels sounds need to be amplified approximately by 45 dB, whereas high-level sounds do not need to be amplified at all

9.2 Formation of Loudness: Normal and Impaired Hearing

A hearing loss can have many different causes. A crude distinction is usually made among *conductive* loss, affecting the sound transmission to the inner ear; *sensorineural* impairment, caused by a dysfunction in the cochlea or auditory neural pathways; and *central* impairment, involving changes in higher perceptual or cognitive functions. Sensorineural impairment is the most common cause of hearing loss and can include several different types of physiological damage. The elevation of the hearing threshold can be caused by a loss or dysfunction of outer hair cells (OHCs), a loss or dysfunction of inner hair cells (IHCs), impaired connections between neurons and IHCs, degeneration or dysfunction of auditory neurons, or by a combination of all these factors. These various types of physiological damage have different effects on loudness perception, as discussed in Sect. 9.2.2. Therefore, the loudness perception can vary greatly among people with hearing loss.

Most listeners with mild or moderate cochlear hearing losses show so-called complete recruitment of loudness, a term used here simply to describe that the listeners perceive high-level sounds as uncomfortably loud at nearly the same sound pressure level as normal-hearing listeners (Fig. 9.1). Listeners with more severe cochlear hearing losses usually show slightly elevated discomfort levels. This is, sometimes called incomplete recruitment of loudness. For a discussion about the effects of peripheral hearing loss on loudness perception, it is necessary to review briefly the sequence of signal transformations involved when an external sound wave is encoded into a stream of auditory-neural impulse patterns.

9.2.1 Peripheral Sound Transformations

Although the encoding of loudness in the auditory nerve is not yet entirely understood, as discussed in Sect. 9.2.2, the following basic physiological auditory functions are reasonably well established and understood, as discussed in detail by Dallos et al. (1996), Robles and Ruggero (2001), and Pickles (2008). Before reaching the brain, the sound is transformed and encoded in the following main steps.

- *External ear transformation*: Sound waves propagate from the sound field into the ear canal and reach the eardrum.
- *Middle ear transformation*: Eardrum vibrations are transferred mechanically to the stapes footplate.
- *Cochlear frequency analysis*: Sound components travel to different depths in the cochlear spiral and cause the highest vibration amplitudes at different places along the basilar membrane (BM) depending on their frequencies.
- *Cochlear compression*: Nonlinear effects of normally active OHCs reduce the range of vibration amplitudes in the cochlea.
- *Inner hair cell transduction*: Excitatory vibrations induce the release of neurotransmitter substances from IHCs to connected dendrites of spiral ganglion nerve cells.
- *Auditory nerve spike generation and transmission*: The stimulus is encoded as impulse patterns in nerve fibers emanating from cell nuclei in the spiral ganglion in the center of the cochlea.

A change in the physiological function of any of these steps will have consequences for the loudness perception of sounds. Each of these steps is discussed in the following sections.

9.2.1.1 External and Middle Ear Transformations

The transformation from the sound field to the eardrum includes wave diffraction around the head and resonances in the ear canal. This changes the power distribution among different frequency components in the sound. These effects can be described as a linear filter with a frequency response that depends on the direction of the sound source. The response has a broad peak around 2.5 kHz dominated by resonances in the ear canal and concha (Shaw 1974). The transformation of eardrum sound pressure signals to vibrations of the stapes can be approximated as another linear filter, with the best transmission in the mid-frequency range (Aibara et al. 2001; Pickles 2008). However, the middle ear muscle reflexes can reduce the transmission efficiency, mainly at low frequencies (Pickles 2008).

A conductive hearing loss reduces the effectiveness of the sound transmission through the external and middle ear. Assuming that loudness is determined mainly by the pattern of auditory nerve activity, it follows that the conductive loss, with a normal inner ear, causes the same reduction in loudness as a corresponding attenuation of the sound spectrum at the source. However, experience with hearing-aid fitting for people with conductive or mixed conductive and sensorineural hearing losses indicates that this might not be entirely accurate, as discussed in Sect. 9.4.4.

9.2.1.2 Cochlear Frequency Analysis

As a result of the mechanical properties of the cochlear structures and the active nonlinear feedback mechanism involving mainly the OHCs, complex sounds are filtered in the cochlea. The strongest response to high frequencies occurs near the basal end of the BM and the strongest response to low frequencies occurs in the apical part (Robles and Ruggero 2001). At the base of the cochlea, this filtering process is highly nonlinear, with sharper frequency resolution and higher gain at low input levels than at high input levels (Fig. 9.3). This nonlinearity is less pronounced near the apex. The frequency causing the strongest response at a given cochlear location is usually called the characteristic frequency of that location.

The frequency response of human cochlear filters can be estimated by psychoacoustic masking experiments (for reviews, see Buus 1997; Moore 2003). Modern estimates are based on tone detection in notched-noise maskers (Patterson 1976). The main characteristics of cochlear filtering can be quantified by the Equivalent Rectangular Bandwidth (ERB) of the filter passbands. The ERB of a band-pass filter is the bandwidth of an equivalent idealized filter with constant frequency response in the passband.

9.2.1.3 Cochlear Compression

If OHCs are functioning normally, a nonlinear feedback mechanism enhances low-level signal components within a narrow frequency range around the characteristic frequency at each location along the BM (Robles and Ruggero 2001). The nonlinearity sharpens cochlear frequency analysis and compresses the amplitude range of the signal (Fig. 9.3).

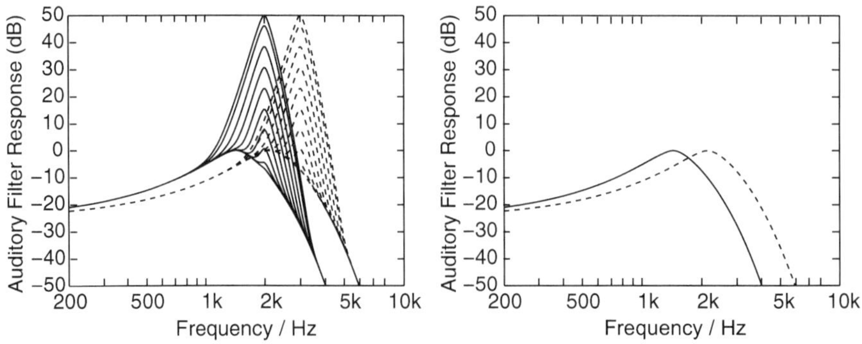

Fig. 9.3 Example of cochlear filter frequency responses (defined as the ratio between basilar membrane vibration amplitudes and input sound pressure amplitudes), for various input signal levels. (Schematically redrawn after Johnstone et al. (1986, Fig. 4) and Robles and Ruggero (2001, Fig. 13).) The decibel scale is arbitrarily chosen with 0 dB gain at high input levels. In the left panel nonlinear filter responses are displayed for a location with the characteristic frequency 2 kHz (*solid lines*), and at a more basal location with characteristic frequency 3 kHz (*dashed lines*). In the right panel, the remaining linear filter responses are displayed, assuming that the nonlinear active gain at low input levels is completely lost because of outer hair cell dysfunction. In this impaired ear, the 2-kHz tone level must be about 50 dB higher than normal to be detected, and the tone will be detected mainly by neural activity originating from cells that normally would be most sensitive at 3 kHz (dashed filter response)

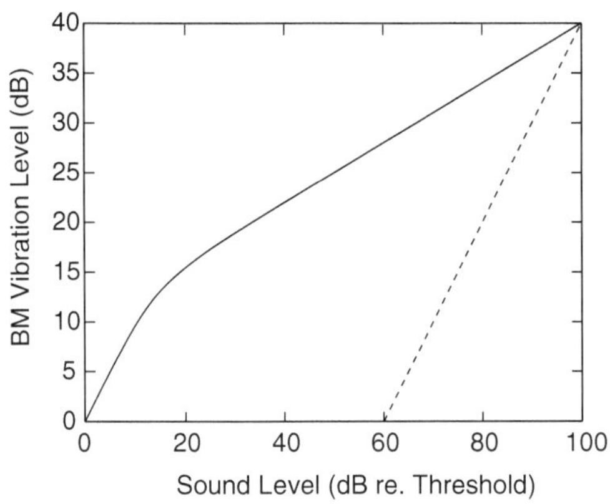

Fig. 9.4 Schematic input–output relation showing an example of basilar membrane vibration amplitude (in decibels with an arbitrary reference level) as a function of ear canal sound level (in decibels re threshold) for a pure tone at 2 kHz. In analogy with Fig. 9.3, the *solid line* indicates the normal compressed response at the location most sensitive to the tone frequency at low levels. The *dashed line* shows the linear response at the location that is normally most sensitive to 3 kHz at low levels, but most sensitive to 2 kHz when the cochlear compression is completely absent because of loss of outer hair cell function

As illustrated in Figs. 9.3 and 9.4, the normal nonlinear amplification is greatest for low-level signals and is gradually reduced for higher input levels. The maximal amplification provided by the nonlinear OHC function is 50–60 dB at low input levels (Robles and Ruggero 2001). Over a wide range of moderate input levels, higher than about 20 dB above threshold, the BM vibration response grows about 0.2–0.3 dB for every 1 dB increase in input sound pressure level (Robles and Ruggero 2001). Within this input-level range, the compression effect can thus be approximated by a power–law relation between input mean-square sound pressure amplitude and the mean-square amplitude of BM vibration. At low input levels, the filter gain remains constant at its maximal value, and the BM response grows about 1 dB for every 1 dB increase in input level (Robles and Ruggero 2001).

This nonlinear relation is fundamental for the growth of loudness with sound intensity and explains about half of the "dynamic-range problem" for intensity perception, discussed by Florentine in Chap. 1 and Epstein in Chap. 4. The physiologically measured power–law relation between mean square sound pressure and BM vibration amplitude is remarkably similar to the psychoacoustic power law often used to describe the relation between sound pressure and loudness at moderate levels. As BM velocity cannot be measured directly in humans, otoacoustic emissions have been used as indicators of BM velocity, showing good agreement with measures of perceived loudness (Buus and Florentine 2001; Epstein et al. 2006, also discussed by Epstein, Chap. 4). Direct vibration experiments in noise-exposed animals have also shown steeper-than-normal growth of BM vibration amplitude with sound intensity (Zhang and Zwislocki 1995). If the normal cochlear gain at low input levels is reduced by OHC dysfunction, the detection threshold is elevated and the growth of loudness will be markedly changed, except at levels near threshold where the OHC compression is not active in the normal ear. The loss of compression of BM vibration amplitudes also reduces the sharpness of the frequency selectivity of BM vibration (Fig. 9.3) (Robles and Ruggero 2001).

The normal cochlear compressor acts nearly instantaneously (Recio and Rhode 2000). In addition to the effects illustrated by the magnitude of frequency responses in Fig. 9.3, the phase response of the cochlear filters has another interesting consequence for loudness: For different complex sounds with identical magnitude spectra, the loudness can be different, because the phase relation between harmonics also influences the BM response. Mauermann and Hohmann (2007) found that an acoustic phase spectrum that produces a highly modulated signal envelope at the BM results in lower loudness than an input signal with identical power spectrum but a phase spectrum that produces less modulation at the BM. They found the effect to be most prominent for modulation rates of about 20–100 Hz. This loudness difference was found to be smaller for listeners with a cochlear hearing loss.

9.2.1.4 Inner Hair Cell Transduction and Neural Spike Generation

The mechanical motion at each longitudinal place along the BM causes an electrochemical response in the IHCs located at that place, and controls the generation of impulses in the auditory nerve cells connected to those IHCs. At frequencies below about 5 kHz, the spike rate varies in synchrony with the sound waveform (e.g., Pickles 2008). The neural representation of sounds can be severely altered by damage to the IHCs or the synapses or by damage to both IHCs and OHCs, for example, in noise-induced hearing loss (Heinz and Young 2004; Heinz et al. 2005a). A loss of synchronization probably disturbs pitch and timbre perception, and the loudness perception might also be affected (Heinz et al. 2005a, b).

Random spontaneous spike activity occurs in auditory nerve fibers in the absence of external sound. This spontaneous background activity limits the detectability of external sound at the absolute hearing threshold, but the spontaneous activity itself does not contribute to any loudness sensation. Morphologically different types of nerve fibers show different spontaneous rate (SR) and different response ranges (Liberman 1978). In a normal ear, the fibers with high SR tend to respond at lower sound levels than low-SR fibers. The spike rate in the most sensitive fibers saturates about 30 dB above threshold, but the activity in less sensitive fibers may still convey loudness information at higher sound levels (Pickles 2008). However, in animals with noise-induced hearing loss, the response ranges of these different fiber populations are overlapping, and the most sensitive medium-SR and low-SR fibers start responding at similar sound levels as the most sensitive high-SR fibers (Heinz et al. 2005a, Fig. 2). If these fiber populations contribute in different ways to the neural encoding of loudness, these physiological effects of the noise-induced hearing loss might also result in altered loudness perception.

9.2.1.5 Auditory Nerve Spike Transmission

Damage to the auditory nerve can affect many aspects of sound perception. An acoustic neuroma reduces the ability of auditory nerve fibers to convey impulse patterns to the brain. In these cases the hearing loss is often unilateral, and the hearing threshold gradually becomes severely elevated in the affected ear. Typically, there are no clinical signs of altered loudness growth, as measured, for instance, by binaural loudness balance tests (for reviews, see Johnson 1977; Brunt 2001). This seems understandable, as the normal compressive function of the cochlea is not affected by the neuroma. Nevertheless, inconsistent test results have been found in many cases and some patients with acoustic neuroma show signs of altered loudness perception (Johnson 1977).

Auditory neuropathy is usually defined clinically as absent or severely distorted auditory brain stem responses combined with normal otoacoustic emissions and cochlear microphonics. These symptoms indicate normal OHC function but abnormal transduction of the inner-ear vibrations into time-synchronous neural patterns, impaired auditory nerve transmission, or damaged functions in the brain stem auditory nuclei. The condition leads to impaired speech recognition, disproportional to the pure-tone audiogram. A literature review shows that auditory neuropathy has a prevalence as high as 8% of newly diagnosed children with hearing loss (Vlastarakos et al. 2008). The condition can probably be caused by several different pathologies that affect the auditory pathways. Common risk factors are neonatal jaundice and hypoxia. The pure-tone hearing thresholds range from normal to profound losses. Psychoacoustic tests indicate that the disrupted neural activity has minimal effects on intensity-related perception, such as intensity discrimination, frequency discrimination at high frequencies, and sound localization using interaural level differences. Instead, it impairs timing-related perception, such as frequency discrimination at low frequencies and temporal integration (Zeng et al. 2005). Thus, the perceptual consequences are the opposite of what is typically found in patients with cochlear impairment.

9.2.1.6 Physiological Interpretation of Altered Loudness Perception

The auditory intensity range is typically reduced to some extent in cases of cochlear hearing loss, regardless of the exact etiology (Miskolczy-Fodor 1960; Pascoe 1988; Hellman and Meiselman 1990, 1993; Kiessling 1995; Launer 1995; Moore and Glasberg 1997), as illustrated in Fig. 9.1. Measured loudness-growth functions can vary greatly across subjects, even if the hearing thresholds are similar, and at least some of this variability is most likely caused by different underlying physiological factors (Buus et al. 1999; Whilby et al. 2006). However, it is impossible to conclude from loudness-growth measurements alone the precise balance between different physiological causes of the impairment. Any discussion of the physiological cause of altered loudness-perception in humans must remain speculative to some extent. The following examples illustrate this difficulty.

Example 1: Assume OHC function is completely normal, and there is a total or nearly total loss of IHCs or afferent fiber connections along some part of the cochlea, as sometimes observed in animals (e.g., Schuknecht and Neff 1952). Such a condition, sometimes called a “dead region,” has consequences for loudness summation, frequency selectivity, and pitch perception (Florentine and Zwicker 1979; Florentine and Houtsma 1983; Florentine et al. 1997; Moore and Glasberg 1997; Moore et al. 2000). In such a case, any test tone normally sensed by cells in the dead region must instead be encoded in fibers innervating other parts of the cochlea, where IHCs and afferent neurons are still functioning. This would give a large threshold elevation for the test tone. Because the tone is perceived by nerve fibers from other locations, where auditory filters behave nearly linearly for the test-tone frequency (Fig. 9.3), the BM vibration amplitude level grows in a one-to-one relation to input sound level. Thus, the growth of loudness with tone level will be steeper than normal, but the total loudness might still not become normal even at high sound levels, because of the missing loudness contributions from the dead region.

Example 2: A loss of some fraction of IHCs or spiral ganglion cells across the whole auditory frequency range might cause only a minor threshold elevation, if OHCs are normal. However, the reduction of the number of independent information-carrying nerve fibers increases the relative variance of the neural spike patterns. This is equivalent to increased internal sensory noise, which reduces discrimination in all perceptual dimensions, including intensity discrimination. Because of the increased internal noise, a higher sound-induced nerve activity than normal might be needed in the remaining fibers to reach the statistical detection threshold for the sound. This can also lead to slightly higher loudness than normal near the threshold, and the loss of neurons might cause the recruitment to be incomplete.

Example 3: Because there are at least three different types of afferent nerve fibers (Liberman 1978), a selective loss or dysfunction of any of these fiber types might cause different changes in loudness perception. For example, the group of fibers with high spontaneous rate, normally having low thresholds, might be more vulnerable to damage than low-SR fibers that normally have much higher thresholds. If the high-SR fibers fail to respond to low-level test sounds, the threshold will be elevated. Further, if the activity in the low-SR fibers that normally code for higher loudness is still interpreted by the brain as indicating high loudness, the loudness at higher input levels might be close to normal; in other words, the recruitment would be complete. As the hearing threshold might be determined partly by activity in low-SR fibers in such a case, loudness at the threshold might also be higher than normal.

9.2.2 From Auditory Nerve Impulse Patterns to Perceived Loudness

It has long been hypothesized that loudness and intensity discrimination might be determined mainly by the overall rate of spikes in the auditory nerve (Fletcher and Munson 1933), but more recent data from cats with noise-induced hearing losses have shown that a loss of BM compression is not directly reflected in a correspondingly steeper growth of the spike rate of auditory-nerve fibers (Heinz et al. 2005a, b). However, some classes of neurons in the ventral cochlear nucleus do show steeper growth of spike rates with sound level after acoustic trauma (Cai et al. 2009; Joris 2009).

Although it is not quite clear how loudness is encoded in the auditory nerve, a large amount of psychoacoustical research has indicated that loudness perception seems to work as if the central nervous system somehow sums loudness contributions across tonotopically organized groups of auditory neurons (Fletcher and Steinberg 1924; Fletcher and Munson 1933; Zwicker 1958). This general concept is fundamental for the design and analysis of psychoacoustic experiments using loudness matching methods to estimate the loudness growth function (e.g., Buus et al. 1998; Buus and Florentine 2002). It is also used in several computational models that aim at predicting loudness from physical sound features by applying signal transformations that mimic some of the physiological facts discussed above (Fletcher and Steinberg 1924; Zwicker 1958; Scharf and Hellman 1966; ISO-532 1975; Florentine and Zwicker 1979; Florentine et al. 1997; Moore and Glasberg 2004; ANSI-S3.4 2007). Other models attempt to predict loudness without explicit reference to physiology, as reviewed by, for instance, Skovenborg and Nielsen (2004). Loudness models are reviewed by Marozeau, Chap. 10.

9.2.2.1 Loudness Summation Across Auditory Neurons

The total loudness of a sound is related to a sum of loudness contributions somehow encoded in the auditory nerve activity in fibers originating from different longitudinal places along the BM. Mathematically, this is most conveniently expressed as an integral (Fletcher and Steinberg 1924). In current loudness-summation models of the Fletcher–Zwicker tradition (Moore and Glasberg 2004; ANSI-S3.4 2007), the integrated quantity represents the loudness contribution per normal auditory ERB. This quantity is usually called *loudness density* or *specific loudness* (Zwicker 1958; Florentine and Zwicker 1979; Moore and Glasberg 2004), and is important for the discussion on hearing-aid fitting in Sect. 9.4.1.

One important effect of loudness summation is the fact that a sound with frequencies spanning across several auditory ERBs is louder than a narrowband sound with the same sound pressure level. The loudness-summation model predicts this loudness difference as a mathematical consequence of the compressed growth of loudness density as a function of the spectral power density of the sound. However, this bandwidth effect on loudness is absent or reduced in cases with cochlear hearing loss (e.g., Scharf and Hellman 1966), which is consistent with a steeper-than-normal growth of loudness density with sound spectral density.

9.2.2.2 Temporal Integration of Loudness

A very brief sound, with duration less than 100 ms, is less loud than a sound with the same amplitude but longer duration. The loudness remains relatively constant for durations greater than about 200 ms. Thus, it seems that loudness contributions are integrated over time, with a time constant on the order of 100–200 ms. The sound level difference required to achieve equal loudness for short and long sounds depends on the presentation level, and the difference is typically reduced in listeners with cochlear impairment (Florentine et al. 1988, 1996; Buus et al. 1997, 1999). By assuming that the ratio between the loudness of short and long tones at equal SPL is independent of the SPL, Florentine et al. (1996) and Buus et al. (1997, 1999) derived individual loudness functions from the observed temporal integration data. The results showed wide variations among listeners with hearing loss, which probably reflect different physiological causes of the impairment (Buus et al. 1999).

9.2.2.3 Binaural Loudness Summation

For a normal-hearing person, a binaurally presented sound is louder than the same sound presented monaurally, but the binaural loudness might be slightly less than the sum of the monaural loudness values. For an average listener with normal hearing, the binaural level difference required for equal loudness of binaural and monaural tones is about 3 dB near the threshold of hearing, up to about 10 dB at comfortable listening levels, and about 5 dB at very high presentation levels

(Whilby et al. 2006). The level differences are typically smaller than normal if the listener has a cochlear hearing loss. The maximal level difference for equal loudness is about 5 dB in listeners with cochlear impairment (Whilby et al. 2006; Marozeau and Florentine 2009). By assuming that the ratio between the loudnesses of equal-level binaural and monaural tones is independent of level, Whilby et al. (2006) and Marozeau and Florentine (2009) derived monaural and binaural loudness functions from the measured binaural summation results. The individual loudness functions varied widely among listeners with a cochlear hearing loss.

9.2.2.4 Loudness near the Hearing Threshold

There is always some spontaneous activity in the auditory nerve, even without an external sound stimulus. Therefore, there exists no definite "threshold" level below which an external sound cannot be perceived at all. A detection threshold can be defined only statistically as the sound level that causes a sufficiently large change in the statistical distribution of neural activity to result in detection with the required degree of probability. Early loudness models (e.g., Zwicker 1958) assigned exactly zero loudness to any sound below the detection threshold. However, if a subthreshold pure tone is presented together with other subthreshold tones, the combined complex sound might be easily detectable and would then also have a subjective loudness clearly greater than zero (e.g., Buus et al. 1998). Therefore, each tone component in the complex sound must contribute greater-than-zero loudness density even at levels a few decibels below threshold. The *detectability* of a sound must, by definition, be the same at the detection threshold, regardless of hearing impairment. However, this does not necessarily imply that the *loudness* at threshold is the same in all listeners.

The function relating log(loudness) to input level has approximately the same slope near threshold in both normal and impaired hearing (Buus and Florentine 2002; Moore and Glasberg 2004; Marozeau and Florentine 2007). This is the expected result if the hearing loss is caused mainly by a loss of OHC compression (Fig. 9.4), because this would change the shape of the loudness-growth function only at higher input levels.

If loudness grows at a normal rate near threshold, how does the loudness catch up to reach normal or near normal values at very high levels, as indicated by the data in Fig. 9.1? Buus and Florentine (2002) fitted a loudness-summation model to individual results of loudness matching between single pure tones and tone complexes with component levels between −3 and 20 dB SL.[2] The fitted model parameters indicated that loudness was higher than normal already at the threshold of impaired hearing, a phenomenon they called softness imperception. Contrary to these results, Moore (2004) presented loudness-matching data that were not consistent with this concept. He concluded instead

[2] A tone level expressed as x dB sensation level (SL) means that the level is x dB above the listener's hearing threshold at the tone frequency.

that the main cause of loudness recruitment is a steeper-than-normal loudness growth at moderate and high sound levels. As discussed in Sect. 9.2.1, if a hearing loss is caused mainly by a loss of OHC compression, loudness would be expected to show rapid growth at moderate sound levels and reach normal values at high sound levels. However, individual differences in loudness growth might indicate different physiological causes of the hearing loss (Buus et al. 1999; Marozeau and Florentine 2007).

9.2.3 Tinnitus

Tinnitus is the sensation of a sound without external cause. This can be perceived as very disturbing by some people. A review of epidemiological studies (Andersson et al. 2005) shows that as many as 40–50% of the population can notice tinnitus in very quiet environments, and about 10–15% of the population are bothered by persistent tinnitus. About 1–2% of the population suffer from "severely annoying" persistent tinnitus, and 0.5–1% report severe effects of tinnitus on their quality of life. Sleep disturbance is the dominant complaint. Tinnitus can be generated by several different physiological mechanisms, at all levels of the auditory pathways. Tinnitus is often associated with hearing loss, but it can also occur without significant hearing threshold elevation (Andersson et al. 2005).

Many attempts have been made to measure the loudness of tinnitus, using either loudness matching or some form of loudness magnitude estimation, as reviewed by Henry and Meikle (2000). Already Fowler (1942) noted that tinnitus loudness was often equal to the loudness of an external tone at only 5 or 10 dB SL. Several later results of loudness matching also show tinnitus to be matched at rather low sensation levels (Reed 1960; Hallam et al. 1985; Jakes et al. 1986; Newman et al. 1994). For a majority of 82 patients in one study, tinnitus was matched to a sound of considerably lower intensity than a sound of comfortable loudness (Hallam et al. 1985). Typically, tinnitus is matched to low levels in decibel SL at frequencies where the hearing loss is large. This is probably related to the altered loudness perception associated with cochlear hearing loss. If loudness grows more steeply with level than normal, a lower sensation level than normal is needed to reach a given loudness value (Tyler and Conrad-Armes 1983; Penner 1986; Tyler 2000).

Attempts have been made to transform the loudness-matching results in decibel SL into other scales designed to represent the tinnitus loudness (Henry and Meikle 2000). However, even with these improved measures that attempt to include effects of the altered loudness perception, the correlation between estimated loudness and psychological measures of tinnitus severity remained low (Henry and Meikle 2000). Hinchcliffe and Chambers (1983) proposed a transformation into "personal loudness units" (PLU), where 1 PLU was defined individually as the most comfortable loudness. The use of an individual PLU scale for tinnitus loudness improved the correlation with subjective suffering. However, the transformation required that the listener produced consistently rank-ordered loudness judgments, and up to 30%

of the participants in one study had to be excluded from the analysis because they had difficulty performing this task (Hallam et al. 1985).

Further, the *annoyance* caused by the tinnitus sound is not necessarily related to the loudness (e.g., Hallam et al. 1985; Newman et al. 1994; Hiller and Goebel 2006, 2007). The severity of tinnitus suffering is strongly related to other psychological dimensions, such as anxiety and depression (Andersson et al. 2005; Hiller and Goebel 2007). Thus, tinnitus loudness and annoyance need to be assessed separately (Hiller and Goebel 2007).

9.2.4 Hyperacusis

Hyperacusis has been defined as "consistently exaggerated or inappropriate responses to sounds that are neither threatening nor uncomfortably loud to a typical person" (Klein et al. 1990). This definition implies that even commonly occurring environmental sounds at moderate levels can elicit strong negative reactions. Hyperacusis is different from the reduced auditory intensity range usually associated with cochlear hearing loss (Fig. 9.1). However, as loudness discomfort measurements show a high within-subject and between-subject variability, there is no consensus on where to draw a line between hyperacusis and the common loudness recruitment. The prevalence varies depending on how hyperacusis is defined (Baguley and Andersson 2007).

The term "phonophobia" is sometimes used to emphasize a strong emotional reaction involving fear of sounds. Abnormal sensitivity to sound has been observed in association with posttraumatic stress syndrome and depression (Andersson et al. 2005). Stansfeld et al. (1985) suggested that 40–50% of highly noise-sensitive people had a recognizable psychiatric disorder. Hyperacusis sometimes disappears when psychological well-being returns to normal (Andersson et al. 2005).

9.3 Exploring the Auditory Dynamic Range

The auditory dynamic range is the usable intensity range of hearing, limited by the hearing threshold level (HTL) and the uncomfortable loudness level (UCL). Roughly midway between the HTL and the UCL is a range of comfortable loudness levels. The most comfortable loudness level (MCL) lies within this range. As described in the preceding text, the auditory dynamic range is reduced for listeners with a cochlear hearing loss. For this reason, various measures of the dynamic range have been used when diagnosing a hearing loss. Knowledge about the auditory dynamic range, measured or estimated, is also important for the rehabilitation of a person with a hearing loss. Several chapters in this book have already covered loudness measurements (Chaps. 2 and 3) and measurements that are correlated with loudness (Chap. 4). The focus in this section will be on measurements that are used for clinical purposes. Generally, psychoacoustic loudness measurements are the most commonly used, and this is true also in the clinical setting. Other methods are mainly used for listeners who cannot participate in psychoacoustic testing, such as infants.

9.3.1 *Psychoacoustic Measurements*

9.3.1.1 Threshold, Most Comfortable Loudness Level, and Uncomfortable Loudness Level

The most important measure of hearing is the HTL. It is a measure of detectability rather than loudness, and constitutes the lower limit of the auditory dynamic range. The second most commonly used measure of the auditory dynamic range is the UCL (e.g., Hawkins et al. 1987). The measure is often called the loudness discomfort level (LDL), sometimes the threshold of discomfort (TD), or (a slightly different measure) the upper limit of comfortable listening (ULCL). The auditory dynamic range, determined as the difference between the UCL and the HTL, is approximately 100 dB for normal-hearing listeners, and it is considerably reduced for listeners with a cochlear hearing loss (Fig. 9.1 and Sect. 9.2.2). UCL measurements have been used in hearing-aid fitting. Most clinicians agree that the maximum output of a hearing aid should not exceed the hearing-aid user's UCL (e.g., Cox 1983; Hawkins et al. 1987; Sammeth et al. 1989). However, because the UCL measurements are influenced by factors such as the psychophysical procedure used and the instructions to the listener, it is somewhat unclear which method should be used to determine the maximum output (Hawkins et al. 1992). This is further discussed in Sect. 9.4.1.2.

The MCL roughly bisects the auditory dynamic range for listeners with a cochlear hearing loss (Fig. 9.1). Measured or estimated MCLs have been used for selecting the gain–frequency response of hearing aids (e.g., Braida et al. 1979; Byrne 1986; Cox 1989; Cornelisse et al. 1995).

9.3.1.2 Loudness Scaling

The previously mentioned psychoacoustic measures together constitute a step toward mapping the whole auditory dynamic range with a more detailed individual loudness-growth curve. Loudness scaling can be performed in a large number of ways (Chap. 2). Clinically, the most commonly used scaling method is categorical loudness scaling, where categories with loudness labels ranging from for instance "very soft" to "uncomfortably loud" are used (e.g., Allen et al. 1990; Launer 1995; Kiessling et al. 1996; Cox et al. 1997; Keidser et al. 1999; Brand and Hohmann 2002). The results of loudness scaling have mainly been used as input to hearing aid prescriptions for loudness normalization (Sect. 9.4.1), and to evaluate hearing-aid fittings.

9.3.1.3 Loudness Balancing

Monaural or binaural loudness balancing techniques used to be a part of a test battery that aimed at determining if a sensorineural hearing loss had cochlear or retrocochlear origin. In the case of a unilateral hearing loss, the alternate binaural loudness balance test (ABLB) can be used to compare the loudness growth in the two ears (Fowler 1936). The level of one tone is held constant and the level of the other tone is adjusted until both tones are equally loud. Figure 9.5 shows a large number of ABLB results for people with cochlear hearing loss, selected to have so-called complete recruitment (Miskolczy-Fodor 1960). Large differences between the presentation levels in the two ears are necessary for equal loudness at low presentation levels, whereas only small differences are needed at high presentation levels.

In the case of a symmetric hearing loss, the balancing can be performed in one ear comparing two frequencies, one with relatively normal and one with elevated thresholds (Reger 1936). Figure 9.6 shows examples of results of monaural loudness balancing for listeners with normal and impaired hearing (Barfod 1978). For the purpose of clinical diagnosis of retrocochlear lesions, the sensitivity of loudness-balance tests appears to be rather low, around 50%

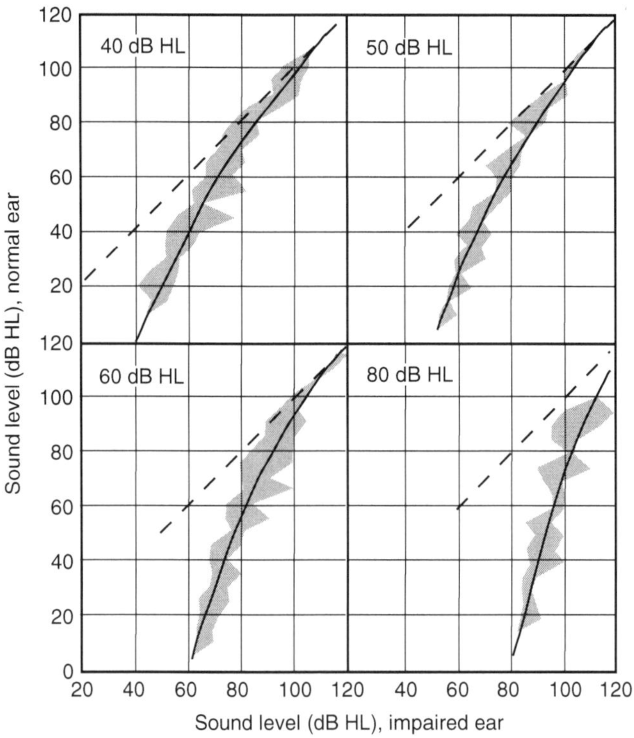

Fig. 9.5 Binaural loudness balance results with pure tones for people with cochlear hearing loss showing complete recruitment (about 300 loudness balance tests). The *four panels* show sound levels at equal loudness for one impaired and one normal ear for hearing losses of 40, 50, 60, and 80 dB HL respectively (schematically redrawn from Fig. 3, Miskolczy-Fodor (1960), with permission from the Acoustical Society of America). The individual balance results are here indicated only by *shaded areas*, where maximum and minimum values for the impaired ears (*horizontally*) are indicated for each 5 dB increment for the normal ears (*vertically*), thus showing the range of individual loudness-balance results

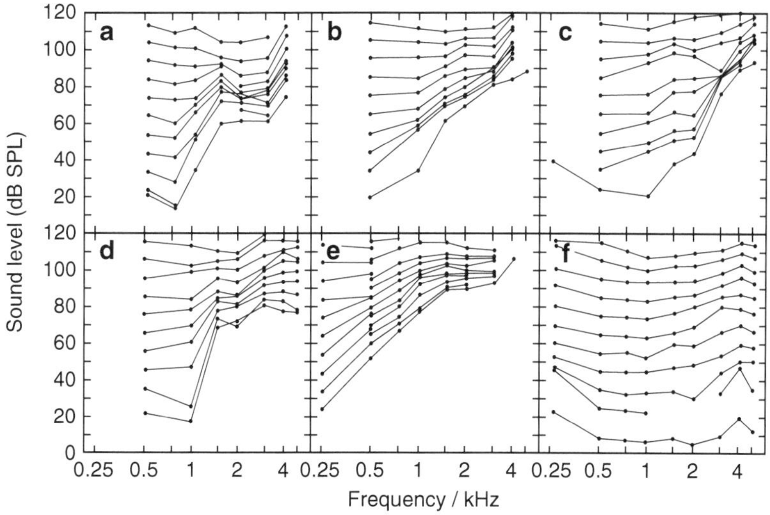

Fig. 9.6 Monaural loudness balancing results (iso-loudness contours) for five listeners with hearing loss (**a–e**) and the mean results for three normal-hearing listeners (**f**). In all cases, the 500-Hz tone acted as the reference for the comparisons (redrawn from Fig. 2, Barfod (1978), with permission of GN ReSound)

(Johnson 1977). This is the main reason why these methods have been replaced by more sensitive diagnostic methods in the clinic. For research purposes, loudness balancing has proved to be a useful tool for exploring the auditory dynamic range (e.g., Figs. 9.5 and 9.6).

9.3.1.4 Measurements for the Pediatric Population

For infants and children, a range of behavioral techniques can be used, mainly to determine the threshold of hearing. These techniques include unconditioned and conditioned behavioral observation audiometry (Ewing and Ewing 1944; Hodgson 2001), visual reinforcement audiometry (Lidén and Kankkunen 1969; Moore et al. 1992), and play audiometry (Dix and Hallpike 1947; Hodgson 2001). As these methods were developed mainly for estimating the hearing threshold, they generally do not provide suprathreshold loudness information.

9.3.2 Otoacoustic Emissions

Otoacoustic emissions (described by Epstein, Chap. 4) are used clinically for hearing threshold screening, particularly for infants (e.g., How and Lutman 2007).

Otoacoustic emissions, measured as a function of level, show correlation with psychoacoustically measured loudness (reviewed by Epstein, Chap. 4).

9.3.3 Electrophysiological Measurements

Electrophysiological tests (described by Epstein, Chap. 4), the auditory brain stem response in particular, are routinely used clinically for instance when examining infants who do not pass hearing screening with otoacoustic emissions (e.g., Hergils 2007) and for adults with a unilateral hearing loss where the clinician wants to investigate if the hearing loss is of cochlear or retrocochlear origin (e.g., Don et al. 2005). In a number of studies, the results of various electrophysiological measurements have shown a correlation with suprathreshold loudness (Eggermont 1977; Thornton et al. 1987; Davidson et al. 1990; Picton et al. 2005; Castro et al. 2008; Menard et al. 2008).

9.3.4 Impedance Measurements

In a meta-analysis of the relation between UCL and acoustic reflex thresholds (ART), Olsen (1999) concluded that a statistically significant correlation exists between the mean HTL, the mean ART, and the mean UCL in impaired hearing, but not in normal hearing. The mean difference between ART and UCL was rather small, 5 dB, but it is an open question whether ARTs can be reliably used for predicting UCLs on an individual basis, because of high between-subject variability (e.g., Olsen 1999; Rawool 2001). Electrically evoked ART have been used when programming cochlear implants to determine a suitable upper limit of electrical stimulation (Gordon et al. 2004; Brickley et al. 2005).

9.3.5 Model-Based Prediction of Loudness

Suprathreshold loudness considerations are central in several applications (Chap. 8). The only way to obtain direct knowledge about an individual's loudness perception is, of course, to perform some form of loudness measurement. However, for many applications, model-based predictions of loudness are important. In hearing-aid fitting, it is still debated if methods that are based on suprathreshold loudness measurements give better results than those that rely on threshold data combined with statistical relations between threshold and suprathreshold measurements (like in Fig. 9.1) or a loudness model where only hearing threshold data is entered. This is further discussed in Sect. 9.4.1.2.

9.4 Loudness and Hearing-Aid Fitting

One of the most prominent features of a cochlear hearing loss is that the cochlear compression associated with normal hearing is reduced (Sect. 9.2). Hearing aids aim at restoring some of the missing nonlinearity by giving low-level input signals higher gain than high-level input signals (Fig. 9.2), that is, the hearing aids provide nonlinear amplification. Dillon (2001) gives a comprehensive description of hearing aids and hearing-aid fitting. This section deals mainly with loudness aspects of hearing-aid fitting for adult hearing-aid users with cochlear hearing losses. A short section on loudness aspects of hearing-aid fitting for people with conductive or mixed hearing losses is found at the end of the section.

9.4.1 Loudness-Based Hearing-Aid Fitting Principles

When an appropriate hearing aid has been selected, the aid needs to be adjusted to fit the hearing-aid user. The initial part of this process is most often carried out in a prescriptive manner (Braida et al. 1979; Byrne 1983). A formula, or "prescription," is used to give the hearing aid its preliminary settings. The aim of the prescription is to suggest gain characteristics that are correct for the average hearing-aid user with the stated hearing loss (or other characteristics that might enter the prescription). This preliminary hearing-aid setting is then evaluated, preferably in the user's home environment. The hearing aid can then be fine-tuned to account for individual loudness growth and individual preferences.

A prescriptive approach to hearing-aid fitting is advocated for the purely practical reason that a large number of hearing aids are available, each with a very large number of potential settings. It would be impossible to evaluate them all to find "the best" hearing aid setting. The prescriptive approach also has another, more theoretical advantage: a consequent use of a prescription based on an explicit theory will increase our knowledge. As formulated by Byrne (1982):

> I want to argue strongly for the adoption of procedures for which the theoretical bases are explicitly stated and therefore accessible to critical examination. This is essential to justify the use of such procedures and to permit the development of better procedures.

Prescriptive methods are based on different rationales. All prescriptions are, to some extent, based on loudness considerations (Braida et al. 1979; Dillon 2001).

9.4.1.1 Rationales for Prescribing Hearing-Aid Gain

A large number of prescriptions are based on *frequency-specific loudness normalization*. The basic idea of this approach is that the perceived loudness and timbre of a sound should be restored to be approximately normal for the hearing-aid user. This is done in a frequency-specific manner so that a broadband sound that is dominated by low-frequency loudness for a normal-hearing person will be dominated by low-frequency loudness also for the hearing-aid user.

A well-defined way to achieve frequency-specific loudness normalization is to apply *loudness density normalization*. As described in Sect. 9.2.2.1, the total or overall loudness of a sound is the integral of the loudness density across frequency. Loudness density normalization implies that each auditory filter band of an input signal should be amplified to restore normal loudness density. This rationale has been described by Villchur (1973) and implemented for instance by Lippmann et al. (1981), Leijon (1991), and Moore (2000). If loudness density normalization is really achieved, then, by the definition of loudness density, the overall loudness of the sound will also be normal. To achieve loudness density normalization over the whole frequency range of speech, the hearing aid needs to have many independent compression channels (with a bandwidth of the order of the auditory filter bandwidth). Other implementations of frequency-specific loudness normalization are based on published data on loudness perception at various input levels (e.g., Killion and Fikret-Pasa 1993) or on individual loudness-growth measurements for narrow-band signals (e.g., Allen et al. 1990; Cox et al. 1997; Valente and Van Vliet 1997).

Another hearing-aid fitting rationale aims at *maximizing speech intelligibility*. Prescriptions based on this rationale allow deviations from normal frequency-specific loudness with the purpose of maximizing speech intelligibility (Braida et al. 1979). The rationale, which in itself does not include a loudness aspect, is often accompanied by a loudness constraint (Radley et al. 1947 cited by Braida et al. 1979; Leijon et al. 1991). If a loudness constraint is not used, very high gain values will be prescribed for low-level speech in a quiet environment. NAL-NL1 (Byrne et al. 2001) is the most well known prescription based on this rationale. This method maximizes the speech intelligibility index (SII, ANSI-S3.5 1997; Ching et al. 2001) for speech in quiet with a constraint that the calculated overall loudness, using a loudness model of Moore and Glasberg (1997), should not exceed normal overall loudness.

Another approach that is focused on speech intelligibility is *loudness density equalization* (Byrne and Tonisson 1976; Byrne and Dillon 1986; Moore et al. 1999). This approach states that the loudness density should be roughly constant across frequency after amplification. In practice it means that also for a normal-hearing person some amplification will be given to low-level high-frequency speech sounds, whereas high-level low-frequency speech sounds will be attenuated to keep the overall loudness normal.

9.4.1.2 Model-Based or Measurement-Based Prescriptions

The prescriptive methods described above are all, to some extent, based on loudness. They are, however, different in how they treat individual loudness data. Consistent with the individual differences seen in loudness-growth data (e.g., Buus et al. 1999; Whilby et al. 2006; Marozeau and Florentine 2009), some methods require measurement of individual loudness-growth functions to tailor the amplification to a hearing-aid user's particular loudness growth (e.g., Allen et al. 1990; Cox et al. 1997; Valente and Van Vliet 1997). However, most of the current prescriptions are threshold based, which means that the only hearing information that enters the prescription is the hearing-aid user's pure-tone detection thresholds. This approach assumes a relation between the threshold of hearing and supra-threshold hearing ability, described for instance in a theoretical loudness model (e.g., Moore and Glasberg 1997) or in statistical data sets like the one presented in Fig. 9.1 (Pascoe 1988). People who advocate this strategy argue that because the hearing aids must be individually fine-tuned anyway, it is more valid and time-efficient to verify, while using the aids, that amplified sound is comfortably loud and that the instruments cannot produce uncomfortably loud sounds.

Elberling (1999) argued that individual categorical loudness scaling has little practical value for clinical prescription of hearing-aid gain–frequency responses. He found that the inverse slope of the loudness function for listeners with impaired hearing generally varied linearly with the hearing threshold, and he suggested that individual differences could be treated in a subsequent fine-tuning process. He also showed that among normal-hearing listeners, there was a very large intersubject variability in the scaling results. Thus, when individual loudness scaling data are used to derive a prescription to achieve "normal" loudness growth, the "normal" reference is not well defined. Dillon and Storey (1998) proposed a procedure to prescribe initial maximum output level settings of hearing aids based only on hearing threshold data. Storey et al. (1998) showed that including individual UCL data into the prescription increased the accuracy by so little that it would not be worthwhile performing the measurements. They concluded that it is still essential to verify, and possibly adjust, the maximum output level setting during the fine-tuning process.

9.4.2 *Prescribed and Preferred Loudness*

Despite the fact that many prescriptions are based on frequency-specific loudness normalization, they differ in the gain–frequency response they prescribe. This is because various loudness data can be used to derive the prescription. When prescriptions based on other rationales are included in the comparison, the prescribed gain differs substantially among prescriptions. An example of the gain prescribed by five generic threshold-based prescriptions for a mild-to-moderate gently sloping cochlear hearing loss is given in Fig. 9.7. If the large number of hearing-aid-specific prescriptions, which are available in hearing aid manufacturers' fitting software, are included in the comparison, the range of measured gain–frequency responses increases even further (Smeds and Leijon 2001). For all input levels used in the study by Smeds and Leijon (speech-like stimuli at 55, 65, and 76 dB SPL), the difference in prescribed gain was substantial, 10–20 dB across frequency. Theoretical estimates of loudness and speech intelligibility for the measured gain curves suggested that the

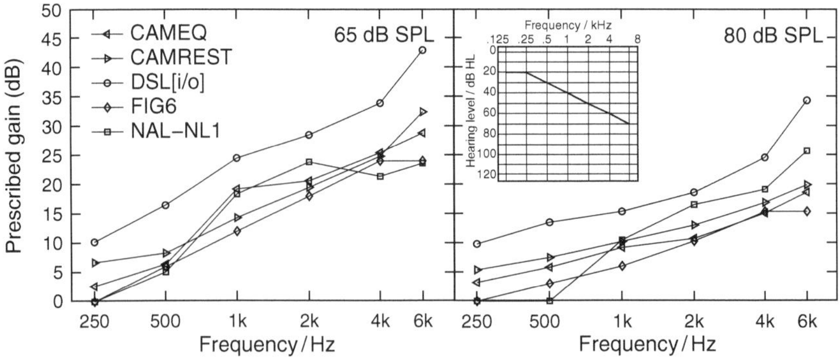

Fig. 9.7 Prescribed gain for speech at two presentation levels, 65 and 80 dB SPL (*left* and *right*), for five prescriptive methods for a person with the illustrated audiogram. CAMEQ is based on loudness density equalization (Moore et al. 1999), and CAMREST is based on loudness density normalization (Moore 2000); both were calculated using the loudness model of Moore and Glasberg (1997). DSL[i/o] is based on fitting an extended range of input signals to the hearing-aid user's dynamic range (Cornelisse et al. 1995). DSL[i/o] prescribes sound pressure levels at the eardrum, rather than the difference in sound pressure levels at the eardrum with and without the hearing aid, and prescribed gain has been calculated using average data on the transfer from sound field to eardrum from the NAL-NL1 software (version 1.28). FIG6 is based on frequency-specific loudness normalization using statistical loudness data (Killion and Fikret-Pasa 1993), and NAL-NL1 is based on speech intelligibility optimization with a constraint that the overall calculated loudness should not exceed normal calculated loudness (Byrne et al. 2001)

gain differences would lead to large differences in perceived loudness but only to small differences in speech intelligibility. The data in the study by Smeds and Leijon were collected in 1998 with the most modern hearing aids available at the time. Repeated measurements 10 years later revealed the same picture.

If the goal of a prescription is to prescribe gain that is correct for the average hearing-aid user with a given audiogram, not all of the prescriptions can be correct, and there is reason to further study what the ideal prescription looks like. In the following, a loudness model of Moore and Glasberg (1997) is used in many of the comparisons between prescribed and preferred loudness. The reason for the choice is that this particular loudness model has been used when deriving a number of prescriptions (see later in the text).

9.4.2.1 Prescribed and Preferred Loudness: Linear Hearing Aids

A large number of studies have aimed at evaluating two prescriptions from the National Acoustic Laboratories (NAL): NAL-R for linear amplification (Byrne and Dillon 1986) and NAL-NL1 for nonlinear amplification (Byrne et al. 2001). The gain prescribed for speech at normal conversational levels for NAL-NL1 corresponds well to the gain prescribed for the NAL-R. If loudness is calculated according to the loudness model of Moore and Glasberg (1997), it is found that

NAL-R prescribes gain that leads to slightly less-than-normal calculated loudness for a mid-level speech input (Smeds and Leijon 2001). The NAL-NL1 prescription for a bilateral fit leads to a calculated loudness level that is approximately 4 phon lower than normal (Smeds et al. 2006a).

A number of studies using linear amplification have shown that hearing-aid users with sensorineural hearing loss prefer 6–9 dB less gain than prescribed by NAL-R (e.g., Leijon et al. 1990; Humes et al. 2000, 2002). At least one study (Byrne and Cotton 1988) has shown good agreement between prescribed and actually used gain when NAL-R was evaluated. It is difficult to explain the discrepancy between these results. One factor that varied between the studies was the time the hearing aids were worn before the evaluation. The participants in the two studies presented by Humes et al. used their hearing aids for the longest period (1 year) and the participants in the study by Byrne and Cotton for the shortest (2–3 weeks). However, the participants in the studies by Humes et al. adjusted their preferred gain down already at their first visit to the clinic 2 weeks (Humes et al. 2000) or 1 month (Humes et al. 2002) after the initial fitting, to about the same gain as they preferred 1 year later.

A general problem with these evaluations of prescriptions for linear hearing aids is that the hearing aids used were seldom able to provide the prescribed gain at 4 kHz and above. When the three studies mentioned above are compared in terms of fitting accuracy, it seems that the participants in the study by Byrne and Cotton were provided with less gain than the participants in the other studies.

In summary, there seems to be a preference for lower gain than prescribed by NAL-R, and, as a consequence, a preference for considerably less-than-normal calculated loudness, according to the loudness model of Moore and Glasberg (1997). This preference might be influenced by the fact that linear amplification, when fitted to give normal loudness for a mid-level input, will give higher-than-normal loudness for a higher-level input for a person with a typical cochlear hearing loss (Fig. 9.2). With the introduction of nonlinear amplification, the picture might look different.

9.4.2.2 Prescribed and Preferred Loudness: Nonlinear Hearing Aids

There are a number of studies where preferred loudness or preferred gain has been studied using nonlinear hearing aids. A majority of these studies have shown results similar to the results obtained for linear amplification. Adults with mild-to-moderate hearing loss prefer amplification that leads to less-than-normal calculated loudness according to the loudness model of Moore and Glasberg (1997).

In a few field studies, gain preferences have been interpreted in loudness terms (Smeds 2004; Smeds et al. 2006b). The results show a preference for gain that led to calculated loudness levels that were, on average, 7–14 phon less than normal. A large number of studies have determined preferred gain relative to the NAL-NL1 prescription for nonlinear hearing aids or the NAL-R prescription for linear hearing aids (e.g., Keidser and Dillon 2006; Peek et al. 2007; Zakis et al. 2007; Keidser et al. 2008). The results show a preference for gain settings that, on average, varied between 3 and 9 dB below the gain prescribed by NAL-NL1. At least one study has determined preferred gain relative to the DSL[i/o] prescription (Laurnagaray and Seewald, reported by Scollie et al. 2005). The results show a preference for gain settings that, on average, varied between 9 and 11 dB below the gain prescribed by DSL[i/o]. One study evaluated the DSL[i/o] prescription as well as two other prescriptive methods for nonlinear hearing aids, CAMEQ and CAMREST (Marriage et al. 2004). This study differs from the rest of the studies in that it concludes that prescriptions like CAMEQ and CAMREST, which lead to normal overall calculated loudness according to the loudness model of Moore and Glasberg (1997), were found acceptable by experienced hearing-aid users, whereas the gain had to be reduced somewhat for inexperienced hearing-aid users. The DSL[i/o] prescription had to be reduced on average by 3.5 dB.

Since CAMEQ is based on loudness density equalization, its prescribed gain–frequency shape is fairly similar to the gain prescribed by NAL-NL1 (Fig. 9.7). Therefore, the results of the study by Marriage et al. (2004) are surprising when compared to studies showing that the NAL-NL1 prescribed gain has to be reduced substantially. One important difference between the studies is the difference in subjective measure used. In the study of Marriage et al., minimal adjustments necessary for *acceptable* loudness were determined as opposed to determining the *preferred* loudness. There is reason to believe that larger gain adjustments would have been seen if preferred gain had been determined. The previously mentioned results for NAL-NL1 point in that direction, as do the results for the DSL[i/o] prescription, which was adjusted down by 9–11 dB on average in the study by Laurnagaray and Seewald (Scollie et al. 2005) and only by 3.5 dB on average in the study by Marriage et al.

A large clinical trial provides some suggestions as to why hearing-aid users might prefer gain values that lead to considerably less-than-normal overall calculated loudness. Shanks et al. (2002) measured speech recognition performance in background babble at three signal-to-noise ratios (+3, 0, and −3 dB), where the 0 dB SNR was defined for each listener as the speech-to-babble ratio that resulted in 50% performance on a sentence test at a conversational speech level (and the other two SNRs were ±3 dB relative to that 0 dB SNR). Participants with mild-to-moderate hearing losses showed *decreasing* mean speech recognition scores when presentation levels *increased* from 52 to 62 dB SPL, and from 62 to 74 dB SPL, using constant SNRs. A plausible explanation of the data is that the speech signal, amplified by NAL-R, was audible at the low presentation levels and the increase in presentation level increased the amount of babble that became audible. Similar results were found by Studebaker et al. (1999) for listeners with normal and impaired hearing. In contrast, participants with more severe hearing losses in the study by Shanks et al. showed better speech recognition scores when the presentation levels increased.

In summary, a large number of studies, using both linear and nonlinear amplification, have demonstrated that the gain prescribed by methods such as NAL-NL1, which actually prescribe gain that leads to less-than-normal overall calculated loudness (using the Moore and Glasberg (1997) loudness model), are judged as "too loud" by listeners with mild to moderate hearing loss. Based on these results, the originators of both the NAL-NL1 and the DSL[i/o] prescriptions have changed, or are about to change, their prescriptions in new versions (Scollie et al. 2005; Keidser and Dillon 2006).

All of the evaluations in this section have been interpreted using a loudness model of Moore and Glasberg (1997). The model has been used when deriving prescriptive methods such as NAL-NL1, CAMEQ, and CAMREST. Despite this, it is natural to first question the model's ability to correctly predict the loudness experienced by listeners with normal and impaired hearing. Laboratory data by Smeds et al. (2006a) indicate that there might be a problem with the loudness model. In that study, listeners with sensorineural hearing losses rated recorded sound samples amplified according to NAL-NL1 as louder than normal-hearing listeners did (without amplification), despite the fact that the NAL-NL1 prescription leads to less-than-normal overall calculated loudness according to the Moore and Glasberg model.

Moore and Glasberg (2004) have published a revised version of the loudness model. In all of the studies presented in the preceding text, preferred loudness levels are presented relative to the corresponding normal loudness level. The new model seems to predict similar differences between loudness calculated for impaired and normal hearing as the 1997 model. The same authors have also published a loudness model that accounts for spectral variation over time (Glasberg and Moore 2002). Further research is needed to see to what extent the loudness level differences presented above would be affected by using the loudness model for time-varying sounds.

9.4.2.3 Hearing-Aid Experience as a Predictor of Preferred Loudness

In some of the studies mentioned in the preceding text, there seemed to be a difference between inexperienced and experienced hearing-aid users in preferred hearing-aid amplified loudness. It has been difficult to find scientific support for the sometimes quoted clinical experience that people fitted with their first hearing aids would like to gradually increase their hearing-aid gain (Convery et al. 2005). Despite this, many hearing aid manufacturers include "acclimatization stages" in their software. The idea is to give a first-time hearing-aid user less gain than the manufacturer thinks is optimal, and over time the hearing-aid user will reach the fully prescribed gain. The gain change implemented in acclimatization stages is usually 5–10 dB (Eberwein et al. 2001).

A small number of studies have shown a significant difference in preferred gain when inexperienced and experienced hearing-aid users are compared (Marriage et al. 2004; Scollie et al. 2005); and at least one study (Keidser et al. 2008) has shown that inexperienced hearing-aid users with a hearing loss described as "more than mild" tend to increase their gain during the first year of hearing aid use. However, the difference in preferred gain between inexperienced and experienced hearing-aid users is small, 2–3 dB, and not nearly as large as the gain reductions implemented in most hearing aid manufacturers' fitting software. This suggests that the acclimatization stages in the fitting software mainly deal with the problem that generic prescriptive methods generally provide adults with a mild to moderate hearing loss with more gain than they prefer or need.

9.4.2.4 Fast Loudness-Restoring Hearing Aid Compression

As discussed in Sect. 9.4.1.1, loudness density normalization over frequency and time would involve fast-acting compression in a large number of channels (of auditory filter bandwidth). Theoretical arguments in favor of fast-acting hearing-aid compression include the fact that cochlear compression is a very fast process. It might therefore be reasonable to apply fast-acting compression in hearing aids to restore short-term loudness perception (e.g., Villchur 1973). Especially for people who have an auditory dynamic range that is smaller than the dynamic range of speech, one might argue that the compressor needs to adapt quickly enough to provide different gain–frequency responses for adjacent speech sounds with different short-time spectra (so-called syllabic compression). On the other hand, fast-acting compression reduces the acoustic temporal and spectral differences between phonemes, especially if the compression is performed in a large number of compression channels (Plomp 1994). A compression system with slow-acting automatic gain control compensates for loudness variations between different sound environments, but it preserves the level and spectral differences between phonemes in an ongoing speech signal.

A number of studies show that compression systems, irrespective of time constants, can result in good speech intelligibility over a wide range of input levels and reduce the need for adjusting a volume control on the hearing aid (summarized by Dillon 1996). The question of whether fast-acting compression will give additional benefit for hearing-aid users has not yet been answered conclusively (e.g., Olsen et al. 2004; Goedegebure 2005; Gatehouse et al. 2006; Kates 2010).

9.4.3 Binaural Loudness and Hearing-Aid Fitting

Generally, the advantages of binaural over monaural listening are the same for listeners with normal and impaired hearing and include benefits of binaural loudness summation, head diffraction, sound localization, and binaural release of masking (e.g., Markides 1982a, b; Dillon 2001). Due to binaural loudness summation (Sect. 9.2.2.3), the gain in each hearing aid can be reduced, which reduces the risk of feedback and increases sound quality. Loudness summation does not seem to lead to a need to reduce the maximum output for bilaterally fitted hearing aids (e.g., Dillon 2001).

9.4.4 *Conductive and Mixed Hearing Losses*

Clinically, the difference between hearing thresholds measured with air-conducted and bone-conducted sound is often interpreted as an indication of the degree of conductive hearing loss. In principle, if a conductive loss is exactly known, the impairment might be perfectly compensated by a linear hearing aid that amplifies sound at each frequency by exactly the same amount as the sound attenuation caused by the conductive loss at the same frequency, as this would restore normal input to the cochlea and thereby normal loudness. However, experience with hearing-aid fitting for people with conductive or mixed losses indicates that it is sufficient to provide hearing aid gain equal to only about 75% of the air–bone threshold gap (Dillon 2001). It is not known if this reflects a real contradiction of the assumed effects of a conductive loss on loudness perception, or if it is a consequence of some other side effects of the hearing loss or the hearing aids. There can be several reasons for this apparent discrepancy:

- The air–bone gap might be an inaccurate indicator of the conductive loss.
- Full compensation of the conductive loss might require very high sound pressure levels at the eardrum, which could cause distortion of the sound in the middle ear.
- The middle ear muscle reflex normally attenuates loud sounds, but this attenuation might be absent or diminished in cases with middle ear conductive loss.
- Technical shortcomings of hearing aids can cause negative side effects when using high amplification.

9.5 Loudness and Cochlear Implants

A CI stimulates auditory nerve cells directly, and can therefore produce auditory sensations, even when cochlear hair cells are totally absent or malfunctioning. The physiological and psychoacoustical bases for the clinical application of CI technology have been extensively reviewed by Zeng et al. (2004) and Clark (2003). The use of CIs can restore remarkably good practical communicative ability for many users with a profound hearing loss, although there are large individual variations in performance (Geers et al. 2003).

A CI system includes an external signal processor that encodes sound into a stream of current pulses emitted from an array of 12–22 electrode contacts, usually placed along the first one or two spiral turns of the cochlear scala tympani. Usually, the acoustic intensity in one frequency band controls the electrical stimulation emitted by one electrode. To the extent that the current from each electrode stimulates only a narrow range of auditory nerve fibers, the neural activity encodes the input sound spectrum tonotopically. However, current CI systems achieve spectral resolution that is far from normal (Henry and Turner 2003; Henry et al. 2005; Molin et al. 2005; Litvak et al. 2007).

The design and clinical fitting of CI systems rely heavily on psychophysical studies using electrical stimulation, reviewed by McKay (2004). One of the most important aspects of CI fitting is the very narrow dynamic range with electric stimulation. The range of currents producing a loudness change between threshold and comfortable loudness is typically only a few decibels (McKay 2004). The loudness sensation caused by electrical stimulation from a single electrode depends on the amplitude, duration, temporal pattern, and rate of electrical current pulses (McKay 2004; Fu 2005). The effect of pulse rate on loudness is highly subject-dependent, and can vary across electrode locations within individual subjects (Fu 2005). There are also individual variations in the effect of current amplitude on loudness (Sanpetrino and Smith 2006). The loudness sensation is also influenced by the summation of loudness contributions from stimulation with multiple electrodes (McKay et al. 2001, 2003). The loudness summation is probably caused by integration of loudness-density contributions across groups of neurons in a similar way as in acoustic hearing, because the integration process is presumed to represent a process that is more central than the transduction from hair cells to auditory neurons (McKay et al. 2001). The loudness sensation caused by bilateral stimulation is approximately equal to the sum of the loudness of each of the two monaural stimuli (van Hoesel and Clark 1997, referred to by McKay 2004), but experimental data for CI users are still sparse.

9.5.1 Exploring the Dynamic Range in Cochlear Implant Fitting

The mapping of instantaneous sound levels to electric pulse rates and amplitudes is usually fitted to the individual dynamic range. Lower and upper limits are defined by estimates of electrical current thresholds (sometimes called T levels) and upper limits of comfortable loudness (sometimes called C levels). If possible, the T and C levels are measured for each electrode. The T level is sometimes defined as the "lowest stimulus level where a response always occurs," and the C level is the "highest level that can be used without causing discomfort" (Clark 2003). The T and C levels might vary from electrode to electrode, and they might also change because of tissue growth after the operation. The CI-user's ability to judge loudness and timbre might also change with use of the implant. Therefore, the mapping must usually be revised several times after the implantation. The use of auditory reflex thresholds has been proposed as a way to estimate C levels and to adjust the electrical stimulation levels in the absence of behavioral measures (Gordon et al. 2004; Brickley et al. 2005).

9.5.2 Loudness Considerations in Cochlear Implant Fittings

Just as with hearing aids, the overall purpose of the CI is to make a wide range of speech sounds audible and to make speech and other environmental sounds comfortably loud, or at least acceptably loud, for the CI user. The narrow electrical dynamic range between T and C levels is mapped to a much wider acoustic range. A mapping similar to a power–law relation between acoustic envelope amplitude and current amplitude has been motivated as follows (Fu and Shannon 1998): Experiments have indicated an expansive power–law relation between current amplitude and loudness. For listeners with normal hearing the loudness growth with sound pressure amplitude can also be described by a compressive power law at moderate levels. Therefore, loudness growth might be approximately normalized for the CI user by a mapping of sound pressure amplitudes to electric stimulation using the power–law form. However, the shapes of loudness functions vary across subjects and also between different electrodes in individual subjects (Sanpetrino and Smith 2006). Experiments have indicated that the best speech recognition was achieved when the shape of the nonlinear mapping restored approximately normal loudness growth, but large deviations from the optimal mapping had only a mild effect on speech recognition (Fu and Shannon 1998). Thus, the exact shape of the mapping function is probably not critical for speech recognition (Fu and Shannon 1998; Boyd 2006).

9.6 Summary

Abnormal loudness perception is an important consequence of the most common types of hearing impairment. Loudness is often normal or near normal at high sound levels, although sounds at lower levels are inaudible or considerably softer than normal. The abnormal loudness perception is determined mainly by physiological changes, primarily in the inner ear. The abnormal loudness perception is often associated with a reduction or loss of the normal cochlear compression resulting from OHC activity. However, given only results of loudness measurements, it is not possible to determine precisely the nature of the physiological damage. An area for future research is how the perceived loudness is affected by various types of physiological damage to the peripheral auditory system.

Another important field where there is still a lack of knowledge is in the habilitation of infants with hearing loss. Very little is known about infants' loudness perception and how it interplays with the particular listening needs these infants have. There is also a strong need to develop methods for exploring an individual infant's auditory dynamic range.

All hearing devices are designed and fitted to compensate to some extent for abnormal loudness perception, and loudness models of normal and impaired hearing are often used to derive the prescription used at the fitting. Studies of prescribed and preferred hearing-aid amplified loudness have shown that hearing-aid users with mild to moderate hearing loss prefer considerably less-than-normal calculated loudness according to the most commonly used loudness models. Existing loudness models seem to underestimate the loudness of amplified sound for these listeners with cochlear hearing loss.

It is questionable whether hearing aids can be fitted more optimally if individual loudness measurements are performed prior to the fitting. The discussion in this chapter on loudness models and statistical relations between hearing thresholds and supra-threshold loudness perception does not reflect neglect for the individual hearing-aid user's loudness perception. Individual variation in loudness perception exists, and the importance of including this variation in the fine-tuning process for an acceptable result is acknowledged. A very important topic for future research is to determine how large the individual variations in loudness perception are, in relation to the model-predicted statistical average, among listeners with hearing losses.

Acknowledgments We thank Stingerfonden for the opportunity to start our work on this chapter in their research facility Forum Auditum in Croatia.

References

Aibara R, Welsh JT, Puria S, Goode RL (2001) Human middle-ear sound transfer function and cochlear input impedance. Hear Res 152:100–109.

Allen JB, Hall JL, Jeng PS (1990) Loudness growth in ½-octave bands (LGOB) – a procedure for the assessment of loudness. J Acoust Soc Am 88:745–753.

Andersson G, Baguley DM, McKenna L, McFerran D (2005) Tinnitus: A Multidisciplinary Approach. London: Whurr.

ANSI-S3.4 (2007) American National Standard Procedure for the Computation of Loudness of Steady Sounds. New York: American National Standards Institute.

ANSI-S3.5 (1997) American National Standard Methods for the Calculation of the Speech Intelligibility Index. New York: American National Standards Institute.

Baguley DM, Andersson G (2007) Hyperacusis. Mechanisms, Diagnosis, and Therapies. London: Whurr.

Barfod J (1978) Multichannel compression hearing aids: experiments and consideration on clinical applicability. In: Ludvigsen C, Barfod J (eds), Sensorineural Hearing Impairment and Hearing Aids (Proceedings of the 8th Danavox Symposium). Scand Audiol Suppl 6:315–340.

Boyd PJ (2006) Effects of programming threshold and map-law settings on acoustic thresholds and speech discrimination with the MED-EL COMBI 40+ cochlear implant. Ear Hear 27:608–618.

Braida LD, Durlach NI, Lippman RP, Hicks BL, Rabinowitz WM, Reed CM (1979) Hearing Aids – A Review of Past Research on Linear Amplification, Amplitude Compression and Frequency Lowering. Rockville, MD: American Speech and Hearing Association.

Brand T, Hohmann V (2002) An adaptive procedure for categorical loudness scaling. J Acoust Soc Am 112:1597–1604.

Brickley G, Boyd P, Wyllie F, O'Driscoll M, Webster D, Nopp P (2005) Investigations into electrically evoked stapedius reflex measures and subjective loudness percepts in the MED-EL COMBI 40+ cochlear implant. Cochlear Implants Int 6:31–42.

Brunt MA (2001) Tests of cochlear function. In: Katz J (ed), Handbook of Clinical Audiology. Baltimore, MD: Lippincott Williams & Wilkins.

Buus S (1997) Auditory masking. In: Crocker MJ (ed), Encyclopedia of Acoustics. New York: Wiley, pp. 1427–1445.

Buus S, Florentine M (2001) Modifications to the power function for loudness. In: Sommerfeld E, Kompass R, Lachmann T (eds), Fechner Day 2001. Proceedings of the 17th Annual Meeting of the International Society for Psychophysics. Berlin, Germany: Pabst Science, pp. 236–241.

Buus S, Florentine M (2002) Growth of loudness in listeners with cochlear hearing losses: recruitment reconsidered. J Assoc Res Otolaryngol 3:120–139.
Buus S, Florentine M, Poulsen T (1997) Temporal integration of loudness, loudness discrimination, and the form of the loudness function. J Acoust Soc Am 101:669–680.
Buus S, Musch H, Florentine M (1998) On loudness at threshold. J Acoust Soc Am 104:399–410.
Buus S, Florentine M, Poulsen T (1999) Temporal integration of loudness in listeners with hearing losses of primarily cochlear origin. J Acoust Soc Am 105:3464–3480.
Byrne D (1982) Theoretical approaches for hearing aid selection. In: Studebaker GA, Bess FH (eds), The Vanderbilt Hearing-Aid Report. Monographs in Contemporary Audiology. Upper Darby, PA, pp. 175–179.
Byrne D (1983) Theoretical prescriptive approaches to selecting the gain and frequency response of a hearing aid. In: Schwarz DM, Bess FH (eds), Monographs in Contemporary Audiology, Vol. 4. Upper Darby, PA, pp. 1–40.
Byrne D (1986) Effects of bandwidth and stimulus type on most comfortable loudness levels of hearing-impaired listeners. J Acoust Soc Am 80:484–493.
Byrne D, Cotton S (1988) Evaluation of the National Acoustic Laboratories' new hearing aid selection procedure. J Speech Hear Res 31:178–186.
Byrne D, Dillon H (1986) The National Acoustic Laboratories' (NAL) new procedure for selecting the gain and frequency response of a hearing aid. Ear Hear 7:257–265.
Byrne D, Tonisson W (1976) Selecting the gain of hearing aids for persons with sensorineural hearing impairments. Scand Audiol 5:51–59.
Byrne D, Dillon H, Ching T, Katsch R, Keidser G (2001) NAL-NL1 procedure for fitting nonlinear hearing aids: characteristics and comparisons with other procedures. J Am Acad Audiol 12:37–51.
Cai S, Ma WL, Young ED (2009) Encoding intensity in ventral cochlear nucleus following acoustic trauma: implications for loudness recruitment. J Assoc Res Otolaryngol 10:5–22.
Castro FZ, Barajas de Prat JJ, Zabala EL (2008) Loudness and auditory steady-state responses in normal-hearing subjects. Int J Audiol 47:269–275.
Ching TY, Dillon H, Katsch R, Byrne D (2001) Maximizing effective audibility in hearing aid fitting. Ear Hear 22:212–224.
Clark GM (2003) Cochlear Implants. Fundamentals and Applications. New York: Springer.
Convery E, Keidser G, Dillon H (2005) A review and analysis: does amplification experience have an effect on preferred gain over time? Aust NZ J Audiol 27:18–32.
Cornelisse LE, Seewald RC, Jamieson DG (1995) The input/output formula: a theoretical approach to the fitting of personal amplification devices. J Acoust Soc Am 97:1854–1864.
Cox RM (1983) Using ULCL measures to find frequency/gain and SSPL90. Hear Instr 34(7): 17–21, 39.
Cox RM (1989) Comfortable loudness level: stimulus effects, long-term reliability, and predictability. J Speech Hear Res 32:816–828.
Cox RM, Alexander GC, Taylor IM, Gray GA (1997) The contour test of loudness perception. Ear Hear 18:388–400.
Dallos P, Popper AN, Fay RR (eds) (1996) The Cochlea. New York: Springer
Davidson SA, Wall LG, Goodman CM (1990) Preliminary studies on the use of an ABR amplitude projection procedure for hearing aid selection. Ear Hear 11:332–339.
Dillon H (1996) Compression? Yes, but for low or high frequencies, for low or high intensities, and with what response times? Ear Hear 17:287–307.
Dillon H (2001) Hearing Aids. Sydney, Australia: Boomerang.
Dillon H, Storey L (1998) The National Acoustic Laboratories' procedure for selecting the saturation sound pressure level of hearing aids: theoretical derivation. Ear Hear 19:255–266.
Dix MR, Hallpike CS (1947) The peep-show: a new technique for pure-tone audiometry in young children. Br Med J 2:719–723.
Don M, Kwong B, Tanaka C, Brackmann D, Nelson R (2005) The stacked ABR: a sensitive and specific screening tool for detecting small acoustic tumors. Audiol Neurootol 10:274–290.

Eberwein CA, Mack HA, Pallett S, Lindley GA (2001) Adaptation managers in hearing aid fitting software. Semin Hear 22:199–207.

Eggermont JJ (1977) Electrocochleography and recruitment. Ann Otol Rhinol Laryngol 86:138–149.

Elberling C (1999) Loudness scaling revisited. J Am Acad Audiol 10:248–260.

Epstein M, Marozeau J, Florentine M (2006) Basilar-membrane activity and loudness. In: Kornbrot DE, Msetfi RM, MacRae AW (eds), Fechner Day 2006. Proceedings of the 22nd Annual Meeting of the International Society for Psychophysics. St. Albans, UK: The International Society for Psychophysics, pp. 181–186.

Ewing I, Ewing A (1944) The ascertainment of deafness in infancy and early childhood. J Laryngol 59:309–333.

Fletcher H, Munson WA (1933) Loudness, its definition, measurement and calculation. J Acoust Soc Am 5:82–108.

Fletcher HF, Steinberg JC (1924) The dependence of the loudness of a complex sound upon the energy in the various frequency regions of the sound. Phys Rev 24:306–317.

Florentine M, Houtsma AJ (1983) Tuning curves and pitch matches in a listener with a unilateral, low-frequency hearing loss. J Acoust Soc Am 73:961–965.

Florentine M, Zwicker E (1979) A model of loudness summation applied to noise-induced hearing loss. Hear Res 1:121–132.

Florentine M, Fastl H, Buus S (1988) Temporal integration in normal hearing, cochlear impairment, and impairment simulated by masking. J Acoust Soc Am 84:195–203.

Florentine M, Buus S, Poulsen T (1996) Temporal integration of loudness as a function of level. J Acoust Soc Am 99:1633–1644.

Florentine M, Buus S, Hellman RP (1997) A model of loudness summation applied to high-frequency hearing loss. In: Jesteadt W (ed), Modeling Sensorineural Hearing Loss. Mahwah, NJ: Lawrence Erlbaum, pp. 187–198.

Fowler EP (1936) A method for the early detection of otosclerosis. Arch Otolaryngol 24:731–741.

Fowler EP (1942) The "illusion of loudness" of tinnitus. Laryngoscope 52:275–285.

Fu QJ (2005) Loudness growth in cochlear implants: effect of stimulation rate and electrode configuration. Hear Res 202:55–62.

Fu Q-J, Shannon RV (1998) Effects of amplitude nonlinearity on phoneme recognition by cochlear implant users and normal-hearing listeners. J Acoust Soc Am 104:2570–2577.

Gatehouse S, Naylor G, Elberling C (2006) Linear and nonlinear hearing aid fittings. 1. Patterns of benefit. Int J Audiol 45:130–152.

Geers A, Brenner C, Davidson L (2003) Factors associated with development of speech perception skills in children implanted by age five. Ear Hear 24:24S–35S.

Glasberg BR, Moore BCJ (2002) A model of loudness applicable to time-varying sounds. J Audio Eng Soc 50:331–342.

Goedegebure A (2005) Phoneme compression – processing of the speech signal and effects on speech intelligibility in hearing-impaired listeners. Rotterdam, The Netherlands: Erasmus University, PhD thesis.

Gordon KA, Papsin BC, Harrison RV (2004) Toward a battery of behavioral and objective measures to achieve optimal cochlear implant stimulation levels in children. Ear Hear 25:447–463.

Hallam RS, Jakes SC, Chambers C, Hinchcliffe R (1985) A comparison of different methods for assessing the 'intensity' of tinnitus. Acta Otolaryngol 99:501–508.

Hawkins DB, Walden BE, Montgomery A, Prosek RA (1987) Description and validation of an LDL procedure designed to select SSPL90. Ear Hear 8:162–169.

Hawkins DB, Ball TL, Beasley HE, Cooper WA (1992) Comparison of SSPL90 selection procedures. J Am Acad Audiol 3:46–50.

Heinz MG, Young ED (2004) Response growth with sound level in auditory-nerve fibers after noise-induced hearing loss. J Neurophysiol 91:784–795.

Heinz MG, Issa JB, Young ED (2005a) Auditory-nerve rate responses are inconsistent with common hypotheses for the neural correlates of loudness recruitment. J Assoc Res Otolaryngol 6:91–105.

Heinz MG, Scepanovic D, Issa J, Sachs MB, Young ED (2005b) Normal and impaired level encoding: Effects of noise-induced hearing loss on auditory-nerve responses. In: Pressnitzer D, Cheveigné AD, McAdams S, Collet L (eds), Auditory Signal Processing: Physiology, Psychoacoustics, and Models. 13th International Conference on Hearing 2003. New York: Springer, pp. 40–49.

Hellman RP, Meiselman CH (1990) Loudness relations for individuals and groups in normal and impaired hearing. J Acoust Soc Am 88:2596–2606.

Hellman RP, Meiselman CH (1993) Rate of loudness growth for pure tones in normal and impaired hearing. J Acoust Soc Am 93:966–975.

Henry BA, Turner CW (2003) The resolution of complex spectral patterns by cochlear implant and normal-hearing listeners. J Acoust Soc Am 113:2861–2873.

Henry BA, Turner CW, Behrens A (2005) Spectral peak resolution and speech recognition in quiet: Normal hearing, hearing impaired, and cochlear implant listeners. J Acoust Soc Am 118:1111–1121.

Henry JA, Meikle MB (2000) Psychoacoustic measures of tinnitus. J Am Acad Audiol 11:138–155.

Hergils L (2007) Analysis of measurements from the first Swedish universal neonatal hearing screening program. Int J Audiol 46:680–685.

Hiller W, Goebel G (2006) Factors influencing tinnitus loudness and annoyance. Arch Otolaryngol Head Neck Surg 132:1323–1330.

Hiller W, Goebel G (2007) When tinnitus loudness and annoyance are discrepant: audiological characteristics and psychological profile. Audiol Neurootol 12:391–400.

Hinchcliffe R, Chambers C (1983) Loudness of tinnitus: an approach to measurement. Adv Otorhinolaryngol 29:163–173.

Hodgson WR (2001) Evaluating infants and young children. In: Katz J (ed), Handbook of Clinical Audiology. Baltimore, MD: Lippincott Williams & Wilkins.

How J, Lutman ME (2007) Transient evoked otoacoustic emission input-output function variation in a large sample of neonates and implications for hearing screening. Int J Audiol 46:670–679.

Humes LE, Barlow NN, Garner CB, Wilson DL (2000) Prescribed clinician-fit versus as-worn coupler gain in a group of elderly hearing-aid wearers. J Speech Lang Hear Res 43:879–892.

Humes LE, Wilson DL, Barlow NN, Garner C (2002) Changes in hearing-aid benefit following 1 or 2 years of hearing-aid use by older adults. J Speech Lang Hear Res 45:772–782.

ISO-532 (1975) Acoustics – Methods for Calculating Loudness Level. Geneva: International Organisation for Standardisation.

Jakes SC, Hallam RS, Chambers CC, Hinchcliffe R (1986) Matched and self-reported loudness of tinnitus: methods and sources of error. Audiology 25:92–100.

Johnson EW (1977) Auditory test results in 500 cases of acoustic neuroma. Arch Otolaryngol 103:152–158.

Johnstone BM, Patuzzi R, Yates GK (1986) Basilar membrane measurements and the travelling wave. Hear Res 22:147–153.

Joris PX (2009) Recruitment of neurons and loudness. Commentary on "Encoding intensity in ventral cochlear nucleus following acoustic trauma: implications for loudness recruitment" by Cai et al. J Assoc Res Otolaryngol 10:1–4. DOI: 10.1007/s10162–008–0142–y.

Kates JM (2010) Understanding compression: modeling the effects of dynamic-range compression in hearing aids. Int J Audiol 49: 395–409.

Keidser G, Dillon H (2006) What's new in prescriptive fittings down under? In: Palmer C, Seewald RC (eds), Hearing Care for Adults. Stäfa, Switzerland: Phonak AG.

Keidser G, Seymour J, Dillon H, Grant F, Byrne D (1999) An efficient, adaptive method of measuring loudness growth functions. Scand Audiol 28:3–14.

Keidser G, O'Brien A, Carter L, McLelland M, Yeend I (2008) Variation in preferred gain with experience for hearing aid users. Int J Audiol 47:621–635.

Kiessling J (1995) Zum überschwelligen Lautheitsanstieg bei Schallempfindungsschwerhörigen - Konsequenzen für die Hörgeräte-Entwicklung und -Anpassung [Loudness growth in sensorineural hearing loss – consequences for hearing aid design and fitting]. Audiologische Akustik 34:82–89.

Kiessling J, Schubert M, Archut A (1996) Adaptive fitting of hearing instruments by category loudness scaling (ScalAdapt). Scand Audiol 25:153–160.

Killion M, Fikret-Pasa S (1993) The 3 types of sensorineural hearing loss: loudness and intelligibility considerations. Hear J 46(11):31–36.

Klein AJ, Armstrong BL, Greer MK, Brown I, Frank R (1990) Hyperacusis and otitis media in individuals with Williams syndrome. J Speech Hear Dis 55:339–344.

Launer S (1995) Loudness perception in listeners with sensorineural hearing impairment. Oldenburg, Germany: Universität, PhD thesis.

Leijon A (1991) Hearing aid gain for loudness-density normalization in cochlear hearing losses with impaired frequency resolution. Ear Hear 12:242–250.

Leijon A, Lindkvist A, Ringdahl A, Israelsson B (1990) Preferred hearing aid gain in everyday use after prescriptive fitting. Ear Hear 11:299–305.

Leijon A, Lindkvist A, Ringdahl A, Israelsson B (1991) Sound quality and speech reception for prescribed hearing aid frequency responses. Ear Hear 12:251–260.

Liberman MC (1978) Auditory-nerve response from cats raised in a low-noise chamber. J Acoust Soc Am 63:442–445.

Lidén G, Kankkunen A (1969) Visual reinforcement audiometry. Arch Otolaryngol 89:865–872.

Lippmann RP, Braida LD, Durlach NI (1981) Study of multichannel amplitude compression and linear amplification for persons with sensorineural hearing loss. J Acoust Soc Am 69:524–534.

Litvak LM, Spahr AJ, Saoji AA, Fridman GY (2007) Relationship between perception of spectral ripple and speech recognition in cochlear implant and vocoder listeners. J Acoust Soc Am 122:982–991.

Markides A (1982a) Reactions to binaural hearing aid fitting. Scand Audiol Suppl 15:197–205.

Markides A (1982b) The effectiveness of binaural hearing aids. Scand Audiol Suppl 15:181–196.

Marozeau J, Florentine M (2007) Loudness growth in individual listeners with hearing losses: a review. J Acoust Soc Am 122:EL81.

Marozeau J, Florentine M (2009) Testing the binaural equal-loudness-ratio hypothesis with hearing-impaired listeners. J Acoust Soc Am 126:310–317.

Marriage J, Moore BC, Alcantara JI (2004) Comparison of three procedures for initial fitting of compression hearing aids. III. Inexperienced versus experienced users. Int J Audiol 43:198–210.

Mauermann M, Hohmann V (2007) Differences in loudness of positive and negative Schroeder-phase tone complexes as a function of the fundamental frequency. J Acoust Soc Am 121:1028–1039.

McKay CM (2004) Psychophysics and electrical stimulation. In: Zeng F-G, Popper AN, Fay RR (eds), Cochlear Implants: Auditory Prostheses and Electric Hearing. New York: Springer, pp. 286–333.

McKay CM, Remine MD, McDermott HJ (2001) Loudness summation for pulsatile electrical stimulation of the cochlea: effects of rate, electrode separation, level, and mode of stimulation. J Acoust Soc Am 110:1514–1524.

McKay CM, Henshall KR, Farrell RJ, McDermott HJ (2003) A practical method of predicting the loudness of complex electrical stimuli. J Acoust Soc Am 113:2054–2063.

Menard M, Gallego S, Berger-Vachon C, Collet L, Thai-Van H (2008) Relationship between loudness growth function and auditory steady-state response in normal-hearing subjects. Hear Res 235:105–113.

Miskolczy-Fodor F (1960) Relations between loudness and duration of tonal pulses III. Response in cases of abnormal loudness function. J Acoust Soc Am 32:486–492.

Molin E, Leijon A, Wallsten H (2005) Spectro-temporal discrimination in cochlear implant users, International Conference on Acoustics, Speech, and Signal Processing, ICASSP, IEEE, Vol. III. Philadelphia, pp. 25–28.

Moore BCJ (2000) Use of a loudness model for hearing aid fitting. IV. Fitting hearing aids with multi-channel compression so as to restore 'normal' loudness for speech at different levels. Br J Audiol 34:165–177.

Moore BCJ (2003) An Introduction to the Psychology of Hearing, 5th ed. London: Academic.

Moore BCJ (2004) Testing the concept of softness imperception: loudness near threshold for hearing-impaired ears. J Acoust Soc Am 115:3103–3111.

Moore BCJ, Glasberg BR (1997) A model of loudness perception applied to cochlear hearing loss. Aud Neurosci 3:289–311.

Moore BCJ, Glasberg BR (2004) A revised model of loudness perception applied to cochlear hearing loss. Hear Res 188:70–88.

Moore BCJ, Glasberg BR, Stone MA (1999) Use of a loudness model for hearing aid fitting: III. A general method for deriving initial fittings for hearing aids with multi-channel compression. Br J Audiol 33:241–258.

Moore JM, Thompson G, Folsom RC (1992) Auditory responsiveness of premature infants utilizing visual reinforcement audiometry (VRA). Ear Hear 13:187–194.

Moore BCJ, Huss M., Vickers DA, Glasberg BR, Alcantara JI (2000) A test for the diagnosis of dead regions in the cochlea. Br J Audiol, 34: 205–224.

Newman CW, Wharton JA, Shivapuja BG, Jacobson GP (1994) Relationships among psychoacoustic judgments, speech understanding ability and self-perceived handicap in tinnitus subjects. Audiology 33:47–60.

Olsen HL, Olofsson A, Hagerman B (2004) The effect of presentation level and compression characteristics on sentence recognition in modulated noise. Int J Audiol 43:283–294.

Olsen SO (1999) The relationship between the uncomfortable loudness level and the acoustic reflex threshold for pure tones in normally-hearing and impaired listeners – a meta-analysis. Audiology 38:61–68.

Pascoe DP (1988) Clinical measurements of the auditory dynamic range and their relation to formulas for hearing aid gain. In: Jensen JH (ed), Hearing Aid Fitting, Theoretical and Practical Views (13th Danavox Symposium). Copenhagen: Stougard Jensen, pp. 129–152.

Patterson RD (1976) Auditory filter shapes derived with noise stimuli. J Acoust Soc Am 59:640–654.

Peek BF, Rosenfeld MA, Bratt GW, Williams DW (2007) NIDCD/VA hearing aid clinical trial and follow-up: coupler and real-ear measurement. J Am Acad Audiol 18:282–291.

Penner MJ (1986) Magnitude estimation and the "paradoxical" loudness of tinnitus. J Speech Hear Res 29:407–412.

Pickles JO (2008) An Introduction to the Physiology of Hearing, 3rd ed. London: Academic.

Picton TW, Dimitrijevic A, Perez-Abalo MC, Van Roon P (2005) Estimating audiometric thresholds using auditory steady-state responses. J Am Acad Audiol 16:140–156.

Plomp R (1994) Noise, amplification, and compression: considerations of three main issues in hearing aid design. Ear Hear 15:2–12.

Rawool VW (2001) Can maximum comfortable loudness levels in hearing impaired listeners be predicted from ipsilateral acoustic reflex thresholds recorded with high frequency probes? Scand Audiol 30:96–105.

Recio A, Rhode WS (2000) Basilar membrane responses to broadband stimuli. J Acoust Soc Am 108:2281–2298.

Reed GF (1960) An audiometric study of two hundred cases of subjective tinnitus. AMA Arch Otolaryngol 71:84–94.

Reger SN (1936) Differences in loudness response of the normal and hard-of-hearing ear at intensity levels slightly above the threshold. Ann Otol Rhinol Laryngol 45:1029–1039.

Robles L, Ruggero MA (2001) Mechanics of the mammalian cochlea. Physiol Rev 81:1305–1352.
Sammeth CA, Birman M, Hecox KE (1989) Variability of most comfortable and uncomfortable loudness levels to speech stimuli in the hearing impaired. Ear Hear 10:94–100.
Sanpetrino NM, Smith RL (2006) The growth of loudness functions measured in cochlear implant listeners using absolute magnitude estimation and compared using Akaike's information criterion. Conf Proc IEEE Eng Med Biol Soc 1:1642–1644.
Scharf B, Hellman RP (1966) Model of loudness summation applied to impaired ears. J Acoust Soc Am 40:71–78.
Schuknecht HF, Neff WD (1952) Hearing losses after apical lesions in the cochlea. Acta Otolaryngol 42:263–274.
Scollie S, Seewald R, Cornelisse L, Moodie S, Bagatto M, Laurnagaray D, Beaulac S, Pumford J (2005) The desired sensation level multistage input/output algorithm. Trends Amplif 9:159–197.
Shanks JE, Wilson RH, Larson V, Williams D (2002) Speech recognition performance of patients with sensorineural hearing loss under unaided and aided conditions using linear and compression hearing aids. Ear Hear 23:280–290.
Shaw EAG (1974) Transformation of sound pressure level from the free field to the eardrum in the horizontal plane. J Acoust Soc Am 56:1848–1861.
Skovenborg E, Nielsen SH (2004) Evaluation of different loudness models with music and speech material, Audio Engineering Society 117th Convention, San Francisco, pp. 1–34.
Smeds K (2004) Is normal or less than normal overall loudness preferred by first-time hearing aid users? Ear Hear 25:159–172.
Smeds K, Leijon A (2001) Threshold-based fitting methods for non-linear (WDRC) hearing instruments – comparison of acoustic characteristics. Scand Audiol 30:213–222.
Smeds K, Keidser G, Zakis J, Dillon H, Leijon A, Grant F, Convery E, Brew C (2006a) Preferred overall loudness. I: Sound field presentation in the laboratory. Int J Audiol 45:2–11.
Smeds K, Keidser G, Zakis J, Dillon H, Leijon A, Grant F, Convery E, Brew C (2006b) Preferred overall loudness. II: Listening through hearing aids in field and laboratory tests. Int J Audiol 45:12–25.
Stansfeld SA, Clark CA, Jenkins LM, Tarnopolsky A (1985) Sensitivity to noise in a community sample. 1. The measurement of psychiatric disorder and personality. Psychol Med 15:243–254.
Storey L, Dillon H, Yeend I, Wigney D (1998) The National Acoustic Laboratories' procedure for selecting the saturation sound pressure level of hearing aids: experimental validation. Ear Hear 19:267–279.
Studebaker GA, Sherbecoe RL, McDaniel DM, Gwaltney CA (1999) Monosyllabic word recognition at higher-than-normal speech and noise levels. J Acoust Soc Am 105:2431–2444.
Thornton AR, Yardley L, Farrell G (1987) The objective estimation of loudness discomfort level using auditory brainstem evoked responses. Scand Audiol 16:219–225.
Tyler RS (2000) The psychoacoustical measurement of tinnitus. In: Tyler RS (ed.), Tinnitus Handbook. San Diego, CA: Singular, pp. 149–179.
Tyler RS, Conrad-Armes D (1983) The determination of tinnitus loudness considering the effects of recruitment. J Speech Hear Res 26:59–72.
Valente M, Van Vliet D (1997) The independent hearing aid fitting forum (IHAFF) protocol. Trends Amplif 2:6–35.
Villchur E (1973) Signal processing to improve speech intelligibility in perceptive deafness. J Acoust Soc Am 53:1646–1657.
Vlastarakos PV, Nikolopoulos TP, Tavoulari E, Papacharalambous G, Korres S (2008) Auditory neuropathy: endocochlear lesion or temporal processing impairment? Implications for diagnosis and management. Int J Pediatr Otorhinolaryngol 72:1135–1150.
Whilby S, Florentine M, Wagner E, Marozeau J (2006) Monaural and binaural loudness of 5- and 200-ms tones in normal and impaired hearing. J Acoust Soc Am 119:3931–3939.

Zakis JA, Dillon H, McDermott HJ (2007) The design and evaluation of a hearing aid with trainable amplification parameters. Ear Hear 28:812–830.

Zeng F-G, Popper AN, Fay RR (eds) (2004) Cochlear Implants: Auditory Prostheses and Electric Hearing. New York: Springer.

Zeng F-G, Kong Y-Y, Michalewski HJ, Starr A (2005) Perceptual consequences of disrupted auditory nerve activity. J Neurophysiol 93:3050–3063.

Zhang M, Zwislocki JJ (1995) OHC response recruitment and its correlation with loudness recruitment. Hear Res 85:1–10.

Zwicker E (1958) Über psychologische und methodische Grundlagen der Lautheit. Acustica 8:237–258.

Chapter 10
Models of Loudness

Jeremy Marozeau

10.1 Introduction

10.1.1 What Is a Model?

Any chapter dedicated to reviewing models should first try to define what a model is. The question is more difficult than it first appears. Whereas most scientists agree that a model should represent a real-world phenomenon, many disagree on the level of complexity that a model should have. In his chapter on pitch models, De Cheveigné (2004) cites Norbert Wiener: "the best model of a cat is another, or preferably the same, cat." De Cheveigné strongly disagrees with this statement, arguing that a cat is not easier to handle than itself. He stressed the need for a model to be simpler than the original. Otherwise, why not use the original?

All models should have some inputs, some outputs, and a specific process in between, a "black box," as engineers like to call it. There are three main goals in creating a model: (1) to describe a complex phenomenon with some simple parameters, (2) to predict the output from known inputs, and (3) to test hypotheses on the mechanisms underlying a phenomenon.

10.1.1.1 Describe

It is often useful to describe a complex phenomenon in simple manner. One of the reasons is that often, at least in experimental data, a part of the complex structure of the data is irrelevant and can be due to some experimental noise. The regression line is the most widely known process to derive a "descriptive model." When data from an experiment are fitted by such a line, the structure of the data is simplified into a usable item. There is no absolute rule to select the number of coefficients

J. Marozeau (✉)
The Bionic Ear Institute, East Melbourne, VIC 3002, Australia
e-mail: jmarozeau@bionicear.org

M. Florentine et al. (eds.), *Loudness*, Springer Handbook of Auditory Research 37,
DOI 10.1007/978-1-4419-6712-1_10, © Springer Science+Business Media, LLC 2011

used in the regression equation. However, the higher the degree, the better the data are described, but also the less handy the model will be. One can fit any data perfectly using a regression line with the same degree as the number of experimental measures. However, this will result in modeling a cat with a cat.

10.1.1.2 Predict

If a model describes the experimental data well, then it can be used to predict the output of a phenomenon. Such use of the model can be found in many fields (seismology, economics, meteorology, etc...) because it is the one with the most commercial potential. A predictor of loudness is also truly valuable. Loudness is a subjective magnitude, and therefore can be assessed only by asking people to rate their sensations. However, it is often useful to be able to predict the loudness of a sound. For example, a car horn sound designer needs to know how loud a new device is perceived. Because it would be time consuming to run a psycho-acoustic experiment for each new prototype, a loudness model will help the designer to create a horn with an appropriate loudness.

10.1.1.3 Test Hypotheses

If a model describes and predicts the experimental data well enough, it can be used to test hypotheses that can lead to a better understanding of the underlying mechanism, or what is inside the black box. This is particularly valuable when studying hearing because the auditory system cannot be dismantled into pieces without serious consequences. A model of the effect of frequency on loudness can, for example, give some insight on how the external and middle ear filters the acoustic input.

This chapter reviews some of the models proposed in the literature in order of their complexity. First, loudness models for steady pure tones will be described. Then the complexity of the model is increased to be able to model the loudness of complex sounds, then time-varying sounds, and finally, hearing loss.

10.2 Modeling the Loudness of a Pure Tone

In psychophysics, the relationship between loudness and intensity of a pure tone was one of the first relationships to be studied. Different equations have been proposed to model this relationship. They might have caused a common misinterpretation between the result of the proposed model and the real-world sensation of loudness. It is often wrongly stated that “a sound has a loudness of X dB SPL,” because the sound pressure level (SPL) is a physical measure that is not a unit of loudness, per se (see Chapter 1).

10.2.1 The Logarithm Model

The first equation that can be considered as a model of loudness was proposed by Fechner (1860). The goal of the equation was to find the relationship between the intensity of a pure tone and corresponding loudness. The frequency of sounds was not taken into account. Fechner's model was based on Weber's law, which states that the minimum amount of intensity increase, ΔI, needed to induce a perceptible loudness difference is proportional to the intensity I.

$$\Delta I = kI \quad \text{or} \quad k = \Delta I / I, \tag{10.1}$$

where k is a constant. ΔI is also known as the Just-Noticeable-Difference or JND (a.k.a. Difference Limen, or DL). Every ΔI can then be associated with an increase on the sensation scale ΔS corresponding to the smallest perceptible sensation increase.

To derive the relationship between the sensation of loudness (S) and the intensity (I), Fechner assumed that each sensation increase would be identical in magnitude (ΔS is independent of S). In other words, for each detectable intensity increase, sensation would change by the same amount. Figure 10.1 illustrates this relationship.

The spacing between horizontal dotted lines represents one ΔS and between vertical dotted lines represents one ΔI. For convenience, the intensity I_0 is assumed to be perceived with a sensation of loudness S_0. Then, to increase the loudness, the

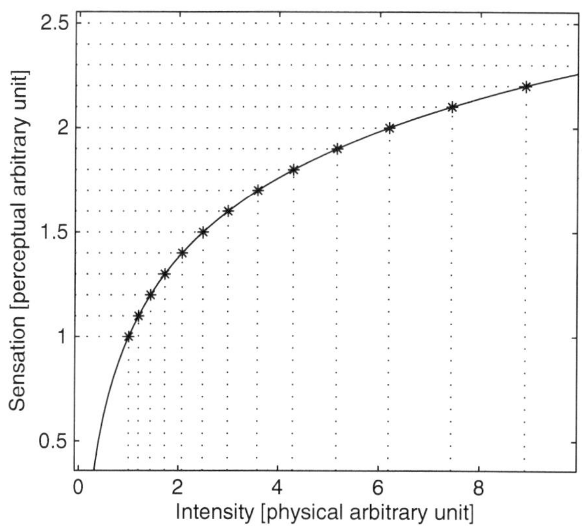

Fig. 10.1 Logarithmic model of the loudness function of a pure tone. Each point *asterisk* represents a new increment of loudness, logarithmically spaced on the physical axis (the intensity), and linearly spaced on the perceptual axis (the loudness). The overall relationship between the two axes has the shape of a logarithmic function

intensity should be raised by ΔI or $k * I_0$. The resulting loudness will correspond to $S_0+\Delta S$. In other words, an intensity of $I_0+k * I_0$, or $I_0 * (1+k)$ should induce a loudness of $S_0+\Delta S$. If the intensity is increased by another ΔI, the total intensity will be equal to $I_0 * (1+K) + k * I_0 * (1+k)$, and should induce a loudness of $S_0+\Delta S+\Delta S$. And so on:

$$I(0) = I_0 => S(0) = S_0;$$

$$I(1) = I_0 + K * I_0 = I_0(1+k) => S(1) = S_0 + \Delta S;$$

$$I(2) = I_0 + k * I_0 + k(I_0 + k * I_0) = I_0 * (1+k)^2 => S(2) = S_0 + \Delta S + \Delta S = S_0 + \Delta S * 2;$$

...

$$I(n) = I_0 * (1+k)^n => S(n) = S_0 + \Delta S * n \tag{10.2}$$

To extract the loudness as function of the intensity, the exponent n is derived from (10.2):

$$n = log_{10}(I / (I_0 * (1+k))); \tag{10.3}$$

So that

$$\begin{aligned} S(I) &= S_0 + \Delta S * log_{10}(I / (I_0 * (I+k))) \\ &= S_0 + C * 10 * log_{10}(I / I_0) \\ &= S_0 + C * L_{dB}, \end{aligned} \tag{10.4}$$

where $C = \Delta S/(10 * log_{10}(1+k))$ is a constant.

Therefore, this model proposes that loudness has a logarithmic relation with intensity (see Fig. 10.1) or a linear relationship with the level of a sound measured in decibels SPL.

10.2.2 The Power Function Model

As described in the preceding text, a logarithmic law is derived from the assumption that each new ΔI will produce a constant increase in sensation. However, another assumption has been proposed by Bretano (as cited by Stevens 1961) that states that each new ΔI will produce an increase of sensation proportional to the original sensation. In other words, it assumes that Weber's law also holds true for sensation: $\Delta S = S * T$ (where T is a constant). Figure 10.2 illustrates this relationship. Again, an intensity of I_0 is defined to evoke a sensation of magnitude S_0. The next differentiable intensity will be equal to $I_0+k * I_0$, corresponding to a sensation $S_0+S_0 * T$ and so on:

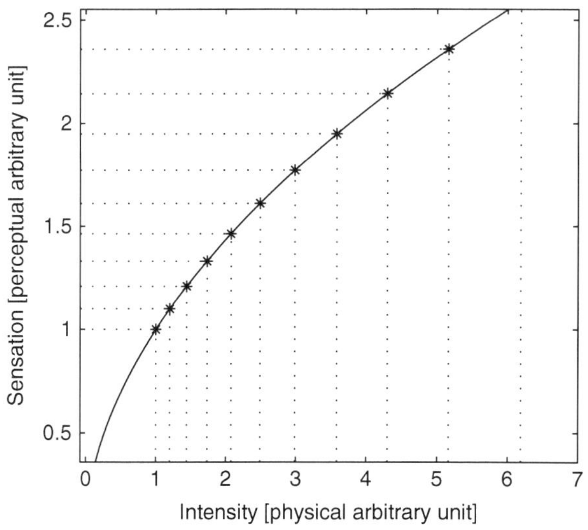

Fig. 10.2 Power model of the loudness of a pure tone. Each point *asterisk* represents a new increment of loudness, logarithmically spaced on both physical (the intensity) and perceptual axis (the loudness). The overall relationship between the two axes has the shape of a power function

$$
\begin{aligned}
I(0) &= I_0; \ \Rightarrow S(0) = S_0;\\
I(1) &= I_0 + k * I_0 = I_0(1 + k); \ \Rightarrow S(1) = S_0 + S_0 * T;\\
I(2) &= I_0 + k * I_0 + k(I_0 + k * I_0) = I_0 * (1 + k)^2; \ \Rightarrow S(2) = S_0 * (1 + T)^2;\\
&\ldots\\
I(n) &= I_0 * (1 + k)^n; \qquad \Rightarrow S(n) = S_0 * (1 + T)^n;
\end{aligned}
\tag{10.5}
$$

To extract the loudness as function of the intensity, the exponent n is derived from (10.5):

$$n = log_{10}\ (I/I_0)\ /\ log_{10}\ (1 + k); \tag{10.6}$$

and then

$$
\begin{aligned}
log_{10}(S / S_0) &= log_{10}(I / I_0) * log_{10}(1+T) / log_{10}(1+k)\\
log_{10}(S / S_0) &= log_{10}(I / I_0) * a\\
S &= S_0 * (I / I_0)^a
\end{aligned}
\tag{10.7}
$$

where $a = log_{10}(1+T)\ /log_{10}(1+k)$. Therefore based on this assumption, the loudness is related to the intensity raised to a power a. The relationship has been strongly supported by Stevens (1961). Therefore it is known now as the Stevens' power law.

The power function can be rearranged to consider the SPL as input instead of intensity:

$$S(I) = S_0 * (I / I_0)^a$$
$$log_{10}(S / S_0) = a * log_{10}(I / I_0) = k_2 * L_{dB} \quad (10.8)$$

where $k_2 = a/10$ is a constant and L_{dB} is the level in decibels SPL. This equation states that level and loudness are related in a logarithmic relationship. This is different from Fechner's law that states that the loudness increases with the SPL in a linear relationship.

To discriminate between these two theories, different psychoacoustic experiments have been performed using the methods of magnitude estimation or magnitude production (see Chapter 2). Most of the literature now supports a power function as a first approximation for levels above 40 dB SPL (see Chapter 5). Different exponents have been proposed, but the value of a = 0.3 is often used because it implies this convenient rule for high levels: Each time the level increases by 10dB, the loudness doubles:

$$S_1 = C * I_1^{0.3};$$
$$S_2 = C * I_2^{0.3};$$

$$if\ S_2 = 2 * S_1;$$

$$C * I_2^{0.3} = 2 * C * I_1^{0.3}$$
$$2 = (I_2 / I_1)^{0.3}$$
$$\log_{10}(2) = 0.3 * \log_{10}(I_2 / I_1) = 0.3 * (L_2 - L_2) / 10;$$
$$\log_{10}(2) \approx 0.3$$
$$10\ dB \approx L_2 - L_1$$

10.2.3 InEx Function Model

The InEx function is now believed to be a better description of the loudness–growth function for tones than the power function (see Chapter 2). Therefore, the InEx function can be used as a descriptive model of the loudness function of a pure tone. Figure 10.3 shows the InEx function as derived by Buus and Florentine (2002) and defined by Florentine and Epstein (2006). The InEx function model differs from a power function in two main ways. At low levels, the slope is close to 1, and the slope gradually declines as level increases until it reaches a minimum at moderate levels, in the 40–60 dB SPL range. At these levels, the slope of the loudness function must be shallower than the standard power function slope of 0.3 in order to explain temporal integration and loudness summation data. At higher levels, the

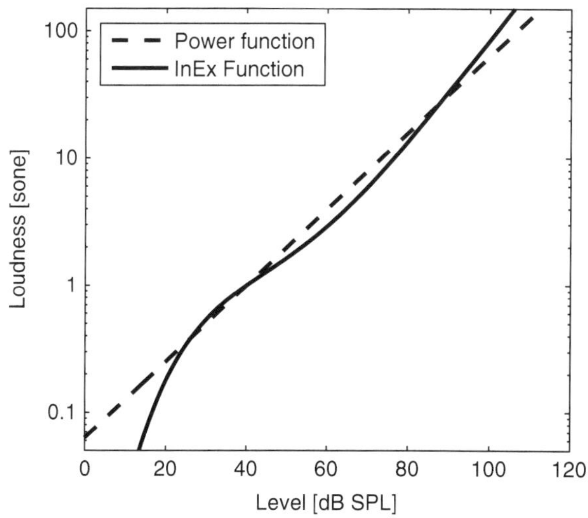

Fig. 10.3 InEx model of the loudness function. The *continuous line* represents the InEx function in comparison with the power function (*dashed line*)

slope gradually increases, but does not reach 1. To account for these variations in the loudness function, the InEx function replaces the constant slope of the power function model with a continuous polynomial, allowing the slope to vary slowly with level.

This function has been derived from a measure of loudness integration (Buus and Florentine 2002) and has been fitted with a five-order polynomial:

$$
\begin{aligned}
Log_{10}(S / S_0) = \ldots \\
&1.7058 * 10^{-9} * L^5 - 6.587 * 10^{-7} * L^4 \\
&+ 9.7515 * 10^{-5} * L^3 - 6.6964 * 10^{-3} * L^2 \\
&+ 0.2367 * L - 3.4831
\end{aligned}
\tag{10.9}
$$

where S is the sensation and L is the level in decibels SPL.

10.3 Loudness of Complex Sounds

The preceding models predict how the loudness of tones will change based on one parameter of a sound: the intensity or level. However, other parameters of sound can influence the loudness such as the frequency or the spectral shape (Chap. 5). Therefore a loudness model with only one input coefficient, as a power function, is simple and easy to implement, but cannot react with precision to the wide range and complexity of environmental sounds.

10.3.1 Loudness as Estimated by Level Meters in dB(A, B, C,…)

To improve the prediction of loudness, most commercial level meters incorporate different filters. Each filter is based on sound sensitivity at a particular SPL, determined using the equal loudness contours, described by Jesteadt and Leibold, (Chap. 5). A-weighting uses a 40-phon reference, B-weighting uses a 70-phon reference, and C-weighting uses a 100-phon reference. Sounds weighted using these references are given levels in units of dB(A), dB(B), and dB(C), respectively. Effectively, the sound is filtered using the inverse of the loudness contour, so that frequency regions that contribute less to loudness are attenuated. These are, however, rough approximations and do not guarantee that loudness is properly balanced across frequency. It is possible to generate sounds with the same level in dB(A) that are distinctly different in loudness. The way loudness increases with bandwidth is not taken into consideration in any of these weightings.

10.3.2 Zwicker's Model

The loudness of two tones with the same level depends on their frequency separation and phase angles (see Chap. 5). If they are in phase and separated by more than a critical band, their loudness will increase gradually with the frequency separation to reach the sum of their individual loudnesses. If they are separated by less than a critical band, their loudness can be modeled as a function of the sum of their intensities. Unfortunately, the story gets more complicated when the relative intensity of each tone varies. If the low-frequency tone has a higher intensity than the high-frequency tone, then the latter can be partially or completely masked by the former. Therefore, the loudness of two tones with different intensities that are separated by more than a critical band can vary from the sum of both loudness (no masking) to the loudness of the louder one (complete masking of the other tone). Such a phenomenon is called spectral masking.

A reliable model of loudness should integrate these spectral interactions between tones of different frequencies. Such a model was developed by one of the most renowned psychophysicists, Eberhard Zwicker (1924–1990), and has been the subject of the standard ISO R532B (Zwicker and Fastl 1990). The standard has been implemented in many different commercial products.

Although the underlying mechanism of coding of loudness is still not clear, this model is based on the assumption that loudness is directly and solely related to the integration of the specific loudness that are based on excitation patterns. The model is divided into three main stages: (1) transformation from the sound's physical property to the level of excitation, (2) transformation to specific loudness, and (3) summation of specific loudness.

Each stage is reviewed with an illustration of a six-tone complex. This sound has been processed by a personal implementation of the model in Matlab according to ISO R532B. The theoretical approach of the model has been detailed in different publications (Zwicker 1961; Zwicker and Scharf 1965; Zwicker and Fastl 1990). However, this section focuses on a practical approach published as a software implementation guideline (Zwicker et al. 1991).

10.3.2.1 Stage I: From Signal to Level of Excitation

The model takes as input the long-term power spectrum of the sound. Therefore, it does not consider the phase relationship between components or any temporal variations. The top panels of Fig. 10.4 show the temporal (Fig. 10.4a) and spectral (Fig. 10.4b) representation of a complex tone composed of a 250-Hz tone at 80 dB SPL, a 500-Hz tone at 73 dB SPL, a 2.3-kHz tone at 80 dB SPL, a 2.5-kHz tone at 65 dB SPL, a 10-kHz tone at 80 dB SPL, and a 10.5-kHz tone at 80 dB SPL.

The result of a discrete Fourier transform (DFT) is typically represented as a histogram, with the frequency on the abscissa and the amplitude (of the pressure or the intensity) on the ordinate. The length of each bin, or bar, is proportional to the amount of energy around a specific frequency inside a constant bandwidth. This bandwidth is related to the number of points of the DFT divided by the sampling frequency, and is constant across each bar. The center frequency of each bin is linearly spaced, and ranges from zero to one half of the sampling rate.

The first stage of the model acts as a modified DFT, but with only 24 bins. The center frequency of each bin has been specifically designed to mimic the presumed frequency selectivity of the cochlea and the bandwidth of each bin reflects the width of the critical bands (Zwicker and Fastl 1990). The values of the center frequencies have been selected so that the 24 bins span without overlapping in a frequency region from 50 Hz to 13.5 kHz (Zwicker 1961), and approximate the Bark scale that models the tonotopic organization of the cochlea. Table 10.1 gives values of the center of frequencies and bandwidths for each of the 24 bands. Figure 10.4c shows the energy summed in the 24 bins for the six-tone complex. The energy is spread into only five bins instead of six in the DFT, because the two highest components, at 10 and 10.5 kHz, are summed into the same critical band.

The effect of the frequency response of the outer and middle ear is modeled as a linear filter. It is implemented by subtracting each bin by a factor, a0 (see Fig. 10.4d, + line). This factor is the inverse of the threshold of hearing above 2 kHz. It amplifies the frequencies around 4 kHz, and progressively attenuates the frequency above 6 kHz. Below 2 kHz, the model assumes that the factor a_0 is constant and equal to unity. The diamond line in Fig. 10.4c shows the energy level in each bin after subtraction of a_0. The high auditory threshold at low frequencies is assumed to be a consequence of a second phenomenon: dependence of the internal noise floor on frequency. The dependence is represented in Fig. 10.4d (o line). It is assumed to be fairly constant from mid-to-high frequencies, and increases significantly at low frequencies. Its effect is modeled in Sect. 10.3.2.2.

When the cochlea is excited by a pure tone, the amount of membrane excited will increase with the level. The first part of the model will model only the effect of the signal in its critical band; the second part models this influence

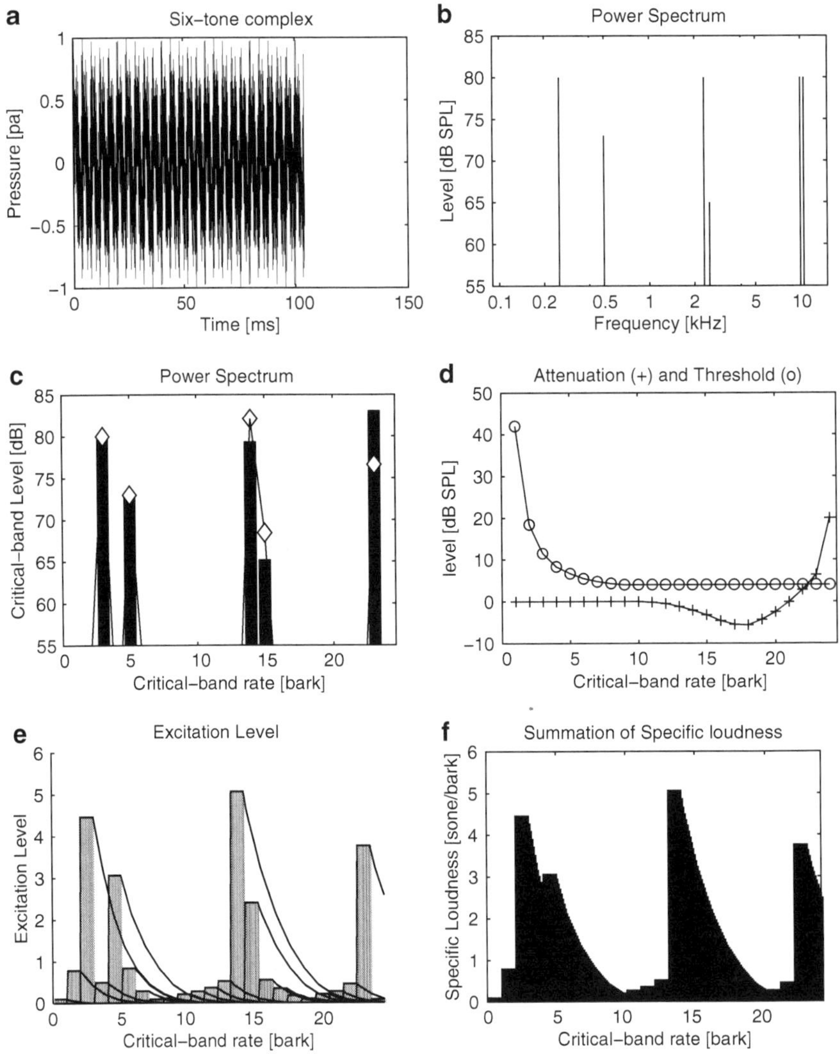

Fig. 10.4 Different stages of Zwicker's model. First, the input signal is shown in the temporal (**a**) and spectral (**b**) domains. Then, the model filters the signal into critical bands (**c**), and attenuates the signal according to two linear filters (**d**, and *diamonds* on **c**). Finally, the signal is transformed into excitation level (**e**) and specific loudness, which is integrated across critical bands (**f**)

outside its critical band. The spread of excitation is greater toward the high frequencies than the low frequencies, and is therefore assumed to be negligible. The spread of excitation toward the high frequencies is given by a table (see ISO R532B).

Figure 10.4e shows the spread of excitation for each tone and that different excitation patterns overlap. If many excitation patterns overlap the same critical

Table 10.1 Summary of the center frequencies and bandwidths of the 24 critical bands (BARK)

Band number	Center frequency	Bandwidth
1	60	80
2	150	100
3	250	100
4	350	100
5	450	110
6	570	120
7	700	140
8	840	150
9	1,000	160
10	1,170	190
11	1,370	210
12	1,600	240
13	1,865	280
14	2,150	320
15	2,500	380
16	2,900	450
17	3,400	550
18	4,000	700
19	4,800	900
20	5,800	1,100
21	7,000	1,300
22	8,500	1,800
23	10,500	2,500
24	13,500	3,500

band, the model will consider only the highest one. For example, the excitation pattern of the 2.5-kHz tone is entirely included within that of the 2.3-kHz tone. This configuration will lead to total masking of the 2.5-kHz tone by the 2.3-kHz tone. Therefore, the 2.5-kHz tone makes no effective contribution. Further, the excitation of the 250-Hz tone is partially included in the excitation pattern of the 500-Hz tone. In that situation, the 500-Hz tone will be partially masked and its contribution to loudness would be diminished. Figure 10.4f shows the excitation pattern.

10.3.2.2 Stage II: From Excitation Pattern to Specific Loudness

The next stage converts the excitation pattern into specific loudness, N', which represents the contribution of each band to the overall loudness. The relationship between excitation pattern and specific loudness is based on the power function and is adjusted in order to correctly predict the empirical results of loudness–growth functions for different types of sounds.

$$N' = \alpha \left(\frac{E_{TQ}}{E_0} \right)^{\beta} \left[\left(0.5 + 0.5 \frac{E}{E_{TQ}} \right)^{\beta} - 1 \right]_{[\text{sone/Bark}]} \qquad (10.10)$$

In this equation, E_{TQ} is the excitation at threshold in quiet as a function of the frequency, produced by the internal noise (see Fig. 10.4d, o line); E is the excitation at the specific frequency; E_0 is the excitation that corresponds to the reference intensity $I_0=10^{-12} W/m^2$; α and β are two constants. When $E >> E_{TQ}$, then N' can be approximated as equal to $\alpha * 0.5 * (E/E_0)^{\beta}$. Therefore β corresponds to the exponent of the loudness function, which has been set in the standard at a value of 0.23. The constant α is related to the arbitrary choice of the scale. It has been set to 0.08, so that a 1-kHz pure tone at 40 dB SPL in free field will induce a loudness of 1 sone.

10.3.2.3 Stage III: From Specific Loudness to Overall Loudness

The last step of the model is to sum across specific loudness.

$$N = \sum_{0}^{24} N' \, dz_{[\text{sone}]} \qquad (10.11)$$

In this equation, N is the total loudness. The model predictions can be transformed to loudness levels Ln in phons, according to these equations:

$$L_n = \begin{cases} N >= 1 & 40 + 10 * \log_2(N) \\ N < 1 & 40 * N^{0.35} \end{cases} [\text{phon}] \qquad (10.12)$$

To illustrate, the model gives a loudness of 38 sones and a loudness level of 92 phons (the complex is as loud as a 1-kHz pure tone at 92 dB SPL). For a comparison, the overall signal level is 85 dB SPL and 83 dB(A). Therefore, the model predicts that this sound will be perceived as having the same loudness as a pure tone with a level 9 dB(A) higher.

10.3.3 Moore and Glasberg's Model

A new ANSI standard (2007) has been published to revise the previous computational model of loudness from Zwicker. This new model, proposed by Moore and colleagues, is strongly based on the previous model (Moore and Glasberg 1997), but each stage is modified. The five stages of the model are: (1) transmission through the outer ear, (2) transmission through the middle ear, (3) transformation of the spectrum to excitation pattern, (4) transformation of excitation pattern to specific loudness, and (5) transformation from specific loudness to

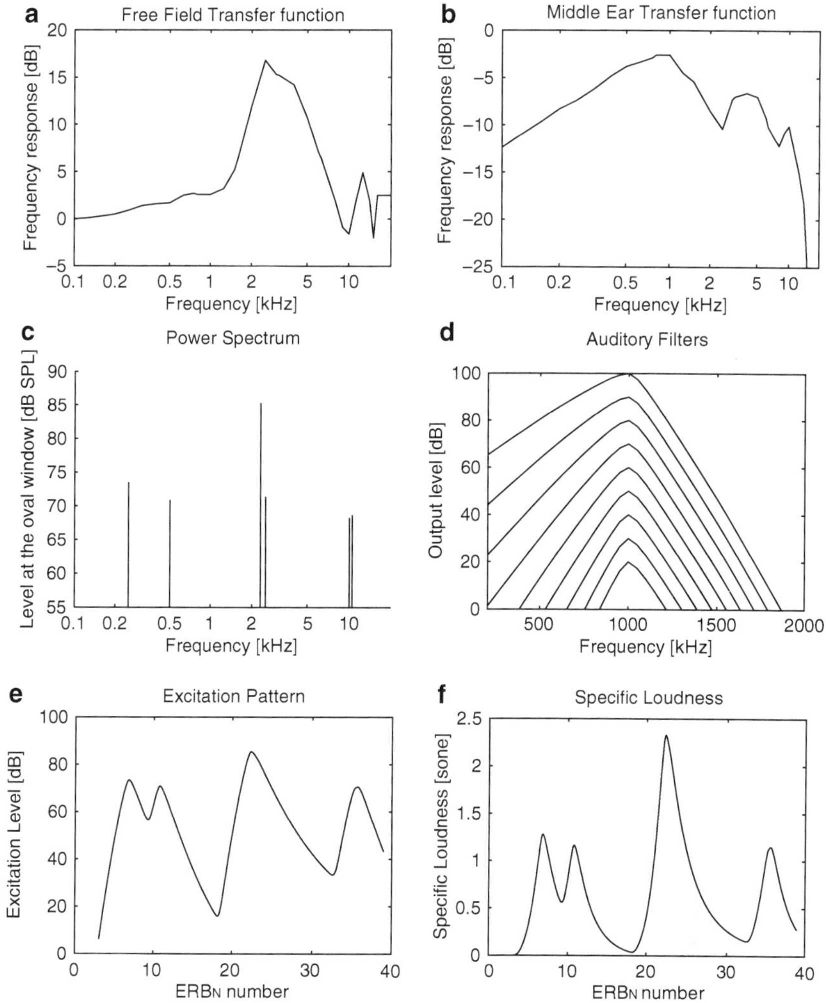

Fig. 10.5 Different stages of Moore's model. The input signal is the same as the one shown in Fig. 10.4. (**a**, **b**) Represent the linear filter that modeled the outer and middle ear; (**c**) shows the spectrum of the signal that will enter the middle ear; (**d**) shows the output level in decibel of the auditory filter centered at 1 kHz for input levels/ ERB_N from 20 to 100 dB in 10-dB steps, plotted as a function of frequency; (**e**) shows the excitation pattern, and (**f**) the specific loudness

overall loudness. To illustrate each stage, the same example as in the previous model will be used. Figure 10.5 is based on the output of a C implementation of the loudness model developed by Brian Moore and colleagues, according to the latest published version (Glasberg and Moore 2006). The software is available from the laboratory website at: http://hearing.psychol.cam.ac.uk. The sound was considered as being presented binaurally in free field with frontal incidence.

10.3.3.1 Stage I. Transmission Through the Outer Ear

The transmission of a sound to the eardrum depends on how it is delivered to the ear. Sounds are usually presented via one of two possible presentation modes: sound field or headphones. Each presentation mode modifies the sound by attenuating or amplifying certain frequencies. Regardless of presentation mode, the ear canal and tympanic membrane resonances are included in this power-spectrum alteration. Provided that a sound is not presented via earphones, resonances of the room, head, body, and pinna shape will also alter the characteristics of a sound. The effects of all of these external processes are modeled in this stage by applying a filter to the power spectrum to characterize the listening condition.

In the present example, the sound is assumed to be presented in free field. Therefore, a filter that modeled the transformation from free-field sound pressure (measured in the absence of the listener at the position corresponding to the center of the listener's head) to eardrum sound pressure is needed. Moore and colleagues specified this filter as in Shaw (1974). Figure 10.5a shows the frequency response of such a filter.

10.3.3.2 Stage II. Transmission Through the Middle Ear

The principal role of the middle ear is to match the impedances between air in the external ear canal and the lymphatic fluid of the inner ear. This system does not have a flat frequency response and, therefore, the amount of acoustic energy transmitted to the inner ear depends on the frequency of the signal. Again, a filter is used to model the middle-ear response. The filter was divided in two parts. Above 500 Hz, the transmission through the middle ear is modeled as the inverted curve of the minimum audible pressure at the eardrum as a function of frequency. Below 500 Hz, the filter is based on measurements of the human middle ear transfer function. Figure 10.5b shows the frequency response of this filter. Figure 10.5c shows the spectrum of the six-tone complex example after both filters. As compared to the spectrum plotted on Fig. 10.4b, mid-frequency components have been amplified and high frequencies have been attenuated.

10.3.3.3 Stage III. Transformation of the Spectrum to Excitation Pattern

Once transmitted to the inner ear, sound vibrations travel along the basilar membrane, which is organized tonotopically. That is, each part of the basilar membrane responds best to a different frequency. A bank of band-pass filters models this stage, with each filter representing the frequency selectivity at a specific place on the basilar membrane. The shapes of these filters were selected using data gathered from notched-noise experiments (Glasberg and Moore 1990), a paradigm designed to examine frequency-specific listening. As in Zwicker's model, the bandwidth of the filter increases with frequency; however, the formula is different:

$$ERB_{\mathrm{N}} = 24.673(0.004368f + 1), \tag{10.13}$$

where f is the center frequency in hertz of the filter and *ERB* stands for Equivalent Rectangular Bandwidth. The model assumes this relationship to hold true only for mid-level sounds (at 51 dB SPL/ERB_N). A correction based on the level is applied to the shape of the filter for low- and high-level sounds. The different equations to derive this filter are beyond the scope of this chapter. However, Fig. 10.5d reproduces an example for a filter centered at 1 kHz at different levels. One can see that the bandwidth of the filter increases with level.

The tonotopic organization of the cochlea is not linear: high frequencies will be compressed compared to the low frequencies. Moore and colleagues propose a scale that represents processing more accurately than the Zwicker model below 500 Hz and called it the ERB_N-number scale.

$$ERB_{\mathrm{N}} \text{ number} = 21.366 \log_{10}(0.004368f + 1), \tag{10.14}$$

Figure 10.5e shows the excitation pattern evoked by the six-tone complex.

10.3.3.4 Stage IV. Transformation of Excitation Pattern to Specific Loudness

Specific loudness is roughly like a loudness density and can be defined as the amount of loudness within a critical band. Psychoacoustical experiments have shown that the magnitude of the excitation on the basilar membrane is not linearly related to the acoustic power. Specifically, changes in basilar membrane vibration do not occur proportionally to level changes, as the system is compressive at moderate levels. The behavior is modeled in this stage by applying a nonlinear, compressive, monotonic function to the excitation pattern to transform it into specific loudness:

$$N' = C[(GE_{sig} + 2E_{Thrq})^{\alpha} - 2E_{Thrq}{}^{\alpha}] \tag{10.15}$$

where N' is the specific loudness, C is a constant, *Esig* is the excitation level, E_{Thrq} is the excitation level at threshold, and G is the low-level gain of the cochlear amplifier. Some modifications of this equation, beyond the scope of this review, occur for the extreme intensity levels, and for frequencies below 500 Hz. However, it is important to note that the specific loudness is modeled as a power function of the excitation pattern. The exponent α will vary between 0.27 and 0.2 depending on the frequency. Figure 10.5f reproduces the specific loudness of the six-tone complex.

10.3.3.5 Stage V. Transformation from Specific Loudness to Overall Loudness

Finally, the overall loudness of the sound is obtained by integrating the specific loudness across the entire ERB_N-number scale. The overall loudness of the six-tone complex is estimated as 45.7 sones or 94.3 phons.

10.4 Loudness Models for Temporally Varying Sounds

Modifications of each of the three models of loudness of complex sounds described above have been proposed to deal with nonstationary sounds. This section reviews each of them.

10.4.1 Time Constants and Sound Level Meters

Commercial sound level meters estimate the level of a sound by measuring the average over a time window of the filtered (A, B, C, D) SPL. The length of the time window depends on the device specification and the user's setup, the most common ones being designated as "fast" and "slow," with nominal exponential time constants of 125 and 1,000 ms, respectively. Therefore, any temporal variation of the amplitude envelope within the time window will be averaged out.

Namba et al. (2008, see also (Chap. 9) showed that the loudness of sounds with a duration shorter than 1s correlates with the total energy, L_{AE}; and the loudness of sounds longer than 1s correlates with the averaged energy, $L_{\mathrm{Aeq,T}}$.

10.4.2 Zwicker's Model for Nonstationary Sounds

Zwicker (1984) proposed a modification of his model to deal with temporal variation. A new final stage was introduced and has been implemented as an analogue electric circuit (see Fig. 10.6). A detailed understanding of the circuit is beyond the scope of this chapter; however, the following gives a brief overview:

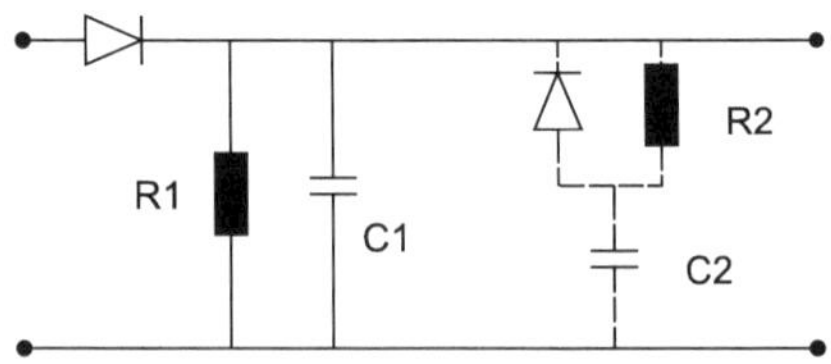

Fig. 10.6 Analogue electric circuit that models the loudness of nonstationary sounds according to Zwicker (1984)

The charge and discharge of the condenser depends on the input and output voltage. If the input signal increases in intensity, the diode, D1, will act as a closed switch and D2 as open switch. Therefore C1 will be instantaneously charged with the signal current, and C2 will be loaded with the time constant T2 = R2 * C2, which is equivalent to 20ms. C2 can be considered as charged at 99% after a duration of 5*T2 = 100ms. This value is within the values proposed by Boone (1973).

After the offset of the signal, two configurations are possible: (1) The tone burst lasted less than 100 ms. The condenser C2 did not have the time to charge fully, and C1 will discharge into R1 and will charge C2 through R2, (2) The tone burst lasted more than 100 ms. The condenser C2 had time to charge fully. Therefore, both condensers will discharge into R1 with a larger time constant ((C1 + C2) * R1), set according to experiments on forward masking.

This stage models how the loudness varies with signal duration. However, to derive a single value of loudness for the overall signal, Zwicker and Fastl (1990) recommend using some statistical descriptor such as N4, N5, or N10, which is the loudness value reached 4, 5, or 10% of the time, respectively.

A more recent adaptation of the Zwicker's model to nonstationary sounds, called the Dynamic Loudness Model, has been described by Chalupper and Fastl (2002). The temporal process of the loudness is modeled within each channel (before the spectral loudness summation) by appending temporal tails to peaks of the specific loudness according to data on forward masking (Zwicker 1984).

10.4.3 The Model of Moore and Glasberg for Nonstationary Sounds

The model of loudness proposed by Moore and colleagues has also been modified to account for nonstationary sounds (Glasberg and Moore 2002). This model is based on three different stages of loudness. The illustrations have been computed with the implementation of the time-varying model of loudness provided by Brian Glasberg (available on their website). The sound example used was the first 4s of a drum roll extracted from the video logo of a famous last century film corporation.

10.4.3.1 Instantaneous Loudness

The first stage of the model consists of deriving the instantaneous loudness by dividing the signal into small segments with a duration that varies with the frequency range. Then, the spectrum of each of the segments is used to extract a loudness value according to the stationary model (see Sect. 10.3.3). This results in a function of instantaneous loudness sampled at 1 kHz (or a period of 1ms). Figure 10.7 shows as a continuous fine line the instantaneous loudness of the drum roll. Glasberg and Moore assume that "*the instantaneous loudness is an intervening variable which is not available for conscious perception.*"

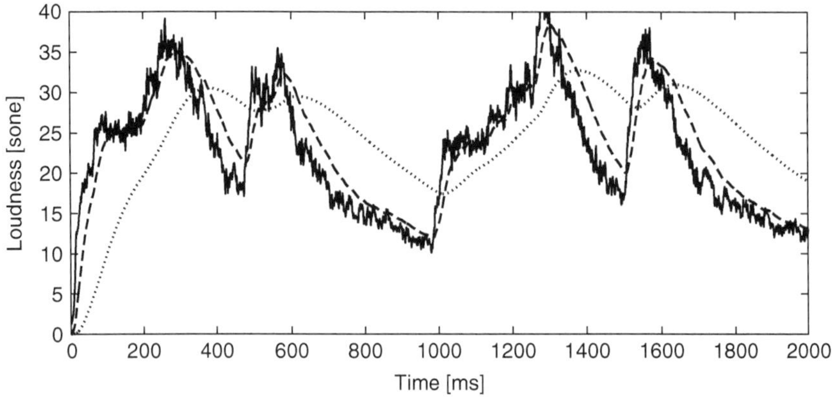

Fig. 10.7 Three different levels of time-varying loudness models: instantaneous loudness (*continuous line*), short-term loudness (*dashed line*), and long-term loudness (*dotted line*)

10.4.3.2 Short-Term Loudness

The second stage consists of modeling the perceived loudness. In the same spirit as Zwicker's model of loudness for nonstationary sounds, this model applies a low-pass filter to the instantaneous loudness. Temporal integration data and forward masking data have two different time constants: the loudness seems to rise more rapidly than it decreases. Therefore, this model proposes two low-pass filters, one for the onset and one for the offset. The instantaneous loudness at the sample n is represented by $Ni(n)$, and short-time loudness by $Nst(n)$.

If $Ni(n) > Nst(n-1)$, then the loudness is increasing. In this case, the value of $Nst(n)$ is modeled as follows:

$$Nst(n) = \alpha_a * Ni(n) + (1-\alpha_a) * Nst(n-1). \tag{10.16}$$

where α_a is a constant that sets the response of the "attack" filter. This constant can be expressed as

$$\alpha_a = 1 - exp(-T_i / T_a), \tag{10.17}$$

where T_i is the sampling period of the instantaneous loudness, i.e. 1ms, and T_a is a constant that set the time response of the filter. If the instantaneous loudness shows a sudden onset like a staircase, then the short-term loudness will reach 99% of the maximum $5{*}T_a$ seconds after the onset of the instantaneous loudness. The model set T_a to 0.027s, which implies a critical duration of 100 ms, and is within the range of value proposed in the literature (see (Chap. 6).

If $Ni(n) < Nst(n-1)$ then the loudness is decreasing. In this case, the value of $Nst(n-1)$ is modeled as follows:

$$Nst(n) = \alpha_r * Ni(n) + (1-\alpha_r) * Nst(n-1), \tag{10.18}$$

where α_r is a constant that sets the response of the "release" filter. This constant can be expressed as:

$$\alpha_r = 1 - exp(-T_i / T_r), \quad (10.19)$$

where T_r is a constant that sets the time response of the "release" filter to 49.5ms. This value is in compliance with measured loudness of amplitude-modulated sounds. Figure 10.7 shows the short-term loudness as a dashed line. Notice the difference between the onset and offset portion of the sound.

10.4.3.3 Loudness Models for Long-Duration Sounds

The short-term loudness models reproduce the perceptual variation of loudness. Even though some variations are perceived, it would be useful to model the perception of the overall loudness impression. For example, a series of pulse trains will be perceived louder than any individual pulse. The sensation is designated as the "long-term loudness," *Nlt*. It is modeled as a low-pass version of the short-term loudness, with two filters (one for onset, one for offset). The equation is the same as for the short-term loudness, except for the time constants of the filters. Figure 10.7 shows the long-term loudness as a dotted line.

10.5 Loudness Models Applied to Hearing-Impaired Listeners

The previous models are based on average results from psychoacoustic experiments performed with listeners with healthy cochleas. As shown by Smeds and Leijon, (Chap. 9, a sensorineural hearing loss can affect loudness in different ways, such as an elevation of threshold, loss of compression, and an increase of the critical bandwidth. This section describes how each type of model can be modified to account for hearing impairment.

10.5.1 Modeling the Loudness of Pure Tones for Hearing-Impaired Listeners

Hearing-impaired listeners (HILs) with sensorineural hearing losses have reduced dynamic ranges. That is, the difference between the highest tolerable sound level and the lowest detectable sound level is less than for normal-hearing listeners (NHLs). This primarily results from the fact that HILs have elevated thresholds and nearly the same (or sometime slightly lower) tolerability at high levels as NHLs. Therefore, the loudness function must be different for HILs than for NHLs. Two models, described below, have been proposed that can explain the reduction of dynamic range.

10.5.1.1 Rapid Growth Model

The primary assumption for the rapid growth (a.k.a. recruitment) model is that loudness at threshold is the same for the HILs and NHLs. If loudness at threshold for a HIL is the same as for a NHL, then loudness functions must grow more steeply for HILs than NHLs for loudness to catch up at high levels. Figure 10.8 shows on the same graph the InEx (see Sect. 10.2.3) function (designated with a star), the power function (dashed line), and the rapid growth model (designated with a circle).

10.5.1.2 Softness Imperception Model

The primary assumption for the softness imperception model is that loudness at threshold is not the same for HILs and NHLs. Softness imperception is based on the concept that loudness at threshold is higher for an HIL than for a NHL and the loudness function shows little-to-no compression at moderate levels, but otherwise is not steeper than the loudness function seen for NHLs. Figure 10.8 also includes the softness imperception model (designated with a square).

The rapid growth model has traditionally been accepted as the foundation for algorithms used in hearing aids to adapt gain as a function of level. Although softness imperception is a more recent concept and is less commonly implemented in hearing aid fittings, many fitting procedures include low-level expansion (reduction of gain

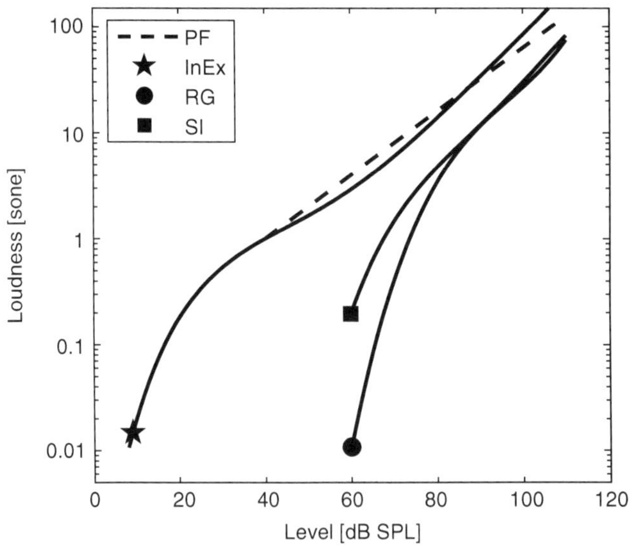

Fig. 10.8 Comparison of the loudness–growth models: power function (*dashed line*), InEx function (*asterisk*); rapid growth (*circle*), and softness imperception (*square*). Loudness in sone is plotted as function of sound pressure level

for very weak sounds) so that low-level sounds are not amplified into the audible range. This fitting is consistent with the idea that low-level sounds amplified to audibility may be disconcertingly loud for HILs and that linear adaptive algorithms such as ADRO (Blamey, 2005) may lead to more comfortable amplification than wide dynamic range compression, WDRC, without reducing speech perception at soft, medium, and loud levels (Blamey and Martin 2009).

There is still significant debate regarding these two models and some evidence supports the idea that some HILs exhibit rapid growth, some exhibit softness imperception, and some exhibit an intermediate behavior (for discussion see Moore and Glasberg 2004; Marozeau and Florentine 2007). Other data support the contention that there are clear individual differences in the loudness functions of HIL (e.g. Marozeau and Florentine 2009).

10.5.2 Zwicker's Loudness Model Applied to Noise-Induced Hearing Loss

10.5.2.1 Modification by Florentine and Zwicker

Zwicker's loudness model was modified to account for noise-induced hearing loss. Different approaches have been proposed. Florentine and Zwicker (1979) first proposed to increase the value of excitation at the threshold of quiet (E_{TQ}) in the formula for the specific loudness (10.10). To understand how this would affect the model, this formula should be restated as function of the level L_E:

$$\begin{aligned} L_E &= 10 * \log_{10}(E / E_0); \\ L_{TQ} &= 10 * \log_{10}(E_{TQ} / E_0); \\ N' &= \alpha * 10^{L_{TQ}\beta/10}[(0.5 + 0.5 * 10^{(L_E - L_{TQ})/10})^{\beta} - 1] sone / Bark \\ \log_{10}(N') &= \log_{10}(\alpha) + L_{TQ}\beta / 10 + \log_{10}[(0.5 + 0.5 * 10^{(L_E - L_{TQ})/10})^{\beta} - 1] \end{aligned} \tag{10.20}$$

This equation shows that the effect of an increase in L_{TQ} will be twofold: first an upward shift ($L_{TQ}\beta/10$) and then a horizontal shift ($L_E - L_{TQ}$). Further, it is worth noting that the derivative: $d(\log_{10}(N'))/d(L_E)$ depends only on the sensation level ($L_E - L_{TQ}$). Therefore, the equation assumes that the slope of loudness growth as a function of sensation level of a HIL is similar to one of normal listener with an inaudible masking noise. However, because the function is shifted upward with E_{TQ}, the growth rate between two fixed loudness will increase with E_{TQ} (as showed in Fig. 8.17 of Zwicker and Fastl 1990).

Florentine and Zwicker (1979) found that this correction was not sufficient to describe experimental data. Therefore, they suggested increasing the bandwidth of the critical band by a factor of 3 to account for the reduced loudness summation in noise-induced hearing loss (see (Chap. 9). With this change they were able to fit the model to empirical data.

10.5.2.2 Modification by Chalupper and Fastl

Chalupper and Fastl (2002) proposed a model where the hearing loss is divided into two components, HLa and HLe, which account for the attenuation and rapid growth phenomenon, respectively. Equation (10.10) is modified as follows:

$$N' = \alpha\left(\frac{E_{TQ} * E_{HLe}}{E_0^{\;2}}\right)^{\beta}\left[\left(0.5+0.5*\left(\frac{E / E_{HLa} * E_0}{E_{TQ} * E_{HLe} / E_0}\right)\right)^{\beta} - 1\right] \text{sone/Bark}$$

$$\log_{10}(N') =$$

$$\log_{10}(\alpha) + (L_{TQ} + L_{HLe})\beta / 10 + \log_{10}[(0.5 + 0.5 * 10^{(L_E - L_{TQ} - L_{HLa} - L_{HLe})/10})^{\beta} - 1] \quad (10.21)$$

If the L_{HLe}= 0, then the function is simply shifted horizontally (as in a conductive hearing-loss). If L_{HLa}= 0, then the model would be similar the one proposed by Florentine and Zwicker (without the increase of critical bandwidth). If L_{HLe} and L_{HLa} are not zero, then these two free parameters of the model will give more flexibility to describe individual hearing-impaired loudness data.

10.5.2.3 Modification by Launer

Another two-parameter approach was proposed by Launer (1995). The first parameter is an attenuation added to each critical band, and the second varies the exponent β. Therefore, the model derived from this approach will vary the slope of the loudness function as described by the phenomenon of rapid growth.

10.5.3 The Loudness Model of Moore and Glasberg Applied to Noise-Induced Hearing Loss

Moore and Glasberg's loudness model (1997) was recently updated to account for hearing-loss (Moore and Glasberg 2004). As in the model of Chalupper and Fastl (2002), the total hearing loss (HL_{TOT}) is divided into two parts: a part due to reduced functioning of outer hair cells (HL_{OHC}) and a part due to reduced functioning of inner hair cells (HL_{IHC}). The two parts add up to give the measured hearing loss:

$$HL_{TOT} = HL_{IHC} + HL_{OHC}. \quad (10.22)$$

First, the HL_{IHC} will primarily affect the response of the inner hair cells. In the extreme case of a dead region of the cochlea, all IHC are damaged so no sounds can be conveyed. This contribution was originally modeled as a simple attenuation of excitation level by the level of HL_{IHC} (Moore and Glasberg 1997), but then was modified to account for the fact that IHCs can respond very poorly to weak sounds while still responding almost normally to high level sounds (Moore and Glasberg 2004). Therefore, the level of excitation was reduced by:

$$HL^{2}_{IHC} / (L_{SIG} - L_{TQ}). \tag{10.23}$$

For example, if HL_{IHC} is 10 dB and L_{TQ} is 40 dB, a level of excitation at 50 dB will be reduced by 10 dB, and a level of 90 dB by 2 dB.

Second, the HL_{OHC} reduces effectiveness of the active mechanism of the cochlea. In a healthy cochlea, the OHCs amplify the low-level sounds with a gain that gradually decreases with level. This phenomenon leads to the compressive aspect of the loudness function. Therefore, a loss of OHC implies an elevation of threshold and a loss of compression. In this model, the low-level sounds' amplification was represented by the factor G (10.15). Therefore, to stimulate HL_{OHC}, G is made more negative, which has the effect of attenuating the input signal. Further, the factor exponent α, which controls the steepness of the slope, is dependent on the factor G. A decrease of G will lead to an increase of exponent α, and result in a loss of compression and will increase the slope in accordance with the rapid growth model.

Finally, a HL_{OHC} will also effectually broaden the critical bandwidth. This effect is modeled by a factor B:

$$B = 10^{0.01 HL_{OHC}} \tag{10.24}$$

10.6 Summary

This chapter shows that models can be designed for different purposes: *to describe*, *predict*, or *test hypotheses*, and can vary in complexity. The simplest models were described first: Fechner's law, Stevens' law, and the InEx function (Sect. 10.2). The last two models are well suited to *describe* and *predict* the loudness functions of pure-tone stimuli. Complex models, such as the loudness model of Zwicker and the model of Moore, have been developed (Sect. 10.3). They both divide the peripheral auditory pathway into different stages. If the output of the model accurately *describes* psychophysical measures, it can be assumed that the model reflects some aspects of the auditory pathway, and could therefore test and validate hypotheses related to those processes. Those complex models were adapted to *predict* the loudness of time-varying sounds (Sect. 10.4), even for the impaired auditory process (Sect. 10.5).

There is no fixed rule to decide which model is better than another. It is important to have simple models (and not to model a cat by a cat); however, cautions should be taken as to not discard important processes that need to be taken into consideration. In the end it depends on the user's needs, the degree of the complexity required and the accuracy needed.

References

ANSI (2007) American National Standard Procedure for the Computation of Loudness of Steady Sounds. J ANSI S3.4–2007.

Blamey PJ (2005) Adaptive dynamic range optimization (ADRO): a digital amplification strategy for hearing aids and cochlear implants. Trends Amplif 9:77–98.

Blamey PJ, Martin, LFA (2009) Loudness and satisfaction ratings for hearing aid users. J Am Acad Audiol 20:272–282.

Boone MM (1973) Loudness measurements on pure tone and broad band impulsive sounds. Acustica 29:198–204.

Buus S, Florentine M (2002) Growth of loudness in listeners with cochlear hearing losses: recruitment reconsidered. J Assoc Res Otolaryngol 3:120–139.

Chalupper J, Fastl H (2002) Dynamic loudness model (DLM) for normal and hearing-impaired listeners. Acustica 88:378–386.

de Cheveigné A (2004) Pitch perception models. In: Plack C, Oxenham A (eds), Pitch. New York: Springer.

Fechner GT (1860) Elemente der Psychophysik. Leipzig: Breitkopf und Heitel.

Florentine M, Epstein M (2006) To honor Stevens and repeal his law (for the auditory system). in Fechner Day, (eds.) DE Kornbrot, RM Msetfi, and AW MacRae, International Society for Psychophysics, St. Albans, UK pp 37–42.

Florentine M, Zwicker E (1979) A model of loudness summation applied to noise-induced hearing loss. Hear Res 1:121–132.

Glasberg BR, Moore BC (1990) Derivation of auditory filter shapes from notched-noise data. Hear Res 47:103–138.

Glasberg BR, Moore BC (2002) A model of loudness applicable to time-varying Sounds. J Audio Eng Soc 50:331–342.

Glasberg BR, Moore BC (2006) Prediction of absolute thresholds and equal-loudness contours using a modified loudness model. J Acoust Soc Am 120:585–588.

Launer S (1995) Loudness Perception in Listeners with Sensorineural Hearing Impairment. Germany: Universitat Oldenburg.

Marozeau J, Florentine M (2007) Loudness growth in individual listeners with hearing losses: a review. J Acoust Soc Am 122:EL81.

Marozeau J, Florentine M (2009) Testing the binaural equal-loudness-ratio hypothesis with hearing-impaired listeners. J Acoust Soc Am 126:310–317.

Moore BC, Glasberg BR (1997) A model of loudness perception applied to cochlear hearing loss. Audit Neurosci 3:289–311.

Moore BCJ, (2004) Testing the concept of softness imperception: Loudness near threshold for hearing-impaired ears, J. Acoust. Soc. Am. 115, 3103–3111.

Namba S, Kuwano S, Fastl H (2008) Loudness of non-steady-state sounds. Jpn Psychol Res 50:154–166.

Shaw EA (1974) Transformation of sound pressure level from the free field to the eardrum in the horizontal plane. J Acoust Soc Am 56:1848–1861.

Stevens SS (1961) To Honor Fechner and Repeal His Law: a power function, not a log function, describes the operating characteristic of a sensory system. Science 133: 80–86.

Zwicker E (1961) Subdivision of the audible frequency range into critical bands (Frequenzgruppen). J Acoust Soc Am 33:248.

Zwicker E (1984) Dependence of post-masking on masker duration and its relation to temporal effects in loudness. J Acoust Soc Am 75:219–223.

Zwicker E, Fastl H (1990) Psychoacoustics: Facts and Models. Berlin: Springer.

Zwicker E, Scharf B (1965) A model of loudness summation. Psychol Rev 72:3–26.

Zwicker E, Fastl H, Widmann U, Kurakata K, Kuwano S, Namba S (1991) Program for calculating loudness according to DIN 45631 (ISO 532B). J Acoust Soc Jpn 12:39–42.

Index

A

Acceptable loudness levels, in communities, 202–203
Acoustic menu method, loudness, 146
Acoustic reflex, loud sounds, 92–93
Acoustic size, 190
Acoustic startle reflex, loud sounds, 89–90
Adaptation level, loudness, 63–64
Adaptation, loud sounds, 95
Adaptive methods, equal loudness, 33ff
Additive summation rule, binaural loudness, 175
Age, loudness, 135ff
Aging auditory system, loudness, 137–138
Air-bone gap, loudness, 249
Alternate binaural loudness balance test, 237–238
Annoyance, and distortion, 200
  and loudness, 200ff
  and tinnitus, 236
  sound source meaning, 200ff
Assimilation effects, loudness, 79–80
Audio-tactile interactions, 210–211
Audio-visual interactions, 208–209
Auditory brainstem response, loudness, 100
Auditory detection, effects of vision, 80ff
Auditory dynamic range, 236–240
Auditory effects, loud sounds, 92ff
  tactile stimulation, 82
Auditory intensity, processing, 59–60
Auditory nerve, 11
  spike generation, 229–230
  spike transmission, 230–231
Auditory periphery, loudness nonlinearity, 116–117
Axiomatic measurement, binaural loudness, 174–175

B

Bandwidth, and loudness, 125
Basilar-membrane velocity, loudness, 100
Beethoven's Ninth Symphony, loudness range of, 213
Binaural equal-loudness-ratio hypothesis, 178
Binaural gain, 171, 176
  headphone investigations, 175ff
Binaural inhibition, 170
Binaural level difference for equal loudness, 171
Binaural loudness, 169ff
  constancy, 189, 212–213
  hearing aids, 191
  summation, 169–170, 212–213, 233–234
Binaural power summation model, 192
Binaural summation, variables affecting, 178ff
Binaural-to-monaural loudness ratio, 173–174
Brain scans, loudness, 102

C

Cardiovascular health, loud noise, 91
Category loudness scales, 38ff
Category scales, 38ff
Childhood, loudness, 135–137
Cochlear compression, loudness, 227ff
Cochlear damage, loud sounds, 95–96
Cochlear frequency analysis, loudness, 227
Cochlear implant fittings, loudness, 250–251
Cochlear implants, dynamic range problem, 250
  loudness, 249–251
Coding of loudness in neurons, 100–101
Cognition and context, loudness, 211ff
Cognitive effects, loud sounds, 91
Color and loudness, 209–210

M. Florentine et al. (eds.), *Loudness*, Springer Handbook of Auditory Research 37, DOI 10.1007/978-1-4419-6712-1, © Springer Science+Business Media, LLC 2011

Commercial Advertisement Loudness Mitigation Act (CALM), 2
Community noise, 202
Complex sound, loudness, 267
Compressive nonlinearity, 116–117
Conductive and mixed hearing losses, loudness, 249
Conductive hearing loss, 225
Context effects, loudness, 57ff
Continuous judgment by category method, loudness, 146, 161ff
Contrast effects, loudness, 79–80
Correlates of loudness, laboratory use, 98ff
Critical band, loudness, 125–127, 269ff
Critical duration of loudness, 158ff
Cross-cultural effects, 203
Cross-modal context effects, 80ff
Cross-modality matching, magnitude production, 45–48

D

Decibel scale, origin, 25–26
Development of loudness, 135–137
Dichotic, definition, 170
Differential context effect, mechanism, 69–73, 77–79
Diffuse field, loudness, 182ff
Diotic, definition, 170
Directional loudness contours, 185ff
Directional sound fields, loudness, 184ff
Discrimination Limen. *See* DL; JND
Distance judgments, anechoic environment, 188
DL (Discrimination Limen), Fechner, 21–23
Duration effects, loudness, 122ff, 158ff
Durlach and Braida's theory of intensity perception, 69
Dynamic range problem, 229
Dynamic range, loudness, 11–12, 100–101, 110

E

Efferent pathway, induced loudness reduction, 75–76
Environments, and loudness, 199ff
Equal loudness matching, 32ff
  adaptive methods, 33ff
  method of adjustment, 33
Equal-loudness-level contours, 181–183
Equivalent rectangular bandwidth, loudness, 275
External and middle ear, loudness, 226–227

F

Fechner, logarithm model, 263–264
  law, derivation, 20–23
  loudness, 19ff
Fechnerian loudness, Principle of Equality, 23–25
Fechnerian measurement, 20ff
Field *versus* laboratory studies, loudness, 164–165
Fletcher and Munson, loudness, 27–28, 110ff, 117, 119
Florentine and Zwicker model, loudness in noise-induced hearing loss, 281–282
Fluctuating noise definition, loudness, 147ff
fMRI, measures of loudness, 102
Fourier time transform (FTT), loudness, 201ff
Free fields, loudness, 181–182
Frequency, 118ff
  contributions to total loudness, 127–129
  effects on loudness, 121–122

G

Gesamtempfindung, 57

H

Habituation to loud sounds, 90
Headphone investigations, binaural gain, 175ff
  binaural loudness, 175ff
Head-related transfer function (HRTF), loudness, 183ff, 190, 192
Hearing aid compression, fast loudness restoring, 248
Hearing aid fitting, binaural loudness, 248
Hearing aid gain, loudness, 241–242, 245
Hearing aid prescriptions, model or measurement based, 242–243
Hearing aids, and loudness, 224, 241ff
  category loudness scales, 38ff
Hearing handicap, 204
Hearing loss, and loudness, 223ff
  loud sounds, 6, 96
  media players, 204
  noise induced (*see* NIHL)
Hearing, cross-modal context effects, 80ff
High sound levels, loudness, 116
Hyperacusis, and loudness, 236

I

Immissions, 207–208
Impulsive noise definition, loudness, 147ff

Individual binaural room impulse response (BRIR), 188
Induced loudness reduction, 73ff, 95
- loudness enhancement, 76–77
- mechanism, 75–76, 77–79
- relation to other sequential effects, 74–75

InEx function model, loudness, 266–267
Infancy, loudness, 135–137
Inner hair cell, dysfunction, 225
- transduction, 229–230

Instantaneous loudness, models, 277
Instantaneous *versus* overall judgment, 163–164
Intensity discrimination, loudness, 154ff, 232
Intensity, loudness, 110ff
- meaning, 6

Interaural level difference, loudness, 185
Interaural time difference, loudness, 185
Intermittent noise definition, loudness, 147ff
International Organization for Standardization. *See* ISO
Intramodal matching, binaural loudness, 170–171
Inverse-square law, 180
Irregular time-varying sounds, loudness, 157–158
ISO 226, loudness function growth, 8–9, 112, 119–120

J

JND (Just Noticeable Difference), Fechner, 21–23

L

Laboratory *versus* field studies, loudness, 164–165
Languages, meaning of "loudness," 7–9
Linear hearing aids, 244–245
Logarithmic law, Fechner, 21–23
Long-duration sounds, loudness, 48
Loudness, aging auditory system, 137–138. *See also* Loud sounds
- and annoyance, 200ff
- and hearing loss, 223ff
- and level meters, 268
- assimilation effects, 79–80
- at high levels, 116
- auditory brainstem response, 100
- balancing, 237–238
- binaural, 169ff
- coding in neurons, 100–101
- compressive nonlinearity, 116–117
- contemporary approaches to measurement, 31ff
- context effects, 57ff
- contours, 118ff
- contrast effects, 79–80
- contributions of frequency, 127–128
- correlates, 89ff
- critical band, 125–127
- cross-modal context effects, 80ff
- defined by International Organization for Standardization, 8–9
- definition, 3ff
- density (specific loudness), 233
- density equalization, 242
- density normalization, 242
- differential context effects, 69ff
- discomfort level, 237
- dynamic range, 11–12, 100–101
- effect of overall level, 129–130
- effects of vision and tactile stimuli, 82
- Fechner's Law, 21ff
- fMRI measures, 102
- function of age, 135ff
- function of bandwidth, 125
- function of duration, 122ff
- function of frequency, 118
- function of intensity, 110ff
- growth near threshold, 234–235
- hearing loss, 6
- importance in society, 1–3
- in daily environments, 199ff
- long-duration sounds, 48
- magnitude estimation, 57–58, 61
- manner of presentation, 5–6
- masking, 4–5, 131ff
- maximizer, 206
- meaning in different languages, 7–9
- measurement methods, 17ff, 57ff
- models, 9ff
- multisensory effects, 80ff
- near threshold, 113ff, 234
- non-steady-state sounds, 145ff
- otoacoustic emissions, 99
- partial masking, 131ff
- physiological correlates, 89ff, 99ff
- physiological models, 9–10
- psychoacoustical models, 9–10
- scale, 27–28, 36ff
- steady-state sounds, 109ff
- Steven's Law, 29–31
- stimulus distribution effects, 60ff, 64ff
- stimulus spacing, 64ff
- summation, 128
- temporally varying sounds, 145ff
- *vs.* intensity, 6
- *vs.* volume, 6

Loudness constancy, source distance, 187ff, 212–213
Loudness enhancement, induced loudness reduction, 76–77
Loudness function, 110–112
 high and low frequency, 121–122
Loudness growth functions, hearing impairment, 231
Loudness judgments, temporal stream, 161ff
Loudness measurement, equal loudness matching, 32ff
 history, 18ff
Loudness models, 261ff
 hearing impaired listeners, 279ff
 long term sounds, 279
 of a pure tone, 262
Loudness perception, physiological interpretations, 231ff
Loudness predictions, physiological models, 240
Loudness summation, across auditory neurons, 233
Loud noise, health, 90–91
Loud sounds, acoustic reflex, 92–93
 adaptation, 95
 auditory effects, 92ff
 cognitive effects, 91
 physiological effects, 89ff
 psychophysical effects, 90–91
 sleep disruption, 91–92
 temporary loudness shift, 93–94
Loudspeaker *versus* headphone, loudness, 189ff

M

Magnitude estimation, loudness, 57–58, 61
 range and spacing of stimuli, 43–44
 scales of loudness, 42ff, 46–48
Magnitude production, cross-modality matching, 45–48
 loudness, 62
Masking level difference, loudness, 179
Masking, of loudness, 4–5, 131ff
Matching, equal loudness, 32ff
Mean energy level, loudness, 160
Mean intensity level, stimulus distribution, 62–64
Measurement, long-duration sound loudness, 48
 loudness, 17ff, 31ff, 57ff
Measuring loudness from judging ratios, 28–29
Measuring loudness from response times, 26–27
Mechanism, differential context effect, 77–79
 induced loudness reduction, 75–79
Media players, hearing loss, 204
Method of adjustment, equal loudness, 33
Method of converging evidence, 4
Methods, loudness measurement, 17ff
Models, definition, 261
 for description, 261–262
 for excitation pattern to loudness, 271–272, 275
 for signal to level of excitation, 269ff, 274–275
 for specific loudness to overall loudness, 272, 276
 loudness, 9ff, 261ff
 loudness in hearing impaired listeners 278ff
 loudness long term sounds, 279
 loudness pure tones, 279
 Parduci's range-frequency, 64ff
 predict, 262
 test hypotheses, 262
Modulation frequency, loudness, 150–151
Monophonic loudness model, 192
Moore and Glasberg model, 272ff
 hearing impaired listeners, 282–283
 nonstationary sounds, 277
Multisensory effects, loudness, 80ff
Multisensory interactions, loudness, 208ff
Music and annoyance, 293ff
Music and loudness, 203–208, 213
Music, hearing loss, 204–205
Musical notation, loudness, 205–206
Music types/styles, loudness, 207–208

N

Neural loss, loud sounds, 96
Neural reorganization, loud sounds, 96
NIHL, importance in society, 1–3
NIHL, loudness model for, 281–282
Noise induced hearing loss *See* NIHL
Noise, exposure limit, 2, 204
Nonlinear hearing aids, 245ff
Nonmetric scaling, binaural loudness, 174–175
Non-steady-state sounds, loudness, 145ff–165

O

Optimal loudness, and environment, 213ff
 in noise, 214
 music halls, 213–214

Otoacoustic emissions, 99, 239–240
Outer hair cell dysfunction, 225
Overall level, effects on loudness, 129–130
Overall *versus* instantaneous judgment, 163–164

P

Parchment-skin illusion, 210–211
Parducci's range-frequency model, 64ff
Partial masking of loudness, 131ff
Pediatric population, loudness, 239
Perceptual experience and stimulus properties, 19
Peripheral sound transformations, 226
  loudness nonlinearity, 116–117
Permanent hearing loss, loud sounds, 95–98
Personal loudness units, 235–236
Phonophobia, 236
Physiological correlates, loudness, 99ff
Physiological effects, loudness, 89ff
Physiological models, loudness, 9ff
Piéron's law, 26–27
Power function, binaural loudness, 172ff
Power function model, loudness, 264ff
Power law, Stevens, 112–113
Preferred listening level, 204–205
Preferred loudness, and hearing aid experience, 247–248
Prescribed *versus* perceived loudness, 243–244
Presentation frequency, stimulus distribution, 64ff
Principle of equality, 46–48
Principle of Equality, Fechnerian loudness, 23–25
Processing, auditory intensity, 59–60
Psychoacoustical models, loudness, 9ff
Psychoacoustics and loudness, 4
Psychological present, 161–162
Psychophysical effects, loud sounds, 90–91
Psychophysical scaling, binaural loudness, 171ff
Psychophysics, loudness measurement, 19–20

R

Rapid growth model, loudness, 280
Reaction time, loudness, 98–99
Recruitment of loudness, 226
Reflections, indirect *versus* direct, 180
Response times, for measuring loudness, 26–27
Rise time, and loudness, 151ff
  and timbre, 151
Roughness, and loudness, 150–151

S

Scale of loudness, 4
  magnitude estimation, 42ff, 46-48
Scales, category, 38ff
Scaling, loudness, 36ff
  headphone investigations, 177–178
Sensorineural hearing loss, 225
Sequential effects, inducted loudness reduction, 74–75
Short term loudness, models, 278–279
Sleep disruption, loud sounds, 91
Softness imperception model, 234, 280–281
Sone scale, Steven's Law, 29–31
Sone scale to measure loudness, origin, 28–29
Sound fields, loudness, 180ff
Sound level meters, loudness models, 276
Sound level meters *versus* sonemeters, 215–216
Sound pressure, conversion to loudness, 110–112
Soundscape, loudness, 2
Sound source meaning, and annoyance, 200ff
Speech intelligibility, maximizing, 242
Steady-state sounds, loudness, 109ff
Stevens Law, sone scale, 29–31
Stevens power law, 112–113
  binaural loudness, 173
Stimulus distribution effects, 60ff
  mean intensity level, 62–64
  presentation frequency, 64ff
  stimulus spacing, 64ff
Stimulus properties and perceptual experience, 19
Stimulus range, loudness, 60f
Stimulus spacing, stimulus distribution, 64ff

T

Tactile stimulation, auditory effects, 82
Temporally varying sounds, loudness models, 145ff, 276ff
Temporal position of intensity increment, loudness, 154ff
Temporary loudness shift, 93–94
Temporary threshold shift, 93–94
Theory of intensity perception, 69
Threshold of discomfort, 237
Threshold, loudness near, 113ff
Time constant, loudness models, 276
Time-varying sounds, ISO classification, 147
Tinnitus, and loudness, 235–236
  loud sounds, 96–97

Total loudness, contribution of individual components, 127–128
Traffic noise, loudness, 157, 159

U
Uncomfortable loudness, hearing loss, 224
Upper limit of comfortable listening, 237

V
Vision, and auditory detection, 80ff
Volume, meaning, 6

W
Weber's law, 21
William Tell Overture, loudness range, 213

Z
Zwicker model of loudness, 127, 268ff
  Chalupper and Fastl modification, 282
  Florentine modification, 281
  hearing impaired listeners, 281–282
  Launer modification, 282
  nonstationary sounds, 276–277